Symbols, Impossible Numbers, and Geometric Entanglements is the first history of the development and reception of algebra in early modern England and Scotland. Not primarily a technical history, this book analyzes the struggles of a dozen British thinkers to come to terms with early modern algebra, its symbolical style, and negative and imaginary numbers. Professor Pycior uncovers these thinkers as a "test group" for the symbolic reasoning that would radically change not only mathematics but also logic, philosophy, and language studies. The book furthermore shows how pedagogical and religious concerns shaped the British debate over the relative merits of algebra and geometry.

The first book to position algebra firmly in the Scientific Revolution and pursue Newton the algebraist, it highlights Newton's role in completing the evolution of algebra from an esoteric subject into a major focus of British mathematics. Other thinkers covered include Oughtred, Harriot, Wallis, Hobbes, Barrow, Berkeley, and MacLaurin.

Symbols, Impossible Numbers, and Geometric Entanglements

Symbols, Impossible Numbers, and Geometric Entanglements

British Algebra through the Commentaries on Newton's *Universal Arithmetick*

HELENA M. PYCIOR
University of Wisconsin, Milwaukee

CAMBRIDGE
UNIVERSITY PRESS

CAMBRIDGE UNIVERSITY PRESS
Cambridge, New York, Melbourne, Madrid, Cape Town, Singapore, São Paulo

Cambridge University Press
The Edinburgh Building, Cambridge CB2 2RU, UK

Published in the United States of America by Cambridge University Press, New York

www.cambridge.org
Information on this title: www.cambridge.org/9780521481243

First published 1997
This digitally printed first paperback version 2006

A catalogue record for this publication is available from the British Library

Library of Congress Cataloguing in Publication data
Pycior, Helena M. (Helena Mary), 1947–
Symbols, impossible numbers, and geometric entanglements : British
algebra through the commentaries on Newton's *Universal arithmetick* /
Helena M. Pycior.
p. cm.
Includes index.
ISBN 0-521-48124-4
1. Algebra – Great Britain – History – 17th century. 2. Algebra –
Great Britain – History – 18th century. 3. Newton, Isaac, Sir,
1642–1727. Arithmetica universalis. I. Title.
QA151.P94   1997
512'.00941'09032 – dc20   96 – 25469
                                                                    CIP

ISBN-13 978-0-521-48124-3 hardback
ISBN-10 0-521-48124-4 hardback

ISBN-13 978-0-521-02740-3 paperback
ISBN-10 0-521-02740-3 paperback

To
R.W. III
&
J.C.L.

Contents

Acknowledgments

A few scholars and institutions will always remain strongly associated with the evolution of this book, at least in my memory. In 1970 the late Henry Guerlac introduced me to the Scientific Revolution and set as my first graduate research paper topic the mathematical lectures of Isaac Barrow. The latter was not my topic of choice, and even after two wonderful seminars on the Scientific Revolution I wrote my dissertation on nineteenth-century British algebra under L. Pearce Williams.

Quite some years and much research later, I discovered that I had fallen into what I sometimes call "the modern historian's trap" – that is, set on finding context for nineteenth-century algebra I had worked my way backward to the eighteenth century and, then, to Barrow's century. The decision to switch the focus of my research from the algebra of the nineteenth century to that of earlier ones seems to have come soon after a research trip to Cambridge and London in 1983–1984. Supported by a grant from the History and Philosophy of Science Program of the National Science Foundation, that trip began with the intent to study Victorian mathematics and mathematicians through 1886. The archives I consulted, however, revealed some unexpected aspects of Victorian mathematics, including fairly explicit links with British mathematical thought of the seventeenth and early eighteenth centuries. In 1985–1987 Robert L. Hall, a professor of mathematics at the University of Wisconsin–Milwaukee (UWM), and I had a grant from the Precollege Science and Mathematics Education Program, National Science Foundation, to apply history to the teaching of mathematics. One of our responsibilities was to offer a history of mathematics course to teachers of grades 7 through 12, and it was for this course that I prepared my first lectures on early modern algebra.

Besides the National Science Foundation, the Center for Twentieth Century Studies at UWM provided key support that helped make this book a reality. I was a fellow at the center during the academic year, 1984–1985. My paper, "Mathematics and Philosophy: Wallis, Hobbes, Barrow, and Berkeley," was first delivered at the center, subsequently

ix

revised there, and submitted to the *Journal of the History of Ideas* in early 1985.

I am grateful to Philip P. Wiener, who wrote a favorable report on the latter paper, as well as to the many other scholars who over the years have shared with me their comments, questions, and, yes, encouragement on various aspects of my algebra studies. Valuable feedback came after I delivered papers at the annual meeting of the Midwest Junto of the History of Science Society (HSS) in 1980 (on William Frend's rejection of negative numbers); the annual meeting of the HSS in 1983 (Berkeley and nineteenth-century British symbolical algebra); the annual meeting of the HSS in 1985 (Wallis, Hobbes, Barrow, and Berkeley); and the XVIIIth International Congress of History of Science in 1989 (Wallis and Sir William Rowan Hamilton). Too, the departments of mathematics at the University of Wisconsin–Madison and at Marquette University let me test some of the book's theses on mathematical audiences. I spoke on Berkeley's *Analyst* at Madison and on Cardano, Viète, and their English disciples at Marquette in 1987 and 1991, respectively.

A great, specific debt is owed to the two referees who read the manuscript of this book for Cambridge University Press during the academic year 1993–1994. Following their advice, I shortened the manuscript but I could neither follow up all their interesting leads nor, when revising the manuscript, do full justice to the literature after 1993. A more general debt for exposing me to the diversity, frustrations, and splendor of the history of mathematics is owed to the many historians of mathematics with whom in the 1980s I worked as the managing editor of *Historia Mathematica*. Friends and colleagues, including Roland Stromberg, prodded me to complete this book by periodically asking about its fate while always seeming to trust that it would eventually appear.

The British Library has graciously granted me permission to reproduce selections from its Thomas Harriot manuscripts. I have been able to quote from the correspondence of Colin MacLaurin with the permission of the University of Aberdeen Library, the Trustees of the National Library of Scotland, and the Department of Special Collections, Stanford University Libraries. Cambridge University Press has granted me permission to quote from *The Correspondence of Isaac Newton*, edited by A. R. Hall and Laura Tilling, and *The Mathematical Papers of Isaac Newton*, edited by D. T. Whiteside. Kent State University Press has given permission to reproduce selections from François Viète, *The Analytic Art: Nine Studies in Algebra, Geometry and Trigonometry . . .* , translated by T. Richard Witmer; and the MIT Press, permission to reproduce selections, including a figure, from Girolamo Cardano, *The Great Art or*

The Rules of Algebra, translated and edited by T. Richard Witmer, and selections from Jacob Klein, *Greek Mathematical Thought and the Origin of Algebra,* translated by Eva Braun. My gratitude also extends to the Johns Hopkins University Press, which has agreed to let me adapt portions of my article, "Mathematics and Philosophy: Wallis, Hobbes, Barrow, and Berkeley," published in the *Journal of the History of Ideas* in 1987.

My biggest debt is, of course, to my husband and dog, who have shared their lives with books-in-progress so long that both have come to regard my summer weekdays and weekends at the computer as almost normal.

Introduction

As a history of algebra in England and Scotland from the early seventeenth century to the mid-eighteenth century, this book considers not only technical algebra but also the personal, philosophical, religious, and institutional factors that affected the introduction, elaboration, and reception of the subject. The first chapter sets the scene for the study by discussing the new algebra that emerged in continental Europe in the sixteenth and early seventeenth centuries. This algebra – which I call "early modern algebra" – included new results (the solutions of the cubic and quartic equations), a new language (the symbolical), an expanded algebraic universe (with negative and, tentatively, imaginary numbers), a new emphasis on analysis as the special method of algebra, and a changing relationship between algebra and geometry. Chapters 2 through 10, the heart of the book, explore British algebra from the introduction of early modern algebra into England by William Oughtred and Thomas Harriot in 1631 through the post-Newtonian formulations of algebra by Colin MacLaurin and Nicholas Saunderson.

There is no book-length history of British algebra of the seventeenth and eighteenth centuries.[1] Historians of British mathematics have con-

1. For an early history of algebra, including British contributions, see Charles Hutton, "Algebra," in *A Mathematical and Philosophical Dictionary*, 2 vols. (1795; reprint, Hildesheim: Georg Olms, 1973), 1:63–97, and, more especially, idem, *History of Algebra*, tract 33 of *Tracts on Mathematical and Philosophical Subjects . . .* , 3 vols. (London, 1812), 2:143–305. Twentieth-century scholarship on early British algebra ranges from Cajori's pioneering studies on Oughtred and Harriot to important articles of the last two decades. Cajori's relevant writings include Florian Cajori, *William Oughtred: A Great Seventeenth-Century Teacher of Mathematics* (Chicago: Open Court, 1916); idem, "A Revaluation of Harriot's Artis Analyticae Praxis," *Isis* 11 (1928): 316–324; idem, "A List of Oughtred's Mathematical Symbols, with Historical Notes," *University of California Publications in Mathematics* 1, no. 8 (1920): 171–186; and (for a larger perspective on the period's mathematical symbols) idem, *A History of Mathematical Notations*,

centrated on later algebra, since it was Great Britain of the late eighteenth and early nineteenth centuries that produced "symbolical algebra," in which algebra began to take its modern, abstract form.[2] While the present book centers on what was in itself as exciting and important an era in the development of algebra, it also provides essential background for the study of the emergence of nineteenth-century symbolical algebra. The book's Epilogue suggests specific ways in which a century and a half of British cultivation primed algebra for development from the early modern stage to the symbolical stage.

Above all, the book reconstructs early modern algebra through its post-Newtonian formulations as a case study in the development of a mathematical discipline from its introduction into a particular country through its fairly wide acceptance. This is a complex and revealing study since the algebra of the seventeenth and eighteenth centuries was more distinguished for its generality and wide applicability than for the clarity of its concepts or the certainty of its methods. British mathematicians from Oughtred and Harriot to MacLaurin and Saunderson were attracted by the power of algebra and by the symbolical language that seemed to hold the secret of that power. This attraction, and the concrete results of algebra, proved enough to sustain their commitment to the subject despite persistent questions about some of its key concepts, mainly the negative and imaginary numbers, and about its methods. This history, then, shows how a group of mathematicians – stretched across more than a century, but sharing a national identity – made and remade their peace with an evolving subject that did not quite adhere to the mathematical standards of their period.

It also analyzes how, especially as they sought to legitimate the fertile negative and imaginary numbers, mathematicians began to undermine the existing mathematics, its rules and its standards. In the privacy of his manuscripts and in informal sessions with his students, Harriot tried to get around the inconsistency involved in imaginary numbers by modi-

2 vols. (Chicago: Open Court, 1929). Among the recent studies are R. C. H. Tanner, "The Ordered Regiment of the Minus Sign: Off-beat Mathematics in Harriot's Manuscripts," *Annals of Science* 37 (1980): 127–158; Chikara Sasaki, "The Acceptance of the Theory of Proportion in the Sixteenth and Seventeenth Centuries," *Historia Scientiarum* 29 (1985): 83–116; and Helena M. Pycior, "Mathematics and Philosophy: Wallis, Hobbes, Barrow, and Berkeley," *Journal of the History of Ideas* 48 (1987): 265–286.

2. "Symbolical algebra" was George Peacock's term. Its use here must be qualified by the recognition that earlier algebra was certainly symbolical if less abstract.

fying the law of signs to read (in certain cases): minus times minus gives minus.[3] In his later defense of the negative and imaginary numbers, which he lumped together as "imaginary quantities," John Wallis somewhat explicitly recognized new criteria for algebraic objects. He emphasized utility, physical applicability, and geometric representation over "strict notions" or traditional definitions. Moreover, he consciously constructed a defense of the negative and imaginary numbers that, with its appeal to precedent, was intended more to persuade than to compel mathematicians to accept an enlarged algebraic universe. As he tried to explain the negatives, MacLaurin suggested a shift in the focus of mathematical foundations from objects to relations and to equations capturing relations. Not only Harriot, Wallis, and MacLaurin but Girolamo Cardano and Nicholas Saunderson as well were algebraic explorers, willing to redefine mathematical "strictness" or rigor in favor of fertility.

The book, then, underscores the importance of individuals and individualistic thought in the making of mathematics. British algebra did not develop in a fundamentally linear fashion, with each mathematician's building carefully on the insights of his predecessors. Studying over a dozen British mathematical thinkers (Oughtred, Harriot, John Collins, John Pell, John Kersey, Wallis, Thomas Hobbes, Isaac Barrow, Isaac Newton, George Berkeley, Robert Simson, MacLaurin, Saunderson, and John Colson), the book uncovers many different twists on algebra, its key concepts, methods, interconnection with geometry, and standing among the mathematical sciences. For example, Wallis was a practitioner of "pure algebra" (his own term) who elaborated an arithmetic approach to algebra and shunned geometric "intanglements," whereas his younger contemporary Newton combined arithmetic algebra with the more geometric, Cartesian algebra. Indeed, British mathematical thinkers referred not only to mathematical "fashions" (Wallis's term) but also to more individualistic "tastes" and "dislikes" (John Robison's terms).

This is not to say that there was no linear development. Although in their published writings Oughtred and Harriot had generally ignored negative roots of equations, Pell and Kersey imported such roots into British algebra from Descartes's *Géométrie* of 1637, and from the 1670s through the late eighteenth century British mathematicians largely accepted negative roots and numbers. Still, each of Pell and Kersey's successors crafted a different approach to legitimation of these numbers.

3. In analyzing Harriot's appeal to this deviationary rule, Chapter 2 builds on an earlier description of Harriot's experiments with the rule of signs given by Tanner, "Minus Sign."

Wallis argued that negative numbers, like numbers involving $\sqrt{-1}$, could be "supposed" to exist by mathematicians; in addition, he showed that these numbers satisfied his new criteria for algebraic objects. Newton and his major commentators MacLaurin and Saunderson defined negative numbers as "quantities less than nothing," a troublesome definition which however fit the prevailing view of mathematics as the science of quantity. Revealing nevertheless some disagreement on exactly how to justify these numbers, no one of the three gave the definition without his own qualification and commentary. Newton made his terse case for the negative numbers with examples of quantities that "may be call'ed" positive and negative, such as assets and debts; MacLaurin, too, offered examples and added a defense of the negatives based on the relation of contrariety; Saunderson gave examples, appealed to contraries, and even blamed the English language for obfuscating the relationship between affirmative and negative quantities.

British mathematical thinkers of the seventeenth and early eighteenth centuries probed not just the negative and imaginary numbers but the methods of algebra and the rationale for symbolical reasoning as well. As the book documents, their concern about the foundations of algebra was fostered by the pedagogical use of mathematics as an instrument for training the human mind. If mathematics was to be a formal academic subject in England (which it was not at the beginning of the seventeenth century), Oughtred realized, the larger merits of the subject had to be carefully elaborated. Oughtred and subsequently Barrow introduced a defense of mathematics that emphasized its value in training the human mind more than its intrinsic worth. They and their successors, in effect, touted mathematics as a logic, worthy of study by all undergraduates. Although highly successful in making mathematics an academic subject (in fact the premier subject at Cambridge), this defense favored the study of Euclidean geometry at the expense of algebra and also put a premium on discussion of foundational issues, whether related to geometry, calculus, or algebra. As long as mathematics was used primarily to train the mind, foundational strictness mattered at least as much as mathematical power and applicability.

The push to explore the philosophy of mathematics (including algebra) came also from sources extrinsic to mathematics and its pedagogy. Chapters 2, 5, and 6, especially and to varying degrees, demonstrate how in England the philosophy of mathematics was affected by some of the major intellectual currents of the Scientific Revolution and by the Civil War. The symbolical style, for example, fit well with Francis Bacon's attack on the "Idols of the Market-place." As Bacon condemned wordy prose, Oughtred and his disciples criticized the "multiplicity of words" involved in the presymbolical or "rhetorical" algebra. Too, the economy of algebra was contrasted with the "pomposity" of classical geometry. As a result of the academic shuffle within and between the English universities, set off in the 1640s by the Civil War, John Wilkins was able to assemble at Oxford a group of scholars whose interests spanned the new science and the new theory of language, including symbolical language. Capitalizing on the period's emphasis on the sensible, Oughtred as well as Wallis and Seth Ward – both members of Wilkins's Oxford circle – depicted algebraic symbols as sensible objects that brought mathematical reasoning "in short Synopsis" before the very eyes of the mathematician. Less favorably to algebra, the empiricist emphasis caused Hobbes and Barrow to argue for the superiority of geometry over arithmetic. As Wallis elaborated arithmetic foundations for algebra, Barrow claimed that numbers had no sensible referents and were mere signs of magnitude.

Barrow's stance was just one manifestation of the competition that existed between arithmetic and algebra, on the one hand, and geometry, on the other, throughout the seventeenth and early eighteenth centuries. Because of the period's struggle with the relative standing of algebra and geometry, and because of the mutual relation of the two sciences, no historian of seventeenth- and eighteenth-century algebra may ignore geometry. The challenge is to capture the texture of British mathematics, with its algebraic and geometric threads, while yet preserving the algebraic focus. For the present author, capturing the texture has necessitated exploring (in varying depth) the personal comfort that Hobbes, Barrow, Newton, and Simson derived from the certainty of classical geometry, Newton's mixed sympathies for geometry and algebra, Berkeley's attack on Newtonian calculus, and MacLaurin's answer to Berkeley, along with more strictly algebraic topics.

As a major theme, which itself adds a new perspective on the history of British mathematics, the book argues that both algebra and geometry were key components of seventeenth- and eighteenth-century English mathematics and, to a lesser extent, Scottish mathematics.[4] The publication of Newton's *Universal Arithmetick* in the early eighteenth century solidified an algebraic tradition in Great Britain despite the favored status conferred on geometry by the defense of mathematics as a logic

4. This theme helps to explain Guicciardini's observation that eighteenth-century British mathematicians did not use exclusively geometric methods in the cultivation of calculus. See Niccolò Guicciardini, *The Development of Newtonian Calculus in Britain, 1700–1800* (Cambridge: Cambridge University Press, 1989), viii.

and by the empiricism associated with the English Scientific Revolution. As the book details, Oughtred, Harriot, Pell, Kersey, and Wallis developed a strong early modern, algebraic tradition to the point where Wallis wrote of a "pure algebra," which was independent of and perhaps superior to geometry. There was, however, a geometric backlash, led by Hobbes and Barrow. From the Lucasian chair, Barrow reduced arithmetic to a subcategory of geometry, dismissed algebra as "yet ... no Science," and emphasized the pedagogical strengths of Euclidean geometry. Although Wallis publicly debated the merits of algebra and geometry with Hobbes, Newton's *Universal Arithmetick* more than anything seems to have signaled to British mathematicians that there were two valid foci of mathematics: the arithmetico-algebraic and the geometric. As the book argues, this Newtonian bifocal mathematical tradition helped to assure that algebra as well as geometry continued to be seriously cultivated in eighteenth-century Great Britain. Thus by the midcentury algebraists seemed firmly entrenched in the Lucasian chair, even though late into the century other Cambridge scholars were still pushing for the centering of the undergraduate curriculum on Euclidean geometry. In early-eighteenth-century Scotland MacLaurin pursued geometry, algebra, calculus (both the geometric and the more algebraic calculus), and geometry again, in roughly that order. Only with MacLaurin's premature death in 1746 and the ascendancy of Simson's mathematical conservatism did the Scots focus more exclusively on geometry.

Berkeley, too, played a role in setting the mathematical priorities of early-eighteenth-century Great Britain. He elaborated a philosophy of mathematics, according to which geometry was the science of perceptible extension, and arithmetic and algebra were sciences of signs. His philosophy of algebra was thus forward-looking and, in fact, his general philosophy inspired British symbolical algebraists of the early nineteenth century. But his philosophy of algebra did not catch on in the first half of the eighteenth century. His ideas on the subject were ignored at least partially because of uneasiness with his antiabstractionism (denial of the existence of abstract general ideas), but also because of his publication of *The Analyst* in 1734. Since he had dared to criticize Newton's calculus and impugn the religious orthodoxy of mathematicians, the British mathematical community was perhaps unable to give his philosophy of algebra a fair hearing. At the practical level, by demanding an immediate, geometric defense of calculus, *The Analyst* helped to concentrate much of British mathematics of roughly the late 1730s into the 1740s on the geometric focus of Newtonian mathematics.

If Berkeley's philosophy of algebra did not immediately influence the development of the subject, it was the case that algebra, perhaps profoundly, affected Berkeley's general philosophy. As another major theme, the book argues that the debate over the foundations of algebra – especially over its symbolical style and its negative and imaginary numbers – forced British thinkers to reflect long and hard on symbolical reasoning. These thinkers – some more known for their mathematics, others for their philosophy – were, so to speak, a "test group" for the symbolical reasoning and later semiotics that would in the nineteenth and twentieth centuries so change not only mathematics but logic, philosophy, and language studies as well.[5] The book is able to capture these fertile interconnections between algebra and philosophy precisely because it rejects as anachronistic any rigid distinction between the mathematical and philosophical elements of the seventeenth-century and early-eighteenth-century discussion of symbolical reasoning. Indeed, from the research through writing stages the book has been informed by the methodological guideline that historians ought to privilege only disciplinary distinctions that were respected by the historical actors themselves.[6]

Early modern algebra called mathematicians to reason on arbitrary symbols that supposedly stood for quantities, but some of the symbols stood for negative quantities that seemed to defy traditional definition and one symbol $(\sqrt{-1})$ seemed to stand for nothing at all. From this perspective, the book is a history of a group of British mathematical thinkers, including the philosophers Hobbes and Berkeley, struggling to come to terms with early modern algebra and the symbolical reasoning that was part and parcel of that algebra but at the same time something greater than it. Hobbes balked at symbolical reasoning, and Barrow worried about its excesses. On the other hand, Oughtred, Harriot, Wallis, and a long line of their disciples reveled in it. Nevertheless, a willingness to reason on symbols and a near mania to coin new symbols did not necessarily entail acceptance of symbols for which there were no ready referents. Oughtred, for example, gave neither negative nor imagi-

5. There has been little historical research on connections between symbolical reasoning in algebra and semiotics. On links between Charles S. Peirce (the American "father of semiotics"), Berkeley, and the nineteenth-century symbolical algebraists, see Helena M. Pycior, "Peirce at the Intersection of Mathematics and Philosophy," in *Peirce and Contemporary Thought: Philosophical Inquiries*, ed. Kenneth Laine Ketner (New York: Fordham University Press, 1995), 132–145. See also Brian Rotman, "Toward a Semiotics of Mathematics," *Semiotica* 72 (1988): 1–35.
6. For my elaboration of this general position, see Helena M. Pycior, "Internalism, Externalism, and Beyond: 19th-Century British Algebra," *Historia Mathematica* 11 (1984): 424–441.

nary roots of equations. Despite an earlier, Cartesian-like acceptance of imaginaries, the Newton of *Universal Arithmetick* accepted "impossible roots" but drew the line at manipulating them. In the seventeenth and early eighteenth centuries, then, symbolical reasoning on idealess symbols was for hardy (or, to use Wallis's term, "venturous") thinkers alone. Wallis made the leap to such reasoning, but only by beginning to rewrite the rules of mathematics. Influenced by Wallis, Berkeley also made the leap but only by beginning to rewrite Western philosophy. As it is described in the present book, Berkeley's *Alciphron* of 1732 was a brilliant explication of a forward-looking theory of language or, more specifically, a theory of signs that was heavily dependent on insights from early modern arithmetic and algebra.

Finally, this is more than a history of major figures of British algebra and philosophy. Second- or third-rate mathematical thinkers contributed to the emergence of the strong algebraic tradition in Great Britain. In the 1660s and 1670s Collins seems to have fueled the algebraic interests of Newton and Wallis by challenging each man in turn to solve the irreducible case of the cubic, in particular, to extract the cube roots of complex binomials. During the same period Collins pressed for a much needed algebra textbook, written in the English language and encompassing early modern algebra through Descartes's *Géométrie*. He involved Newton in a textbook project, which (more than any other of Newton's endeavors) brought out his pedagogical talents and led circuitously to the publication of the first Latin edition of *Universal Arithmetick* in 1707. Collins perhaps inspired Wallis to begin his *Treatise of Algebra,* which appeared in 1685. With more immediate success, he assisted Pell – by many accounts, a first-rate mathematician whose potential was never realized – in publishing his English edition of Rahn's *Algebra* in 1668, and Kersey, a solid textbook author, in publishing his algebra in 1673. Indicative of the lack of a consensus on algebra into the eighteenth century as well as the importance of what historians of mathematics might be tempted to ignore as the lesser algebras, Pell's and Kersey's textbooks appeared alongside those of Oughtred, Harriot, Wallis, and Newton on Cambridge reading lists of the period, and Kersey's textbook continued to be reissued through 1741.

In short, from the early seventeenth century British mathematical thinkers of the first and not-quite-first orders, some more mathematically inclined, others more philosophically inclined, worked and reworked algebra. By the mid-eighteenth century they had published, among others, the eight textbooks covered in the present book. In consecutive stages, they had accepted the symbolical style, largely come to terms with the negative numbers, and begun to reach (what Chapter 10 describes as) a pragmatic détente with the imaginaries. Due to their efforts, there was an established arithmetico-algebraic tradition in England and a somewhat less secure tradition in Scotland.

Still, even as MacLaurin and Saunderson tried to perfect Newtonian algebra in the 1720s to 1740s, they left room for lingering discontent with the subject. What their writings showed was that after more than a century of discussion of the negative and imaginary numbers, British mathematicians could not quite agree on their defense. In his *Elements of Algebra* of 1740, Saunderson wondered publicly if he ought to call the imaginaries "quantities"; in the abstract of the *Treatise of Fluxions* that he wrote for the *Philosophical Transactions* of 1742–1743, Mac-Laurin declined to apply the term "quantity" to such "expressions." It was not, however, only the admission of these lingering difficulties that by the midcentury hinted at the need for a major reform of algebra. In his *Elements*, Saunderson argued persuasively for the ideal of an algebra that was a demonstrative science on par with, if not superior to, geometry. In the late eighteenth century, there were enough algebraically inclined British mathematicians to take his suggestion seriously and thereby to begin the transformation of algebra from its Newtonian stage into a new, deductive, and more abstract stage, known to all historians of mathematics as the British symbolical algebra of the early nineteenth century.

1

Setting the Scene
The Foundations of Early Modern Algebra

> In this book, learned reader, you have the rules of algebra (in Italian, the rules of the coss). It is so replete with new discoveries and demonstrations by the author – more than seventy of them – that its forerunners [are] of little account or, in the vernacular, are washed out.[1]

Thus the very advertisement to Girolamo Cardano's *Great Art* (*Ars magna*) suggested the dawning of a new epoch in the algebra of the Western world. Although less original than implied here, *The Great Art* – published in 1545, just two years after Copernicus's *De revolutionibus* and Vesalius's *De fabrica* – was an exciting scientific classic. Exciting in the mathematical way, the work announced the solution of two hitherto unsolved problems, finding the roots of cubic and quartic (or biquadratic) equations. As many other classics, it helped to redefine its field. In particular, it fostered an expanding universe of algebraic objects through its consistent acknowledgment of negative roots as well as its brush with imaginary roots.

Still, *The Great Art* was not solely responsible for the major reconstruction that Western algebra underwent in the sixteenth and early seventeenth centuries. Algebra's reconstruction involved new results, objects, language, methodological justification, and a changing relationship with geometry. It was as much due to François Viète's *Analytic Art* as to Cardano's *Great Art*. Cardano's acceptance of negative numbers and his tentative recognition of imaginary numbers, along with Viète's use of arbitrary symbols to denote algebraic entities, his appeal to analysis as the method of algebra, and his eschewing of geometry as the basis for algebra – all led to a substantially new kind of algebra, so new that the present book will hereinafter refer to the algebra that emerged from the

1. Girolamo Cardano, *The Great Art or The Rules of Algebra,* trans. and ed. T. Richard Witmer (Cambridge, Mass.: MIT Press, 1968), [1]. The name "great art" referred to algebra in contradistinction to the "lesser art" of arithmetic.

sixteenth and early seventeenth centuries as "early modern algebra." The consequences of this new algebra came in time to rival the effects of the astronomical and anatomical classics of the same period. Not only did early modern algebra contribute to the development of calculus, but its objects, symbolism, methods, and relationship to geometry became focal points of a far-reaching and long-running debate on the very foundations of mathematics as well as on the nature of language, general terms, and human reasoning. This was a debate in which British thinkers of the sixteenth through eighteenth centuries, the protagonists of this book, were major participants.

I

Algebra was imported from the Islamic world to western Europe through Latin translations of al-Khwārizmī's *Kitāb fī hisāb al-jabr wa'l muqābala,* made in the twelfth century, and Leonardo of Pisa's *Liber abbaci* of 1202.[2] The subject, which took its name from a European corruption of the Arabic term "al-jabr" in al-Khwārizmī's title, concentrated on the solution of equations by essentially arithmetic algorithms, that is, series of basically arithmetic steps taken to reach the values of the unknowns. The new equation theory caught on quickly in the Western world, principally in Italy where a commercial revolution led in the late thirteenth century to the founding of abacus schools, where young boys from the expanding merchant and commercial class studied the methods and principles of elementary arithmetic.[3]

By the next century, the instructors in these schools, the *maestri d'abbaco,* were not only transmitting the Arabic algebra but also adding original contributions to the subject. In their manuscripts, these *maestri* tackled many different equations, including specific cubic and quartic

2. My intent is to highlight elements of late medieval and early modern algebra that are essential to an understanding of British algebra, the focus of the present book. For more detailed expositions of early Western algebra, see, e.g., Cynthia Hay, ed., *Mathematics from Manuscript to Print, 1300–1600* (Oxford: Clarendon Press, 1988), and Karen Hunger Parshall, "The Art of Algebra from al-Khwārizmī to Viète: A Study in the Natural Selection of Ideas," *History of Science* 26 (1988): 129–164. On the transmission of Arabic algebra to Europe, see Warren van Egmond, "How Algebra Came to France," in *Mathematics from Manuscript to Print,* ed. Hay, 127–144, esp. 127–128.
3. On the abacus mathematics, see van Egmond, "How Algebra Came to France," 128–129.

equations, equations that had hitherto ranked among the unsolved problems of mathematics. Generally, the equations related to commercial problems, the *maestri* paid more attention to cubic than quartic equations, they succeeded in solving certain types of cubic and even quartic equations, and they gave positive solutions only.[4]

The algebraic tradition of the *maestri d'abbaco* culminated in the work of Girolamo Cardano (1501–1576), an Italian physician, professor of medicine, and polymath. The beneficiary of two centuries of Italian equation theory, he was also an exploiter of Europe's new mode of communication, the printed text. He collected, synthesized, and supplemented the early results on cubic and quartic equations, and published the whole for the scrutiny and enlightenment of mathematical thinkers across western Europe. Although somewhat derivative, his *Great Art* impressed readers with algebra's power and potential, and, according to Markus Fierz, the "book made him truly famous."[5]

Whereas the principles for solving quadratic equations had been known since ancient times, Cardano announced the principles for solving cubic and quartic equations. His solutions (whereby roots were expressed in radicals, as in the case of quadratic equations) were neither completely original nor general in the modern sense. In addition to solving cubic equations immediately reducible to linear or quadratic ones ($ax^3 = bx$, e.g.), some *maestri d'abbaco* of the fourteenth century had arrived at a correct solution of a more difficult case of the cubic equation – $ax^3 + bx^2 + cx = d$, where $b^2/3a = c/a$ – which they recorded in their unpublished manuscripts.[6] Cardano himself admitted that his path to solution of the cubic equation began as Niccolò Tartaglia (ca. 1506–1557), a mathematics teacher, revealed to him the solution of the equation,[7]

$$x^3 + ax = b. \qquad (1)$$

4. Raffaella Franci and Laura Toti Rigatelli, "Fourteenth-Century Italian Algebra," in *Mathematics from Manuscript to Print,* ed. Hay, 11–29. On early Italian algebra, see also R. Franci and L. Toti Rigatelli, "Towards a History of Algebra from Leonardo of Pisa to Luca Pacioli," *Janus* 72 (1985): 17–82.

5. Markus Fierz, *Girolamo Cardano, 1501–1576: Physician, Natural Philosopher, Mathematician, Astrologer, and Interpreter of Dreams,* trans. Helga Niman (Boston: Birkhäuser, 1983), 6.

6. Four manuscripts from the late fourteenth century offer a correct solution of this type of cubic equation (Franci and Toti Rigatelli, "Fourteenth-Century Italian Algebra," 18–21).

7. Cardano, *The Great Art,* 8.

Even Tartaglia, however, had not been the first to solve this case of
the cubic equation; Scipione del Ferro (1465–1526), a professor of
mathematics at the University of Bologna, had solved it around 1515,
and then passed his solution on to his students, including Antonio Maria
Fiore. Twenty years later, Fiore had challenged Tartaglia to a public
contest centering on equations that could be reduced to form (1). Such
contests made or broke mathematicians of the period by gaining or
losing for them crucial patronage. Under the pressure of "solve, or
perish," Tartaglia had independently solved equation (1). When Cardano learned of the contest, he asked Tartaglia for the solution. Tartaglia
initially refused to share his result. The solution was his private property,
according to the mathematical mores of the time, and as such would
give him an edge in future mathematical contests; furthermore, he had
told Cardano that he intended to publish a book expanding on the
result. Subsequently, however, Tartaglia visited Cardano in Milan, and
divulged the secret. According to a later account by Tartaglia, he had
done so only when Cardano took an oath on the Gospels, and gave his
word as a gentleman, never to publish the solution. Once armed with
Tartaglia's result for equation (1), Cardano and his secretary, Lodovico
Ferrari (1522–1565), solved the remaining cases of the cubic. All the
results on the cubic – those of del Ferro, Tartaglia, Cardano, and Ferrari – were then published in *The Great Art*.[8]

The result was a complete solution of the cubic equation, but not a
general one in the modern sense. Cardano worked within the mathematical bounds of his period. He wrote and reasoned in prose, not in the
symbolism that would become a hallmark of algebra within the next
century, and, because of the traditional qualms about negative numbers,
he used only positive coefficients. Thus, for him, the modern-day general
quadratic, cubic, and quartic equations did not exist. There was no
consistent symbolism in which to express even the modern, general
quadratic equation:

$$ax^2 + bx + c = 0.$$

Rather Cardano and his contemporaries referred to the unknown as the
"thing" (*res* in Latin, *cosa* in Italian, and *coss* in German) and subse-

8. Cardano's admission that "Scipione del Ferro of Bologna ... solved the
    case of the cube and first power equal to a constant ... [and] my friend
    Niccolò Tartaglia of Brescia, wanting not to be outdone, solved the same
    case when he got into a contest" (ibid.) did not assuage Tartaglia's fury at
    seeing his solution in print. On this feud, see Oystein Ore, foreword to *The
    Great Art*, ed. Witmer, ix–xii.

quent multiples of the "thing" as the "square" and the "cube"; they wrote of such specific situations as "the square and the thing equal to the number."

Along with the lack of algebraic symbolism, abhorrence for the negative numbers stood in the way of the emergence of general equations. Entertaining no negative or zero coefficients, mathematicians of the period saw the equations,

$$x^2 = ax + b$$

and

$$b = x^2 + ax,$$

for example, as distinct equations rather than equations that could be reduced to a common form. For Cardano, then, there were five basic types of quadratic equations:

1. the square is equal to a number,
2. the square is equal to the first power,
3. "the square is equal to the first power and constant,"
4. "the number is equal to the square and first power,"
5. "the first power is equal to the square and number."

He directed his readers that equations of type 1 had two roots, and he gave a specific algorithm for each of the other specific types.[9] For the solution of an equation of type 3, for example, he instructed readers to "add the square of one-half the coefficient of the first power to the constant of the equation and take the square root of the whole. To this add one-half the coefficient of the first power, and the sum is the value of x." The rules for the fourth and fifth cases were similar, all three being but special cases of the present-day quadratic formula.[10]

Cubic equations came in many more types, over thirteen in all, which Cardano solved one by one. Thus, in chapter 11 of *The Great Art*, he tackled the case of "the cube and first power equal to the number, generally," the cubic equation solved by del Ferro and Tartaglia. "Cube one-third the coefficient of x," he began,

> add to it the square of one-half the constant of the equation; and take the square root of the whole. You will duplicate this, and to one of the two you add one-half the number you have already squared and from the other you subtract one-half the same. You

9. Cardano, *The Great Art*, 10, 15, 28–29, 33–47. Witmer has substituted some modern terminology for Cardano's, e.g., translating *res* (thing) as the "first power" (preface to *The Great Art*, xxii).

10. Ibid., 36–39 (quotation from p. 36).

will then have a *binomium* [a sum of two terms] and its *apotome* [the difference of the terms]. Then, subtracting the cube root of the *apotome* from the cube root of the *binomium*, the remainder [or] that which is left is the value of *x*.

So went chapters 11 through 23, as Cardano stated largely rhetorically (that is, in prose interspersed with some abbreviations and numerals) algorithms for the solutions of thirteen types of cubic equations.[11]

Although awkward and prolix from a modern perspective, Cardano's theory of cubic and quartic equations was extremely impressive in its own time. Defying complete solution for so long, the cubic and quartic equations had come to seem rather intractable. In *The Great Art,* Cardano claimed that even the fine algebraist Luca Pacioli (ca. 1445–1517) had intimated that the solution of some cubic and quartic equations was impossible. Moreover, as he announced the work of del Ferro and Tartaglia, Cardano wrote: "this art surpasses all human subtlety and the perspicuity of mortal talent and is a truly celestial gift and a very clear test of the capacity of men's minds."[12]

2

Although Cardano was obliged to share credit with others for the solutions of the quadratic and cubic equations offered in *The Great Art,* he claimed the bulk of the credit for demonstrating these solutions. In chapter 1 he wrote that: "The demonstrations, except for the three by Mahomet [that is, al-Khwārizmī] and the two by Ludovico [Ferrari], are all mine." In including demonstrations, Cardano evidenced his intent that algebra be more than a collection of rules or algorithms for solving equations. He thereby fostered a commitment to the theoretical as well as practical development of algebra, which would be shared by the best algebraists of early modern Europe. As he put it: "Now it is meet that we should show these very wonderful things by a demonstration . . . , so that, beyond mere experimental knowledge, reasoning may reinforce belief in them." "To know by demonstration is to understand."[13]

But how did mathematicians get to the "reasons" behind algebraic

11. Ibid., 98–99 (quotation from p. 99). Earlier chapters covered simpler cubic equations, such as $x^3 = a$ and $x^3 = x$ (ibid., 11, 28).

12. Ibid., 8. In his *Summa de arithmetica* of 1494, Pacioli stated only that certain cubic and quartic equations had so far defied solution (Franci and Toti Rigatelli, "Towards a History of Algebra," 64–65).

13. Cardano, *The Great Art,* 9, 20, 28.

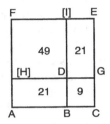

Figure 1. Cardano's geometric demonstration of the solution of $x^2 + 6x = 91$ (reproduced, by permission, from Girolamo Cardano, *The Great Art or The Rules of Algebra,* trans. and ed. T. Richard Witmer [Cambridge, Mass.: MIT Press, 1968], 33). ©1968 by The Massachusetts Institute of Technology.

rules? Cardano was dissatisfied with the argument that roots obtained through application of a particular algorithm to a particular equation could be shown through substitution to satisfy the equation. Following al-Khwārizmī, he offered geometric demonstrations leading to the algorithms for solutions of quadratic and cubic equations.[14] For example, chapter 5 began with a geometric demonstration that 7 is the root of $x^2 + 6x = 91$. Cardano drew "the square FD" (figure 1) as the square whose side was to be found, and continued: "Then I produce DB and DG, which are 3, one-half of 6, the coefficient of *x,* and complete the square DGBC and then, having produced CG and CB, I perfect the square AFEC, just as in II, 4 of the *Elements.*" He observed in succession that: the area HDAB is $3x$; the area IEDG is $3x$; the two aforementioned areas plus the square FD equal 91 (according to the conditions of the problem); but the square DGBC is 9; "therefore the square of AC is 100 and its side, AC, is 10"; and finally, AB (or *x*) is AC − BC = 10 − 3 = 7. Following this and two other geometric demonstrations, he "formulated" the algorithms for quadratic equations of types (3)–(5).[15]

14.   According to one interpretation, al-Khwārizmī had united "aspects of two previously distinct varieties of algebraic thought, namely, the calculationally oriented Babylonian approach to algebra and Euclid's formal geometrical interpretation of algebra" (Parshall, "The Art of Algebra," 134). For varying opinions on the hypothesis of a Greek geometric algebra, see "Historians Debate Geometrical Algebra," in *The History of Mathematics: A Reader,* ed. John Fauvel and Jeremy Gray (London: Macmillan Education, 1987), 140–147.

15.   Cardano, *The Great Art,* 33–36 (quotations and diagram from p. 33).

However convincing such demonstrations of algebraic results were to earlier mathematicians and, indeed, to Western mathematicians of the sixteenth century, Cardano himself admitted to gaps in the geometric defense of algebra. Most blatantly, there were no geometric counterparts of quartic powers, and therefore no geometric demonstrations for solutions of irreducible quartic equations – the very solutions that ranked among the gems of *The Great Art*. As he wrote:

> For as *positio* [the first power] refers to a line, *quadratum* [the square] to a surface, and *cubum* [the cube] to a solid body, it would be very foolish for us to go beyond this point. Nature does not permit it. Thus, it will be seen, all those matters up to and including the cubic are fully demonstrated, but the others which we will add, either by necessity or out of curiosity, we do not go beyond barely setting out.[16]

In short, Cardano set the high standard of a demonstration for every major algebraic result. But, here as elsewhere in *The Great Art*, he exhibited the prescience (or was it just daring?) that seemed to be a special attribute of his genius as he exempted parts of his work from adherence to his own standards. "Necessity or . . . curiosity" compelled him to sketch the theory of quartic equations, even if it was not yet justified.

3

Besides extending the power of mathematics over hitherto unsolved equations and trying to preserve the theory of algebra as he did so, Cardano initiated a fundamental change in Western algebra by expanding the universe of objects accepted as legitimate algebraic entities. Through the fifteenth century algebra was primarily the study of equations with whole number or fractional solutions. The ancient Egyptians, for example, had sought solutions to such practical problems as the division of a certain number of loaves of bread among a certain number of workers. Whole numbers as well as fractions constituted the universe of acceptable solutions for such everyday problems. But, at various points, mathematicians had found that some equations present solutions lying outside this universe. Even some simple quadratic equations (such as $x^2 + x = 2$) produce negative roots ($x = -2$); others (such as $x^2 + 4 = 0$) produce imaginary roots ($x = +2\sqrt{-1}$ and $x = -2\sqrt{-1}$). Nevertheless, early Western algebraists largely ignored negative and

16. Ibid., 9.

imaginary numbers. They did not use negative coefficients, generally avoided equations with a negative root, or, if they did consider such equations, cited only their positive roots.

From the thirteenth century on, however, a few Western mathematicians experimented with negative numbers. According to Kurt Vogel, among these pioneering algebraists was Leonardo of Pisa, who calculated with negative numbers in his *Liber abbaci* and gave some negative roots in his *Flos,* which survives in a manuscript of 1225. In his manuscripts, Nicolas Chuquet (ca. 1440–1487/1488) employed negative coefficients and exponents, and admitted a negative root.[17] The increased willingness of European mathematicians to consider negative numbers was perhaps linked to an appreciation of the newfound practical usefulness of these numbers. By the fourteenth century Italian merchants were using double entry bookkeeping, where numbers on one page (or in one column) stood for assets and numbers on the other page (or in the other column), for debts. Correspondingly, early modern algebra textbooks, including *The Great Art,* used practical examples involving debts to illustrate negative roots.

Cardano was one of the most consistent of the early supporters of the negative numbers. He not only divided numbers into two kinds, positive and negative, and regularly gave the negative solutions of equations, but even tentatively explored the consequences of working with imaginary roots, which he considered "sophistic" kinds of negative numbers. In chapter 1 of *The Great Art,* he wrote: "[It will be remembered also that] 9 is derivable equally from 3 and − 3, since a minus times a minus produces a plus." He then generalized that there are always two roots (one negative and the other, positive) of equations of the form $x^{2n} = a$, where $n = 1, 2, 3, \ldots$ and a is a positive rational number. Yet, showing how difficult it was, even for a bold adventurer like himself, to move beyond the traditional universe of algebra (that is, the universe of whole and fractional numbers), Cardano referred to a positive root as a "true

17. Kurt Vogel, "Leonardo Fibonacci, or Leonardo of Pisa," *Dictionary of Scientific Biography,* ed. Charles C. Gillispie, 18 vols. (New York: Scribner's Sons, 1970–1990), 4:604–613, on 607, 610; Graham Flegg, "Nicolas Chuquet – An Introduction," in *Mathematics from Manuscript to Print,* ed. Hay, 59–72, on 63. Although more detailed research is needed on the history of negative numbers and roots, the standard histories of mathematics conclude that, unlike Western mathematicians, early Chinese and Indian mathematicians accepted negative numbers but not necessarily negative roots. Brahmagupta (fl. 628) gave negative and positive roots of quadratic equations (see, e.g., Carl Boyer, *A History of Mathematics* [New York: John Wiley, 1968], 223, 242, 252).

solution" and a negative root as "a fictitious one (for such we call that which is *debitum* or negative)."[18]

In chapter 37 (On the Rule for Postulating a Negative), he elaborated on negative roots, covering those cases in which a mathematician "assumes a negative [that is, a simple negative root], or seeks a negative square root [an imaginary root], or seeks what is not." The least troublesome case, from Cardano's perspective, was the first. Here negatives were "solutions for problems which can at least be verified in the positive." He meant, in modern terms, that a negative root of an equation could be shown to correspond to a positive (or "true") root of a second equation, which differed from the first equation only in the signs attached to the coefficients of the odd powers of the unknown. Lacking negative coefficients and the general form of equations, Cardano made the latter point through specific examples. Here he cited:

$$x^2 = 4x + 32,$$

the positive root of which is $x = 8$. But he then considered

$$x^2 + 4x = 32,$$

with the positive root $x = 4$. If $+4$ solves the latter equation, he wrote, "-4 is the solution for" the former. "Hence you change the equations. . . . If it is a true [problem] with a positive solution in one, it will be a true [problem] with a negative solution in the other."[19]

There followed three practical problems involving debts. The first was: "The dowry of Francis' wife is 100 *aurei* more than Francis' own property, and the square of the dowry is 400 more than the square of his property. Find the dowry and the property." The solution is that the groom is in debt ("has -48 *aurei*, without any capital or property") and the bride will bring 52 *aurei* to the marriage. In modern terms, the solution would proceed as follows. Let x denote Francis's property; then the dowry is $x + 100$, and $(x + 100)^2 = x^2 + 400$; and

$$x^2 + 200x + 10{,}000 = x^2 + 400 \qquad \text{(A)}$$
$$200x = -9{,}600$$
$$x = -48.$$

The solution offered in *The Great Art* however differed in a significant way. Clearly already knowing the answer, Cardano translated the problem into a prose equation that has a positive solution. To do this, he "assume[d] that Francis has $-x$." The assumption led to the equation

18. Cardano, *The Great Art*, 9–11.   19. Ibid., 217–218.

$$x^2 + 400 + 200x = 10{,}000 + x^2 \qquad \text{(B)}$$

(written in prose of course), with the solution $x = 48$.[20]

And so Cardano indirectly made it clear that negative roots did not enjoy quite the same status as positive ones; they were dependent on "true problems" (ones with positive roots) for their verification. Thus, his route to solution of the problem of Francis's property centered around the somewhat artificially constructed equation (B). Since the problem described by (B) was true, the problem described by (A) was also true – or, in other words, the negative solution of -48 was verified.

In his discussions of the second case of "negatives," that of the "negative square root," Cardano evidenced more hesitation. Early in *The Great Art,* he conservatively dismissed equations with complex roots as "false," and in chapter 37 he recognized such roots only as "sophistic." The earlier allusion to imaginary roots came in chapter 5 where he discussed the solution of the quadratic equation of type (5), that is, $ax = x^2 + b$:

> If the first power is equal to the square and number, multiply as before one-half the coefficient of the first power by itself and, having subtracted the number from the product, subtract the root of the remainder from one-half the coefficient of the first power or add the two of them, and the value of x will be both the sum and the difference.

In modern terminology,

$$x = \frac{a}{2} \pm \sqrt{\left(\frac{a}{2}\right)^2 - b},$$

where x is real only for $(a/2)^2 \geq b$. Temporarily choosing to stay within the traditional universe of algebra, Cardano warned: "If the number [b] cannot be subtracted from the square of one-half the coefficient of the first power [$(a/2)^2$], the problem is itself a false one and that which has been proposed cannot be." "It must always be observed as a general rule . . . ," he added, "that, when those things which have been directed cannot be carried out, that which is proposed is not and cannot be."[21]

In chapter 37, however, he defied the latter "general rule," and proceeded to solve an equation of the preceding type involving an "impossible" subtraction. Thus rule II of chapter 37 stated:

> The second species of negative assumption involves the square root of a negative. I will give an example: If it should be said, Divide 10 into two parts the product of which is 30 or 40, it is clear that this case is impossible. Nevertheless, we will work thus: We divide 10 into two equal parts, making each 5. These we square, making 25.

20. Ibid., 218.   21. Ibid., 38–39.

Subtract 40, if you will, from the 25 thus produced, as I showed you in the chapter on operations in the sixth [*sic*] book, leaving a remainder of -15, the square root of which added to or subtracted from 5 gives parts the product of which is 40. These will be $5 + \sqrt{-15}$ and $5 - \sqrt{-15}$.

"It is clear that this case is impossible." Here Cardano first merely reiterated the obvious: the problem has no solution in the traditional universe of algebra. But then he, ever the algebraic explorer, pressed on into the uncharted territory of the imaginaries. Ignoring his earlier proscription, he solved the equation $10x = x^2 + 40$ by using the formula given in chapter 5:

$$x = \frac{10}{2} \pm \sqrt{\left(\frac{10}{2}\right)^2 - 40} = 5 \pm \sqrt{25 - 40} = 5 \pm \sqrt{-15}.^{22}$$

Still he remained uneasy as he faced the dilemma of the imaginaries: these numbers followed the rules of algebra but seemed to defy any kind of rational explanation. The roots $5 + \sqrt{-15}$ and $5 - \sqrt{-15}$ arose naturally from the algorithm of chapter 5 for equations of the type $ax = x^2 + b$, and both roots satisfied the equation posed in rule II of chapter 37. But an imaginary number remained an arithmetic absurdity. As Cardano noted in *Ars magna arithmeticae*: "$\sqrt{9}$ is either $+3$ or -3, for a plus [times a plus] or a minus times a minus yields a plus. Therefore $\sqrt{-9}$ is neither $+3$ nor -3 but is some recondite third sort of thing."[23] The imaginaries seemed also to defy geometric representation. Cardano's geometric demonstration of rule II of chapter 37 broke down in the middle as he admitted that the reader would "have to imagine $\sqrt{-15}$." At this point, he urged readers to put "aside the mental tortures involved" and note that the complex roots did indeed solve the given problem. At the conclusion of the demonstration, he remarked: "So progresses arithmetic subtlety the end of which, as is said, is as refined as it is useless." Finally, toward the end of chapter 37, he tellingly described the imaginary number as a "sophistic negative" – a term capturing the dilemma of the new number which was correct in algebraic form but still perhaps invalid.[24]

4

Behind Cardano's tentative exploration of operations with imaginary quantities there lay perhaps more than an adventurous spirit that was

22. Ibid., 219.   23. Witmer makes this point (*The Great Art*, 220 n. 6).
24. Cardano, *The Great Art*, 219–221.

willing to follow the prose algorithms wherever they led, even to imaginary roots. The solution of the cubic equation placed the imaginary numbers in a new light, which both dazzled and confused mathematicians. The solution of some cubic equations, early modern mathematicians realized, involves imaginary numbers that cancel out on the way to finding the equations' roots, which are in turn all real numbers. Put dramatically, from that which was seen as recondite and sophistic came something real. Such is the situation with some equations of the form $x^3 = ax + b$, including $x^3 = 15x + 4$. The latter equation has, for example, a root

$$x = \sqrt[3]{2 + 11\sqrt{-1}} + \sqrt[3]{2 - 11\sqrt{-1}}$$

that can be reduced to $2 + \sqrt{-1} + 2 - \sqrt{-1} = 4$.

In chapter 12, where he gave the rule for "the cube equal to the first power and number" – an important type of cubic equation since, through substitution, all cubic equations can be reduced to this type[25] – Cardano alluded to, but did not pursue, the special case involving imaginary numbers. His algorithm began with a call for the subtraction of the cube of one-third the coefficient of the first power from the square of one-half the number, followed by the taking of the square root of the result – or, in succinct modern terms,

$$\sqrt{\left(\frac{b}{2}\right)^2 - \left(\frac{a}{3}\right)^3}.$$

His final formula was:

$$x = \sqrt[3]{\frac{b}{2} + \sqrt{\left(\frac{b}{2}\right)^2 - \left(\frac{a}{3}\right)^3}} + \sqrt[3]{\frac{b}{2} - \sqrt{\left(\frac{b}{2}\right)^2 - \left(\frac{a}{3}\right)^3}}.$$

But this algorithm was not a truly general one, for Cardano seemed to exclude the case where "the cube of one-third the coefficient of x is greater than the square of one-half the constant of the equation." This is the case (later called "irreducible") in which the equation has three real roots, but roots that cannot be found by Cardano's formula without introducing imaginary expressions. "The solution of this [case]," he tersely suggested in the original edition of *The Great Art,* "can be found by the aliza problem which is discussed in the book of geometrical

25.   In the general cubic equation, $x^3 + a_1x^2 + a_2x + a_3 = 0$, substitute $x = y - 1/3a_1$. For the mathematical details and a brief history of the cubic equation, see H. W. Turnbull, *Theory of Equations* (Edinburgh: Oliver and Boyd, 1939), 117–124.

problems." In the second edition he referred his readers to "the Aliza book," which he now appended to *The Great Art*.[26]

As in rule II of chapter 37 Cardano defied his earlier proscription against calculating imaginary roots of quadratic equations, so in chapter 13 he proved unable to restrain himself from giving a root of an irreducible cubic equation. Without any explanation, he wrote: "Solving $y^3 = 8y + 3$, according to the preceding rule, I obtain 3."[27] Attentive readers of the first edition of *The Great Art* may have been frustrated with the application of this rule to the very case Cardano had excepted. But they were probably fascinated with a real solution to an equation for which the general rule led to imaginary expressions. No one could deny that 3 satisfied the given equation, since $3^3 = 8(3) + 3$. What Cardano had implied was that a trip through the realm of the imaginaries was a legitimate, perhaps algorithmically necessary, path to a real solution of a certain cubic equation. In leading to real numbers, the imaginaries became all the more mysterious as they became the more useful.

Whereas Cardano seemed almost to hide his solution to an irreducible cubic in a single sentence of *The Great Art*, his countryman Rafael Bombelli (1526–1572) soon tackled the irreducible case head-on. An engineer-architect and probably a nonuniversity man, Bombelli praised Cardano's algebraic work even as he complained about the obscurity of some of his explanations.[28] If uneasiness about the "sophistic negatives" lay behind the inconsistency of Cardano's restricting certain rules to the real cases and later applying them to cases involving imaginaries, Bombelli somehow managed to throw off such inhibitions. Writing of his direct and successful assault on the irreducible case of the cubic equation, he explained: "It was a wild thought in the judgment of many; and I too for a long time was of the same opinion. The whole matter seemed to rest on sophistry rather than on truth. Yet I sought so long, until I actually proved this to be the case."[29] Indeed, in chapter 2 of his *Algebra* of 1572, Bombelli took the equation

26. Cardano, *The Great Art*, 103. See Witmer's note 5 for information on the edition of 1570.
27. Ibid., 106.
28. S. A. Jayawardene, "Unpublished Documents Relating to Rafael Bombelli in the Archives of Bologna," *Isis* 54 (1963): 391–395; idem, "Rafael Bombelli, Engineer-Architect: Some Unpublished Documents of the Apostolic Camera," *Isis* 56 (1965): 298–306; idem, "The Influence of Practical Arithmetic on the *Algebra* of Rafael Bombelli," *Isis* 64 (1973): 510–523; and idem, "Rafael Bombelli," *Dictionary of Scientific Biography*, 2:279–281.
29. Quoted in David M. Burton, *The History of Mathematics: An Introduction* (Boston: Allyn and Bacon, 1985), 314.

$$x^3 = 15x + 4,$$

brashly applied Cardano's general rule, and thus found

$$x = \sqrt[3]{2 + \sqrt{-121}} + \sqrt[3]{2 - \sqrt{-121}}.$$

Although labeling the imaginary roots "sophistic," he observed that, despite these sophistic roots, the given equation was not impossible, since it had a real root, $x = 4$. He went on to find the cube roots of the complex numbers $2 + \sqrt{-121}$ and $2 - \sqrt{-121}$, and in the process began elaborating algorithms for the imaginary numbers. Assuming that

$$\sqrt[3]{2 + \sqrt{-121}} \text{ and } \sqrt[3]{2 - \sqrt{-121}}$$

are complex numbers, he reasoned at length to the conclusions:

$$\sqrt[3]{2 + \sqrt{-121}} = 2 + \sqrt{-1}$$

and

$$\sqrt[3]{2 - \sqrt{-121}} = 2 - \sqrt{-1},$$

and thus $x = 4$.[30]

5

Despite Cardano's endorsement of negative roots, his mention of a pair of complex roots, and Bombelli's explicit solution of the irreducible case of the cubic, Western algebraists did not immediately or, over the next century, unequivocally embrace the expanding algebraic universe. Written in the vernacular, Bombelli's *Algebra* remained in the shadow of *The Great Art*. And Cardano himself championed the expanding universe only with reservations: after all, in *The Great Art* he referred to negative numbers as "fictitious," appealed to equations with positive roots to verify those with negative roots, described imaginary roots as "sophistic," and excluded the irreducible case of the cubic equation from the general rule of chapter 12.

Furthermore, as shown by R. C. H. Tanner, he hinted at deep concerns about the negative and imaginary numbers in his *De aliza regula liber* ("the Aliza book"), which appeared with *The Great Art* in 1570. *De aliza* revealed a Cardano who was so troubled by the imaginary numbers

30.   For a more detailed explication of Bombelli's calculations, see B. L. van der Waerden, *A History of Algebra: From al-Khwārizmī to Emmy Noether* (Berlin: Springer-Verlag, 1980), 60–62.

that he considered abandoning the traditional rule of signs to resolve the problem and, in doing so, he implicitly recognized the conventional element of mathematics.[31] Although *De aliza* began with the standard rules of algebra, chapter 22 was entitled "On the contemplation of plus and minus, and that *minus into minus makes minus*. . . ." The statement "minus into [times] minus makes minus" contradicted the standard rule of signs, for which the period's mathematicians had a proof, and a geometric one at that. In chapter 22, Cardano demolished the proof, assumed his new version of the rule ("minus into minus makes minus"), and pursued its consequences.[32]

In presenting the new rule, he epitomized early modern mathematicians struggling to make sense of an algebraic universe that was expanding in strange directions and seemingly by forces beyond their control. He showed that he was unable to put either the negatives or the imaginaries into the same intellectual category as the whole numbers. "It is necessary," he confessed, "to lay down that it [minus] is as it were not of the same kind as plus; it is in fact something alien." Following this line of reasoning, he maintained that, whereas plus multiplied by plus gives plus, "minus into minus, that is alien into alien, and minus into plus, or plus into minus, that is that which exists into the alien thing, or the alien thing into that which exists, all produce only minus, that is the alien thing." There were some new, corresponding rules of division. Plus divided by plus is of course plus, and minus divided by plus is minus, he began. But the remaining cases of division took on new twists, including "plus divided by minus gives nothing." Suppose it gives minus; then minus times minus makes plus – a contradiction of the new rule of signs. Neither can it give plus, for that too leads to a contradiction. Minus divided by minus gives both plus and minus, since minus times minus, as well as minus times plus, makes minus.[33]

Alluding to the purpose behind the new rule of signs was a brief paragraph, speaking of "the true nature of multiplication by minus and division by minus and of taking a root, both square and cubic (for as regards the cube root there is no doubt, it is minus), things hitherto

---

31. My discussion of *De aliza* is based on R. C. H. Tanner, "The Alien Realm of the Minus: Deviatory Mathematics in Cardano's Writings," *Annals of Science* 37 (1980): 159–178. The word "aliza" is possibly "a faulty transcription of an Arabic word which combines the meanings of risky, doubtful, burdensome" (162).

32. Tanner, "Alien Realm of the Minus," 165–166.

33. All quotations are taken from Tanner, who translated the original Latin into English ("Alien Realm of the Minus," 166–167).

not known." In short, Cardano seems to have changed the rules of multiplication in order to accommodate imaginary numbers. Given his new rules, the square root of a negative number was a negative. Thus numbers such as $\sqrt{-4}$ were no longer "sophistic"; $\sqrt{-4}$ was simply -2. Cardano had thus managed "to elude the imaginary in the square root of a negative number."[34]

On the expanding universe of algebra, as on other topics, Cardano eschewed consistency in order to share his evolving thoughts with his readers. As his ideas about the negative and imaginary numbers changed, he recorded and published them, without revising his earlier statements on these numbers.[35] His algebraic publications document growing concern about the expanding algebraic universe as he aged and perhaps reflected more on the topic. In *The Great Art,* he had regularly used negative roots and even briefly considered imaginary roots, but with reservations. Yet, in *De aliza* he wrote of the minus sign as "something alien" and, in his determination to rid algebra of the "sophistic" imaginaries, even rewrote the rule of signs. But from a more positive perspective, in *De aliza* Cardano showed himself truly committed to the power and generality associated with algebra's expanding universe. Indeed, Tanner has argued that he rewrote the rule of signs precisely "to retain all the new techniques of algebra that brought in not only 'minus' quantities but also their square roots, and to escape the 'impossible' status of these last."[36] In his own way, Cardano spoke for coming generations of algebraists who would find that they could not work without negative and imaginary numbers, but that working with them was philosophically confusing and somewhat insidious to the established canons of mathematics.

Still, the actual influence of *De aliza* would be limited. Through the nineteenth century few commentators mentioned the work and some misinterpreted it. Thomas Harriot, one of the two original English symbolical algebraists of the seventeenth century, understood and possibly built on *De aliza.*[37] In his *History of Algebra* of 1812, on the other hand, Charles Hutton simply corrected the title of Cardano's chapter 22 to read "minus times minus makes plus."[38] Most later commentators, in fact, knew Cardano the algebraist exclusively through *The Great Art* and thus as an oversimplified algebraic hero who had been among the

34. Ibid.
35. Fierz speculates that Cardano purposely chose to so document his evolving ideas (Fierz, *Girolamo Cardano,* 57).
36. Tanner, "Alien Realm of the Minus," 159 (abstract).
37. Ibid., 164–165.   38. Ibid., 163.

first mathematicians to support the negative and, to a certain extent, the imaginary numbers. In short, for later generations Cardano's was a story of algebraic novelty, stripped of much of the philosophical questioning that had in fact surrounded his experiments with an expanding algebraic universe. Viewed retrospectively, Cardano as author of both *The Great Art* and *De aliza* – as an uneasy proponent of the negatives and imaginaries, who began to tamper with existing mathematical canons to accommodate these numbers – would have been a more appropriate model for algebraists through the late eighteenth century.

6

Whereas Cardano's algebraic novelty concerned new results and the expanding universe of algebraic objects, François Viète's concerned the language and methods of algebra. In his *Introduction to the Analytic Art* (*Isagoge in artem analyticem*)[39] of 1591, Viète substituted arbitrary symbols for the words and abbreviations that had formed the language of earlier algebra, and he also elaborated a theory of algebra as an art based on the method of analysis.

Viète's new symbolism came after centuries of sporadic experimentation with the language of algebra. The earliest algebra, which historians of mathematics have described as "rhetorical," was algebra expressed in complete prose statements. With the work of Diophantus of Alexandria of the third century, however, algebra began the passage from the rhetorical to the "syncopated" stage. Tiring of multiple references to the "unknown," the "square," the "cube," and so on, Diophantus took the major step of developing a shorthand for common algebraic quantities and even for the operation of subtraction. For example, he wrote the abbreviation ς (probably derived from a combination of the first two letters of the Greek word *arithmos* for "number") for the unknown; the abbreviation Δ^Y (the first two letters of the Greek word *dynamis* for

39. Most references to this work are taken from François Viète, *The Analytic Art: Nine Studies in Algebra, Geometry and Trigonometry . . . ,* trans. T. Richard Witmer (Kent, Ohio: Kent State University Press, 1983). The "Translator's Introduction" gives bibliographical details on the *Introduction to the Analytic Art* and the other eight works included in this translation. A few references are taken from J. Winfree Smith's translation of the *Introduction to the Analytic Art,* appended to Jacob Klein, *Greek Mathematical Thought and the Origin of Algebra,* trans. Eva Braun (Cambridge, Mass.: MIT Press, 1968), 315–353.

"power") for the square of the unknown; and the abbreviation K^Y (the first two letters of the Greek word *kybos* for "cube") for the cube of the unknown.

Diophantus's *Arithmetic*, in which he used the syncopated algebra, was among those works lost to western Europe in the early Middle Ages. Before its recovery at the end of the sixteenth century, Western mathematicians invented some algebraic abbreviations of their own. Some fourteenth-century *maestri d'abbaco* wrote "*co*," "*ce*," and "*cu*" for *cosa* (thing), *censo* (square), and *cubo* (cube).[40] Later, Cardano used abbreviations, including "*p*" (plus), "*m*" (minus), and "℞" (the square root of). For example, he checked his solution to the imaginary problem of chapter 37 of *The Great Art* as follows:

$$5p:℞m:15$$
$$5m:℞m:15$$
$$\overline{25m:m:15 \ \bar{q}d \ est \ 40.}^{41}$$

Viète, however, took a bolder step in his *Analytic Art* and introduced the new "symbolical" algebra. Symbolical algebra, which was more abstract than syncopated, consisted of symbols with little or no connection with the entities they represented. Whereas Diophantus provided an abbreviation for a single unknown, Viète used the vowels *A, E, I, O,* and *U* for multiple unknowns. These symbols were arbitrary, chosen from the alphabet without concern for resemblance to the quantities for which they were to stand or the words used to express those quantities. Underlining this arbitrariness, Viète remarked almost casually that "symbolic logistic . . . employs symbols or signs for things as, say, the letters of the alphabet."[42]

Although the initiator of the symbolical style, Viète was not its final arbiter, not even in his own period. His "analytic art" offered a hybrid of the syncopated and symbolical styles rather than a purely symbolical algebra. For example, he used abbreviations to denote powers. He wrote the square of the unknown variously as *A quadratum, A quad,* or, in the extreme, *Aq.* A later generation of algebraists would develop the more consistently symbolical notation for powers that is used in modern algebra. For example, the English mathematician Thomas Harriot would let *aa* stand for the square of the unknown, *aaa* for the cube, and so on;

40.   Franci and Toti Rigatelli, "Fourteenth-Century Italian Algebra," 15.
41.   Florian Cajori, *A History of Mathematical Notations*, 2 vols. (Chicago: Open Court, 1929), 2:126.
42.   Viète, *The Analytic Art*, 17.

in 1637 Descartes would introduce the nearly modern notation for exponents, writing xx, x^3, and so on.

7

More important, Viète himself elaborated a theory of the symbolical logistic. That logistic not only changed the language of algebra but facilitated its generality as well, since, in addition to denoting unknowns or variables by vowels, he used consonants to denote known quantities or parameters.[43] He thus made it "possible to build up a general theory of equations – to study [for example] not cubic equations, but *the* cubic equation."[44] Whereas Cardano had built his theory of equations on careful studies of specific equations with numeral coefficients, the new symbolism enabled Viète and his successors to work directly with general equations. An early problem in Viète's *Five Books of Zetetica* gave a simple demonstration of the new kind of general, economical argument made possible by literal coefficients. "Given the difference between two roots and their sum," Viète proposed, "to find the roots." He let B (a consonant) stand for the difference between the two roots; D (another consonant) for the sum of the roots; and A (a vowel) for the unknown or the smaller root. The greater root, he continued, is $A + B$; the sum of the roots is $2A + B = D$, and, by transposition, $2A = D - B$, or $A = \frac{1}{2}D - \frac{1}{2}B$.[45] Thus, without looking at particular cases, Viète had arrived at a formula for determining any two numbers with a given sum and a given difference.

Symbolical notation was not essential to the solution of the preceding problem. Mathematicians of Cardano's generation had found complete solutions for equations of much greater complexity. These earlier mathematicians would have pursued either of two nonsymbolical paths to solution of the given problem. First, a general, rhetorical solution could have been constructed.[46] But the rhetorical procedure would have been

43. Ibid., 24.
44. Carl B. Boyer, *History of Analytic Geometry* (New York: Scripta Mathematica, 1956), 59–60.
45. Viète, *Five Books of Zetetica,* in *The Analytic Art,* trans. Witmer, 83–84. The *Five Books* was published in 1591 or 1593.
46. A rhetorical argument would go as follows: the smaller root plus the greater root is equal to the sum. But the greater root equals the smaller root plus their difference. Therefore, the smaller root taken twice plus their

longer than the symbolical, and perhaps more mentally taxing. Alternatively, a mathematician could have solved specific cases of the given problem and then tried to generalize. Viète's message, however, was that mathematicians, armed with literal coefficients, could prove a general result first and then apply it to relevant cases. Viète followed his solution of the given problem with a specific example where the difference is 40 and the sum, 100. Given the general formula, he quickly found the smaller root, 30, and the larger root, 70.[47]

Trying to work from specific cases to the general result – that is, in a direction opposite to Viète's – illustrates the power of literal coefficients. Suppose that he first solved the specific case of the given problem where the difference is 40 and the sum, 100. In doing so, he would have let A be the smaller root. Then $A + 40$ equals the greater root. Since the sum of the two roots is 100, $2A + 40$ equals 100, or $2A$ equals $100 - 40$, or A equals $\frac{1}{2}(100) - \frac{1}{2}(40)$, and thus A equals $50 - 20$, or 30. But then the numeral coefficients (40 and 100) have vanished by the final solution, leaving Viète with a veiled solution to the general problem of any sum and any difference. In his symbolical solution, on the contrary, the literal coefficients persist to the end and thus express the general solution.

The symbolical logistic laid the basis for early modern equation theory not only by facilitating general solutions but also by directing attention to the very structure of equations. The new symbolism seems to have focused the mathematician's eyes and mind on the relationships existing between the coefficients and the roots of an equation. Viète himself wrote of the structure (*constitutio*) of equations and asked rhetorically: "Will an analyst attempt [the solution of] any proposed equation without understanding how it is composed so that he can avoid the rocks and crags?" As his first theorem of the *Two Treatises on the Understanding and Amendment of Equations,* which centered on the structure of equations, he gave: "If $A^2 + AB = Z^2$, there are three proportionals the mean of which is Z and the difference between the extremes is B, making A the smaller extreme."[48] (Expressed in modern terms, for the equation $x^2 + bx = c,$ there is a relationship between its roots and its coefficients,

difference equals the sum, and the smaller root taken twice equals the sum minus the difference. Thus, the smaller root equals one-half the sum minus one-half the difference.

47. Viète, *Five Books,* 84.
48. Viète, *Two Treatises on the Understanding and Amendment of Equations,* in *The Analytic Art,* trans. Witmer, 160–162. These treatises were originally published in 1615.

in particular, $x + b : \sqrt{c} :: \sqrt{c} : x$.) Although the preceding and like theorems were not new major algebraic results, they did represent an important "shift from direct problem-solving to the theoretical investigation of the structure of equations." The "art of the coss" soon became the "doctrine of equations."[49]

8

Viète also sketched a new approach to the justification of algebra. He abandoned the Arabic name "algebra," and wrote instead of the "analytic art." A new name and "an entirely new form" for algebra were essential, as he vividly wrote, "lest it [algebra] should retain its filth and continue to stink in the old way."[50] Under his new approach, algebra was no longer dependent on geometry for confirmation. Rather algebra was subsumed under the analytic art, which had a method and justification of its own. The analytic method was Viète's primary concern; it gave his major work its title, and it was the very first thing to which he referred in that work: "There is a certain way of searching for the truth in mathematics that Plato is said first to have discovered. Theon called it analysis."[51]

The Analytic Art opened with references to classical mathematicians, and attested to Viète's view of his work as renovation of the ancient mathematics rather than revolution. He was familiar with the classical mathematical tradition, including the writings of Pappus and Diophantus of Alexandria, which had just been rediscovered.[52] Through Pappus he learned of geometric analysis; and in Diophantus he found a wealth of early arithmetic algebra. According to Jacob Klein, Viète's "renovation" involved the conjoining of the ideas on analysis, which Pappus had presented in relation to the theorems and problems of geometry only,

49. Michael Sean Mahoney, *The Mathematical Career of Pierre de Fermat (1601–1665)* (Princeton: Princeton University Press, 1973), 38, 36.
50. Viète, dedication to Princess Mélusine, in Klein, *Greek Mathematical Thought*, appendix, 318.
51. Viète, *The Analytic Art*, 11.
52. According to van Egmond, "Diophantus . . . provided the chief inspiration for Viète's conceptual revolution in algebra, as he tried to fit the work of the classical arithmetician into the context of the contemporary algebraic tradition." Van Egmond suggests that Viète learned of Diophantus either through a Latin translation of the *Arithmetic*, made in 1575, or from the Greek manuscripts. See van Egmond, "How Algebra Came to France," 143.

with Diophantus's procedure of solving arithmetic questions by operating on an unknown as if it were known. Although the analysis of Pappus thus had a "purely geometric character,"[53] Viète now adapted it as the method of a new "analytic art," under which all of algebra from Diophantus's *Arithmetic* through Cardano's *Great Art* was to be subsumed.

Viète defined analysis (in the traditional way of Theon) as "assuming that which is sought as if it were admitted [and working] through the consequences [of that assumption] to what is admittedly true." According to him, mathematicians solved algebraic problems by analysis and not by the rival method of synthesis, or by "assuming what is [already] admitted [and working] through the consequences [of that assumption] to arrive at and to understand that which is sought."[54] For example, when faced with the problem of finding two roots with a given difference and a given sum, Viète assumed the smaller unknown root as if it were known, calling it *A,* and then worked through the consequences of this assumption until a formula for *A* was found. Algebraists, he stressed, reasoned from the unknown to the known, and such reasoning was perfectly acceptable. Some mathematicians erred, he implied, in forgetting that there were two valid mathematical methods, synthesis and analysis.

Viète thus offered Western mathematicians the rudiments of a new foundational approach to algebra, which was so influential that through the nineteenth century some mathematicians would use the terms "algebra" and "analysis" interchangeably while others would write of an analysis that included calculus as well as algebra. The approach suggested the possibility of an algebra on equal footing with, if not superior to, geometry. Unlike Cardano's *Great Art,* Viète's *Analytic Art* offered no geometric diagrams or proofs to confirm algebraic results.[55] When necessary, he explained, analytic results could be "confirmed" by synthesis.[56] But this synthesis was not geometric; it was a symbolical demonstration that the results actually satisfied the given problem. Thus the solutions $A = \frac{1}{2}D - \frac{1}{2}B$ and $E = \frac{1}{2}D + \frac{1}{2}B$ could be shown to satisfy the requirements that they differ by B and sum to D. $E - A = \frac{1}{2}D + \frac{1}{2}B - (\frac{1}{2}D - \frac{1}{2}B) = B$, and so on.

Although thus freeing algebra of dependence on geometry for justifi-

53. Klein, *Greek Mathematical Thought,* 157.
54. Viète, *The Analytic Art,* 11.
55. "Viète does not use ... [geometric] diagrams even as illustrative matter except where he is dealing with triangles" (Witmer, "Translator's Introduction" to *The Analytic Art,* 6). Here Viète followed Diophantus.
56. Viète, *The Analytic Art,* 28.

cation, Viète's elaboration of the analytic art was no clear-cut step toward an independent discipline of algebra. Viète not only distorted the classical notion of analysis, he also muddied existing distinctions between algebra and geometry, and even arithmetic and geometry, and thereby placed a new obstacle in the way of the emergence of an independent early modern algebra. As Michael Mahoney has argued, "algebra and classical Greek geometry represent two substantially different approaches to mathematics and reflect different demands on mathematical knowledge." For the Greeks, analysis was a method of geometry, applied to that subject alone. Viète extended the method of analysis to cover algebra, and made analysis algebraic. That is, he identified algebra and analysis. To do so, he rewrote some of the elements of the classical definition of analysis.[57] Whereas Pappus had delineated two kinds of analysis, zetetic (theorem-proving analysis) and poristic (problem-solving analysis), Viète explained that there were three kinds: (1) "zetetics by which one sets up an equation or proportion between a term that is to be found and the given terms"; (2) "poristics by which the truth of a stated theorem is tested by means of an equation or proportion"; and (3) the new "exegetics [or rhetics] by which the value of the unknown term in a given equation or proportion is determined."[58] In short, equations and equation theory – once the domain of algebra – were now at the very heart of the analytic enterprise. According to Mahoney, with Viète's *Analytic Art* "analysis has become algebra. More accurately stated, analysis has become the analytic art. . . . Viète and his successors found it appropriate to drop the name 'algebra' altogether."[59]

Viète's exclusive use of the term "analytic art" in place of algebra reflected the reality of an art that subsumed not only traditional algebra but arithmetic, geometry, and trigonometry as well. The analytic art was powerful, Viète stressed, precisely because the language used in zetetics was new and that language could stand for magnitude or continuous quantity (traditionally, the subject of geometry) as well as number or discrete quantity (the subject of arithmetic). "Zetetics," he wrote, ". . . no longer limits its reasoning to numbers, a shortcoming of the old analysts, but works with a newly discovered symbolic logistic which is far more fruitful and powerful than numerical logistic for comparing magnitudes with one another." Following the setting up of an equation or proportion (that is, the zetetical part of analysis), rhetics or exegetics

57. A careful discussion of the relationship between algebra and Viète's analytic art appears in Mahoney, *Fermat*, 33–34 (quotation from p. 33).
58. Viète, *The Analytic Art*, 11–12.
59. Mahoney, *Fermat*, 34.

performs its function. It does so both with numbers, if the problem to be solved concerns a term that is to be extracted numerically, and with lengths, surfaces or bodies, if it is a matter of exhibiting a magnitude itself. In the latter case the analyst turns geometer by executing a true construction after having worked out a solution that is analogous to the true. In the former he becomes an arithmetician, solving numerically whatever powers, either pure or affected, are exhibited.[60]

Examples in the *Five Books of Zetetica,* published the same year as *The Analytic Art,* illustrated the cross-disciplinary nature of the symbolical logistic. In Zetetic VII of book II, Viète solved the problem: "Given the difference between the roots and the difference between their squares, the roots will be found." His result was: "If the difference between the squares is divided by the difference between the roots, the quotient is the sum of the roots." (In modern terms, given $x - y = a$ and $x^2 - y^2 = b$, $x + y = b/a$.) Now, from the difference and the sum of the roots, one could easily calculate the roots, as Viète did for the case where the difference between the roots was 8 and the difference between their squares, 96. In book III Viète turned to geometric problems, including Zetetic III: "Given the perpendicular of a right triangle and the difference between the base and the hypotenuse, to find the base and the hypotenuse." Highlighting the power of the symbolical logistic, he noted that the result of Zetetic VII of book II solved the given geometric problem. By the Pythagorean theorem, the square of the perpendicular of a right triangle equals the square of the hypotenuse minus the square of the base. But the perpendicular is given, and so Viète had its square as well. In short, he had the given difference between the hypotenuse and base as well as the difference between their squares, and so he could simply apply Zetetic VII of book II.[61]

With the analytic art now producing solutions whose applicability spanned arithmetic, geometry, and trigonometry, its power seemed limitless to its inventor. In the (much quoted) concluding sentence of *The Analytic Art,* Viète declared that "the analytic art, endowed with its three forms of zetetics, poristics and exegetics, claims for itself the greatest problem of all, which is *To solve every problem.*"[62]

In constructing an art that was supposed to solve every mathematical problem Viète had abandoned the traditional demarcation between arithmetic and algebra, on one hand, and geometry, on the other. The

60. Viète, *The Analytic Art,* 13, 28–29.
61. Viète, *Five Books,* 104, 114–115.
62. Viète, *The Analytic Art,* 32.

analytic art was clearly more than traditional algebra. It was applicable to geometry and trigonometry in ways that perhaps no earlier mathematical practitioner could have dreamed of. This applicability helped assure a strong, continued interest in equations and equation theory. It was this applicability that was soon exploited by René Descartes and Pierre de Fermat, and later by Isaac Newton, in analytic geometry. But, as Descartes's *Géométrie* would show, as algebra became firmly entrenched under the guise of the analytic art, it became more and more entangled with geometry. Analysts, including Descartes and Newton, promoted algebraic solutions for geometric problems, but they also promoted geometric solutions for algebraic equations. To a large extent, then, Viète's legacy was that of an algebra whose ties and applicability to geometry were exploited before the subject's own independent status was ever established.

Furthermore, an algebra liberated from geometric justification through appeal to the analytic method was not necessarily the equal of the subject Euclid had brought to such perfection in the *Elements*. Western mathematical thinkers were influenced by not only Plato but also Aristotle, who had exalted deduction as the reasoning of science. Although (as we shall see) many of Viète's English successors would no longer turn to geometry for confirmation of algebra, some would nevertheless acquiesce in ranking synthesis over analysis and hence Euclidean geometry over algebra. From the perspective of Viète and his French and English followers, however, analysis was at least the equal of synthesis. Viète admitted that synthesis was "rated the most rational method of demonstration" and, in the same paragraph, promoted analysis as "the inventive art." Indeed, analysis produced results that were "confirmed by . . . [synthesis], this being a great miracle" of analysis. But, he added, that the retracing of "the footsteps of analysis" – traditionally labeled synthesis – was "itself [a form of] analysis and, thanks to the introduction of symbolic logistic, . . . no longer difficult."[63] For the next two centuries, Western mathematicians would concern themselves with not only the relationship between algebra and the analytic art but also the relative standings of the analytic art and classical geometry.

9

Along with the theory of the analytic art, Viète enunciated its "rules" or "laws," some of which were presented as self-evident, others of which

63. Ibid., 28.

were demonstrated. It was conformance of mathematics to rules that seemed to attract Viète, a lawyer by vocation, to the subject. In dedicating his *Analytic Art* to Princess Mélusine, Catherine of Parthenay, he explained that "it is not the same in mathematics as in other studies, that everyone's opinion is free and free his judgment. Here things are done by rule and effort, and neither the persuasions of rhetoricians nor the pleadings of lawyers are of use."[64] Like Cardano's call for the justification of algebraic results, Viète's stress on rules helped to establish the ideal of proof in early modern algebra.

In chapter 2 of his *Analytic Art,* Viète listed "the well-known fundamental rules of equations and proportions that are given in . . . [Euclid's] *Elements,*" which he here "accept[ed] . . . as proven."[65] The next chapter stated additional laws for analysis, including the law of homogeneous terms, which evidenced a lingering geometric influence on algebra. According to this law, "homogeneous terms must be compared with homogeneous terms." For example, addition could involve only homogeneous terms: lengths could be added to lengths only, squares to squares, cubes to cubes, and so on.[66] This law compelled him to assure the homogeneity of each term of every equation; thus he wrote the first proposition of *The Analytic Art* as:

$$A \text{ square} - D \text{ plane} = G \text{ square} - B \text{ in } A,$$

in order to indicate that every term of the equation was a square or of the order two.[67]

Chapter 4 presented the basic "rules of symbolic logistic" – that is, of symbolical addition, subtraction, multiplication, and division. The rules were patterned on those for the corresponding arithmetic operations. Viète wrote that "There are four basic rules for symbolic logistic just as there are for numerical logistic," although he gave no particular justification for extending the rules from arithmetic to the analytic art. According to rule I, the symbol + was to be used to indicate addition of

64. Viète, dedication to Princess Mélusine, in Klein, *Greek Mathematical Thought,* appendix, 319.
65. Viète, *The Analytic Art,* 14–15. These included: "If equals are added to equals, the sums are equal."
66. Ibid., 15–16. Whereas most historians see this law simply as a geometric remnant in Viète's algebra, Klein offers a different interpretation (Klein, *Greek Mathematical Thought,* 173–178).
67. Viète, *The Analytic Art,* 25. See also *Introduction to the Analytic Art,* in Klein, *Greek Mathematical Thought,* appendix, 342–343. The modern equation is: $x^2 - d = y^2 - bx$ (ibid., 343 n. 36).

homogeneous terms. Rule II dealt with subtraction, which, he explicitly stated, involved the taking of a smaller homogeneous magnitude from a greater, and was usually indicated by the symbol $-$. More complicated subtractions are, of course, possible, and Viète gave rules for $A - (B + D)$ and $A - (B - D)$. For the first case, the result is $A - B - D$, "the terms B and D having been subtracted individually." For the second, he explained, "the remainder will be A minus B plus D, since in substracting the magnitude B, more than enough, to the extent of D, has been taken away and compensation must therefore be made by adding it." Viète was concerned about questionable subtractions as well as complicated ones. In the former cases, algebraists are not sure which of two magnitudes, say A and B, is the greater, and therefore they do not know if A is to be subtracted from B or vice versa. To avoid an illegitimate subtraction (where the greater is taken from the smaller), he introduced a second symbol, $=$ (which had not yet come to stand for equality), as "the sign of difference" between two unordered magnitudes.[68]

In rule III, he used the words "times" and "by" to indicate multiplication, stated the distributive property, discussed the rule of signs governing multiplication, and gave rules for the multiplication of powers. His rule of signs, significantly, concerned terms of polynomial expressions affected by the subtraction sign rather than isolated negatives. In chapter 3, he had explained that a power "is affected when it is associated [by addition or subtraction] with a homogeneous term." Such affection was the source of "positive terms" (terms to be added to others) and "negative terms" (terms to be subtracted from others). Of these terms he wrote in rule III:

> If a positive term of one quantity is multiplied by a positive term of another quantity, the product will be positive and if by a negative the result will be negative. The consequence of this rule is that multiplying a negative by a negative produces a positive, as when $A - B$ is multiplied by $D - G$.

Thus taking the rule "positive times negative gives negative" as a given, he proved "negative times negative gives positive" – the very rule that Cardano had exposed as unproven. Viète noted that $+A$ times $-G$ gives a negative. The latter, however, "takes away or subtracts too much." Similarly, taking $-B$ times $+D$, he arrived at the negative product, "which takes away too much since D is not the exact magnitude to be multiplied." But "the positive product when $-B$ is multiplied by $-G$ makes up for this."[69]

This last part of the rule of signs was not the only rule for which Viète

---

68. Viète, *The Analytic Art*, 18–19.   69. Ibid., 17, 19–20.

offered a somewhat detailed proof. In chapter 5 of *The Analytic Art*
("On the Rules of Zetetics") he demonstrated the rules of transposition,
depression, and reduction – all algebraically and symbolically. Thus he
proved that an equation was not changed by transposition, or "a removal of affecting or affected terms from one side of an equation to the
other with the contrary sign of affection." He took the equation:

$$A^2 - D^p = G^2 - BA.$$

He then added $D^p + BA$ to both sides. "Then by common agreement
[the Euclidean axiom that 'when equals are added to equals, the sums
are equal']," he continued,

$$A^2 - D^p + D^p + BA = G^2 - BA + D^p + BA.$$

"The negative affection on each side of this equation cancels a positive:
on one side the affection D^p vanishes, on the other the affection BA."
He now had the desired result, or

$$A^2 + BA = G^2 + D^p.^{70}$$

10

The statements and proofs of rules II and III of chapter 4 of *The Analytic
Art* highlight key differences between Viète and Cardano, differences
that would separate Western algebraists through the early nineteenth
century. Cardano was basically an algebraic explorer. He was inclined
to let problems dictate the development of his subject. Viète was a more
traditionally philosophical thinker. He wanted to deal with the rules of
the analytic art first, and only then move to its applications.

For Viète rules limited results; for Cardano results could lead to a
broadening of rules. Viète's rules II and III set the boundaries for his
use of "negative terms." Rule II assured that impossible subtractions,
generating negative numbers, would not slip into the analytic art. Rule
III sanctioned negative coefficients and even a rule of signs, but subject
to the limitation that negative terms were terms of polynomials. Correspondingly, he almost completely confined the examples of *The Analytic
Art* to equations with positive roots only, and he avoided any discussion

70. Ibid., 25–26. The rule of depression permitted division of the equation by
the highest power of the unknown contained in each term, and the law of
reduction provided for division by the coefficient of the highest power of
the unknown (ibid., 26–27).

of negative roots.[71] Cardano, on the other hand, worried about negative and, indeed, imaginary numbers but he still used them. These very different responses foreshadowed the twists and turns that would mark the history of algebra's expanding universe into the nineteenth century. To a certain extent, Cardano and Viète themselves served as models for two basic types of later algebraists: algebraic explorers, whose primary concern was with results; and philosophical algebraists, whose foundational qualms sometimes obstructed their full participation in the algebra of their generations.

71. On the last point, see Witmer, "Translator's Introduction" to *The Analytic Art,* 8.

2

William Oughtred and Thomas Harriot

"Inciting, Assisting, and Instructing Others"
in the Analytic Art

Viète's symbolical style so captured the imaginations of the Englishmen William Oughtred and Thomas Harriot that at least the former seems to have taken as the mission of his advanced years the "inciting, assisting, and instructing others" in the analytic art.[1] Both produced textbooks on the subject, which brought early modern algebra, under the guise of the analytic art, to England. Published in 1631, these symbol-laden books struck a responsive chord in an English scientific community that was beginning to favor, among other things, plain prose. By the third quarter of the century, then, there was a growing school of English analytic mathematicians; there were also the beginnings of a related drive toward the acceptance of mathematics, including algebra, as a scholarly pursuit.

Oughtred, Harriot, and their disciples shared first and foremost a deep and irrevocable commitment to symbolical reasoning. As the two English algebraic pioneers fostered Viète's symbolical style, they also perpetuated his hesitancy about the expanding algebraic universe. They and most of their immediate successors reasoned symbolically, wrote algebraic equations for geometric problems, but largely ignored negative and imaginary roots or – in the most daring of cases, including Harriot's speculations in unpublished manuscripts – struggled to make some sense of them. In short, the algebra that was imported into England through Oughtred's and Harriot's textbooks was Viète's, not Cardano's. In England, a deep appreciation of the symbolical style came first; openness to the expanding universe of algebra followed slowly.

---

1. William Oughtred, *To the English Gentrie, and All Others Studious of the Mathematicks* (1632?), 9. (I have numbered the pages of this unpaginated pamphlet.) Although Oughtred's description here relates to his spreading of general mathematical knowledge during his early Cambridge years, there seems no better description of his later efforts on behalf of the analytic art.

I

The first major English proponent of symbolical algebra, William Oughtred (1575–1660), neither learned early modern mathematics as part of his formal university studies nor earned his livelihood through the teaching or application of the subject. Perhaps significantly, Oughtred – like John Wallis and Nicholas Saunderson after him – was first exposed to mathematics in the form of arithmetic, taught to him by a family member. Indeed, Oughtred's father, a teacher of writing at Eton, instructed him in basic arithmetic, and only later did Oughtred study some Greek mathematics as a scholar at King's College, Cambridge University. Moreover, what the young Oughtred learned of early modern mathematics while at Cambridge did not come from his tutors or professors. As he later recounted, "the time which over and above those usuall studies [at Cambridge] I employed upon the Mathematicall sciences, I redeemed night by night from my naturall sleep, defrauding my body, and inuring it to watching, cold, and labour, while most others tooke their rest."[2]

For Oughtred, mathematics mastered was mathematics to be shared. Even in his early days as a fellow of King's College, "by inciting, assisting, and instructing others, [he] brought many into the love and study of those Arts."[3] His mathematical teaching continued throughout his life, although between 1603 and his death he pursued mathematics as an avocation only and earned an ecclesiastical living, serving as the rector of Albury in Surrey for the last fifty years of his life. Flourishing in this environment, he found time for mathematics, pupils, and mathematical writing; according to his son, he did mathematics late into the night, kept writing instruments at his bedside, and slept little. His critics charged him with slighting his clerical obligations, and he was said to have applied himself to theological studies and preparation of sermons only when threatened with sequestration for his staunch royalism during the Civil War.[4]

2. Oughtred, *To the English Gentrie*, 8–9. For an important survey of major developments in basic arithmetic in England during 1500–1700 (developments underlying the period's algebra), see Keith Thomas, "Numeracy in Early Modern England," *Transactions of the Royal Historical Society* 37 (1987): 103–132.
3. Oughtred, *To the English Gentrie*, 9.
4. John Aubrey, *"Brief Lives," Chiefly of Contemporaries, Set Down by John Aubrey, between the Years 1669 & 1696*, ed. Andrew Clark, 2 vols. (Oxford: Clarendon Press, 1898), 2:107, 111.

Oughtred fostered the study of the new mathematics in England through private lessons given in his rectory as well as an extensive correspondence with students of mathematics drawn from both the university community and the mathematical practitioners of his time. Some university men, caught up in the academic shuffle occasioned by the Civil War, turned to Oughtred for private instruction. Numbered among his resident and correspondent pupils of the 1640s and 1650s were such major figures of English science as Seth Ward, Christopher Wren, and John Wallis.[5] Ward had already been elected a fellow of Sidney Sussex College, Cambridge, when in the early 1640s he and Charles Scarburgh, both royalists, went to study at Albury.[6] Although having made some progress in the mathematical sciences (including astronomy) at Cambridge, Ward seems to have learned early modern mathematics from Oughtred, with whom he lived a half year.[7]

Wren and Wallis, on the other hand, were Oughtred's pupils in a more removed way. In the late 1640s Wren, another royalist, resided for a time with Scarburgh, by then an established physician. Perhaps first his patient and later his pupil, Wren was directed by Scarburgh to Oughtred's mathematical writings. Wren wrote to Oughtred and, with Scarburgh's encouragement, prepared a Latin translation of Oughtred's treatise on dialing.[8] Unlike Wren, Wallis, a parliamentarian, seems to have learned of Oughtred directly from his writings, which were said to have first awakened Wallis's mathematical interests. The two men corresponded on mathematical points, and Wallis, in turn, recognized an intellectual debt to Oughtred. In 1654/1655 he sent him an observation of a solar eclipse, which he presented as "a trifle of my own . . . not worthy of your acceptance, yet such as the respect, which I owe you, commands me to present unto you, having nothing better to

5. Florian Cajori, *William Oughtred: A Great Seventeenth-Century Teacher of Mathematics* (Chicago: Open Court, 1916), 58–60.
6. W. W. Rouse Ball, *A History of the Study of Mathematics at Cambridge* (Cambridge: Cambridge University Press, 1889), 37. Oughtred invited the men to Albury after the parliamentarians stripped Ward of his fellowship.
7. Aubrey, *Brief Lives,* 2:108. Ward "learned all his mathematiques of" Oughtred (ibid.). Feingold, however, has stressed Ward's prior exposure to the mathematical sciences at Cambridge (Mordechai Feingold, *The Mathematicians' Apprenticeship: Science, Universities and Society in England, 1560–1640* [Cambridge: Cambridge University Press, 1984], 88–90). Oughtred perhaps taught Ward all his analytic mathematics.
8. J. A. Bennett, *The Mathematical Science of Christopher Wren* (Cambridge: Cambridge University Press, 1982), 16.

tender."[9] In a public gesture of gratitude, he dedicated his *Arithmetica infinitorum* of 1656 to Oughtred.[10]

2

In 1631 the publication of *Clavis mathematicae,* subsequently translated as *The Key of the Mathematicks,*[11] extended Oughtred's influence to students of mathematics throughout England and even some on the continent. This treatise, which Oughtred originally composed for a student, Lord William Howard, the son of the earl of Arundel, has been described as "the most influential mathematical publication in Great Britain which appeared in the interval between John Napier's *Mirifici logarithmorum canonis descriptio,* Edinburgh, 1614, and the time, forty years later, when John Wallis began to publish his important researches at Oxford."[12] *The Key* – which went through five Latin editions and two independent English editions[13] – was essentially a textbook covering arithmetic, algebra, and some geometry as well. It was a textbook whose author made no concessions to weak students. It was written in a somewhat cramped style (the original edition contained eighty-eight small pages) and filled with the new symbols of algebra.

The Key introduced English students to the methods and symbols of

9. Wallis to Oughtred, 5 Feb. 1654/1655, *Correspondence of Scientific Men of the Seventeenth Century,* ed. Stephen Jordan Rigaud, 2 vols. (Oxford, 1841; reprint, Hildesheim: Georg Olms, 1965), 1:79–80. Begun by Stephen Peter Rigaud, the editing of the *Correspondence* was completed by his son, Stephen Jordan Rigaud.

10. Wallis wrote here that: "I thought that I should look not only for a great man, but rather a great mathematician to whom I should inscribe my work." The translation from the Latin is by J. F. Scott, *The Mathematical Work of John Wallis, D.D., F.R.S. (1616–1703)* (London: Taylor and Francis, 1938), 27.

11. William Oughtred, *The Key of the Mathematicks, New Forged and Filed* . . . (London: Thomas Harper, 1647). The *Clavis* was originally published in Latin in 1631; references herein are to the first English edition of 1647, unless otherwise indicated.

12. Cajori, *William Oughtred,* 57.

13. On the various editions, see Florian Cajori, "A List of Oughtred's Mathematical Symbols, with Historical Notes," *University of California Publications in Mathematics* 1 (1920): 171–186, on 171–172; or, idem, *William Oughtred,* 17–19.

the new algebra of the continent as developed principally by Viète.[14] The preface to the first Latin edition declared that the book's concern was not with "the so-called practice [of mathematics], which is in reality mere juggler's tricks with instruments, the surface so to speak, pursued with a disregard of the great art," but rather with the "most beautiful science" of mathematics.[15] Concern for science meant concern for methods – in particular, the analytic method. Like Viète, Oughtred wrote not of algebra but of "the Analyticall art . . . in which by taking the thing sought as knowne, we finde out that we seeke."[16] His *Key* was "not written in the usuall syntheticall manner . . . , but in the inventive way of Analitice."[17]

Indeed, the theme of *The Key* was that mathematics ought to be pursued analytically, which meant for Oughtred as for Viète that mathematical problems, including geometric ones, ought to be translated into symbolical equations and those equations solved algebraically. Whereas Viète had written philosophically of the analytic art, Oughtred tried to bring the art down to the level of young mathematical scholars. According to him, the art would permit those scholars not only to solve new problems, as Viète had emphasized, but also to improve their understanding of the mathematical classics. Thus, from its first introduction into English mathematics, the analytic art was characterized as both forward and backward looking.

As explained in the preface ("To the Reader") to the first English edition of *The Key*, Oughtred originally wrote the book to give Lord Howard "a method of precepts for the more ready attaining, not a superficial notion, but a well-grounded understanding of those mysterious [mathematical] Sciences, and of the ancient Writers thereof." He claimed that he himself had often been puzzled as to how the ancient mathematicians, including Euclid, Archimedes, and Apollonius, had arrived at their results:

> [W]ith wonder [I] observed their most witty demonstrations, so skilfully framed out of principles, as one would little expect or thinke, but laid together with divine Artifice: I was even amazed, whence possibly any power of imagination should be able to sustaine so immense a pile of consequences, and cause that so many things, so far asunder distant, could be at once present to the minde,

14. Although Oughtred failed to mention Viète, *The Key* features enough of Viète's ideas, terminology, and analogies to make the debt clear (Cajori, *William Oughtred*, 39).

15. Quoted in translation by Cajori, *William Oughtred*, 20.

16. Oughtred, *The Key*, 4.

17. Ibid., preface, 1–2. Pagination of the preface is mine.

and as with one consent joyne and lay themselves together for the
structure of one argument.

Only when he rewrote classical propositions in the symbolical style and
applied the method of analysis did he understand mathematical invention. It was this analysis, this "inventive way," that permitted "more
easie and full understanding of the best and antientest Authors." Analysis enjoyed this privileged position, he claimed, because it was the very
method by which "those antient Worthies have beautified, enlarged, and
first found out this most excellent Science."[18] In short, Oughtred argued
forcefully that analysis was the inventive method of not only the present
and future but also the past. In retrospect, his was a very effective
argument, with wide appeal to English mathematical thinkers of varying
degrees of loyalty to classical and early modern mathematics.

3

But the analytic art of the moderns was not the same as that of the
ancients. For Oughtred as for Viète, the analytic art was inextricably
linked to the symbolical style. Oughtred's personal enthusiasm for the
new symbols informed a spirited defense of the style, which went beyond
Viète's and was perhaps the most important element in the favorable
reception of symbolical algebra in seventeenth-century England. As he
explained, he embraced the new style because, first, symbols permitted
the mathematician to grasp a mathematical situation more quickly and
clearly than corresponding prose statements and, second, symbols fostered algebraic inventiveness and generality. He attributed the power of
symbols to their involving the human eye as well as reason in the
mathematical process.

Symbols eliminated "verbous expressions." Furthermore, as Oughtred
enthused in the preface to *The Key*, "the specious and symbolicall manner, neither racketh the memory with multiplicity of words, nor chargeth
the phantasie with comparing and laying things together; but plainly
presenteth to the eye the whole course and processe of every operation
and argumentation." Symbols could replace a "multiplicity of words,"
and thus do the memory a service. But Oughtred suggested, more fundamentally, that symbols could also change the process of human thought.
They brought a hitherto unknown clarity to the algebraic process. Given
symbols, he seemed to maintain, the "phantasie" no longer needed to
compare the ideas or the things for which the symbols stood. The

18. Ibid., preface, 1, 3–4.

symbols were visual objects with a sort of independence, which "plainly presenteth to the eye the whole course and processe of every operation and argumentation." "Wherefore that I might more clearly behold the things themselves," he wrote, "I uncasing the Propositions and Demonstrations out of their covert of words, designed them in notes and species appearing to the very eye."[19] In short, from Oughtred's perspective, symbols were visual objects that brought the "things themselves" of algebra – that is, its propositions and demonstrations – right before the senses. In his *Circles of Proportion* of 1632 he stated explicitly that the "manner of setting downe Theoremes, whether they be Proportions, or Equations, by Symboles or notes of words ... is easie, being most agreeable to reason, yea even to sence."[20]

The time was ripe for such a defense of the symbolical style, especially in England where Francis Bacon had outlined his basically inductivist approach to science. The move to concise, symbolical algebra fit well with Bacon's attack on the "Idols of the Market-place." Writing of science rather than mathematics in his *Great Instauration* of 1620, Bacon had postulated that "all depends on keeping the eye steadily fixed upon the facts of nature and so receiving their images simply as they are." According to his *New Organon,* there were however certain obstacles or "Idols" that "beset men's minds" and obstructed the path to understanding nature, "the most troublesome" of which were the "Idols of the Market-place" or distortions caused by "the ill and unfit choice of words." Good science, he argued, depended upon "let[ting] all those things which are admitted be themselves set down briefly and concisely, so that they may be nothing less than words."[21] Choice of words was thus crucial, and the fewer words in a scientific account the better.

In the later seventeenth and early eighteenth centuries, some key English scientists, including Robert Hooke and Robert Boyle, as well as the historian of the Royal Society Thomas Sprat, set the symbolical style as a model for terse, scientific expression. In his *History of the Royal-Society* Sprat claimed that the society "exacted from all their members a close, naked, natural way of speaking ..., bringing all things as near the Mathematicall plainness as they can."[22] Hooke specifically referred

19. Ibid., preface, 1–2, 4.
20. William Oughtred, *The Circles of Proportion and the Horizontall Instrument,* trans. William Forster (London, 1632), 20.
21. *The Works of Francis Bacon,* ed. James Spedding et al., vol. 4 (London: Longman, 1860), 32, 53, 60, 54–55, 254. The last quotation is from *Parasceve,* which Bacon appended to the *New Organon* of 1620.
22. Thomas Sprat, *The History of the Royal-Society of London, for the Improving of Natural Knowledge* (1667; reprint, ed. J. I. Cope and H. W. Jones, St. Louis: Washington University Press, 1958), 113.

to the symbolical style as a model for scientific prose. In his "General Scheme, or Idea of the Present State of Natural Philosophy," he wished for scientific reports that "are [the] shortest and express the Matter with the least Ambiguity, and the greatest Plainness and Significancy." This led to his suggestion that these reports be written in "some very good Short-hand or Abbreviation . . . , as in Geometrical Algebra, the expressing of many and very perplex Quantities by a few obvious and plain Symbols."[23] In brief, although the symbolical style, which owed so much to Viète, was not spawned by Bacon's attack on the "Idols of the Market-place," the style served and was served by the English quest for plain scientific prose.

Symbols fostered not only brevity and clarity – the special concerns of English scientists – but generality as well, a thesis developed earlier by Viète. Oughtred reiterated the thesis, even using Viète's analogy of "the footsteps of analysis."[24] Literal, but not numerical, coefficients left their prints in proofs:

> [I]n the Numerous [arithmetic], the numbers with which we worke, are so, as it were, swallowed up into that new which is brought forth, that they quite vanish, not leaving any print or footstep of themselves behind them. But in the Specious, the species remaine without any change, shewing the process of the whole worke; and so doe not onely resolve the question in hand; but also teach a generall Theoreme for the solution of like questions in other magnitudes given.[25]

Reflective of Oughtred's strong commitment to the symbolical style, *The Key* was replete with symbols, at least 150 different ones.[26] Some of Oughtred's specific symbols were simply adaptations of Viète's. For example, in place of Viète's *A quadratum, A cubus,* and *A quadrato-quadratum,* he used the shorter versions *Aq, Ac,* and *Aqq.*[27] But *The Key* also offered a myriad of new symbols peculiar to Oughtred, who was charmed by mathematicians' freedom to coin symbols "at pleasure" so long as they kept "in mind for what magnitude every Species is

23. Robert Hooke, "A General Scheme, or Idea of the Present State of Natural Philosophy, and How Its Defects May Be Remedied," in *The Posthumous Works of Robert Hooke* (London, 1705; reprint, with an introduction by Richard S. Westfall, The Sources of Science, no. 73, New York: Johnson Reprint, 1969), 63–64. On the relationship between science and language in seventeenth-century England, see Brian Vickers, ed., *English Science, Bacon to Newton* (Cambridge: Cambridge University Press, 1987), 8–22.
24. Viète, *The Analytic Art,* 28.
25. Oughtred, *The Key,* preface, 4.
26. This is Cajori's count (Cajori, *William Oughtred,* 28).
27. See Cajori, "Oughtred's Mathematical Symbols," 176.

set."[28] Oughtred replaced the symbol =, used by Viète to denote the difference between two quantities when their relative values were unknown, with ∽. He used ⊏ for "greater than" and ⌐ for "less than."[29] He also coined symbols that eventually became elements of the standard language of modern mathematics. In place of the words "by" and "times" used by Viète to denote multiplication, he introduced the St. Andrew's cross ×; he also introduced :: as the symbol for proportion.[30]

As Viète before him, Oughtred proved no final arbiter of algebraic symbolism. He wantonly coined symbols: thus he used three (slightly different) variants of a capital Z to denote $A + E$, $Aq + Eq$, and $Ac + Ec$,[31] and Zq, to denote the square of $A + E$. Some of his symbols confused his students and typesetters, if not himself, and these symbols soon gave way to more commodious replacements. Examples of the latter were his symbols for inequalities, which proved difficult to remember. Oughtred himself used them inconsistently; his student Ward wrote ⊏ for "less than"; and Wallis, who regularly employed the symbols > and <, incorrectly reproduced Oughtred's symbol for "greater than" when describing the latter's work in his *Treatise*.[32] As the history of the signs for inequalities shows, symbolical standardization came slowly, largely by trial and error.

That many of Oughtred's English mathematical successors worried about which algebraic symbols to adopt, rather than about the legitimacy of the symbolical style, was a tribute to *The Key*. Oughtred's rigorous defense of the style did not immediately convince all mathematical students. In the preface to the English edition of 1647 of *The Key*, he admitted that the first edition of the textbook had "seemed unto many very hard; though indeed it was but their owne diffidence, being scared by the newnesse of the delivery; and not any difficulty in the thing it selfe."[33] Still, by the midcentury key English mathematicians had

28. Oughtred, *The Key*, 4.
29. Cajori, "Oughtred's Mathematical Symbols," 177–178, 184–185. The "difference" symbol appeared not in *The Key* proper, but in a tract added to the editions of *The Key* of 1652, 1667, and 1693.
30. Ibid., 175, 183–184, 181–182. "Oughtred's notation A.B :: C.D is the earliest serviceable symbolism for proportion" (ibid., 181–182). Although the symbol x was used for multiplication in an appendix to the English translation of 1618 of Napier's work on logarithms, Cajori conjectures that Oughtred wrote this appendix (ibid., 183, and Cajori, *William Oughtred*, 27, 54–56).
31. Cajori, "Oughtred's Mathematical Symbols," 177.
32. Ibid., 184. See also Wallis, *Treatise of Algebra*, 127.
33. Oughtred, *The Key*, preface, 2.

imbibed so much of Oughtred's symbolical brew that most were addicted to the new symbolism. From this point on, thanks largely to Oughtred but also (as we shall see) somewhat to Harriot, there would be few serious English challenges to the symbolical style.

4

On the other hand, Oughtred's ambivalence toward the expanding universe of algebra – he accepted negative numbers but typically only positive roots of equations – would help to slow the development of the subject in seventeenth-century England. Going beyond Viète, he introduced and sanctioned negative numbers from the very first section of his textbook. But his treatment of these numbers was far from clearcut, and their introduction seemed to owe more to the emerging theory of logarithms than to the theory of equations. Already in his early-sixteenth-century work on powers, which helped prepare the way for logarithms, Michael Stifel had accepted negative exponents, which were then incorporated into the first specifically logarithmic work, Napier's *Mirifici logarithmorum canonis descriptio* of 1614.[34] Oughtred mentioned "Logarithmes" (as "artificiall" numbers) in the very first sentence of *The Key*. The sentence was followed immediately by a table giving "Indices or Exponents of termes continually proportionall from an unit both wayes":

$$9\ 8\ 7\ 6\ 5\ 4\ 3\ 2\ \overline{1}\ 0\ \overline{1}\ \overline{2}\ \overline{3}\ \overline{4}\ \overline{5}\ \overline{6}\ \overline{7}\ \overline{8}\ \overline{9}.$$

These exponents, Oughtred quickly (and still on page 1 of *The Key*) explained, were "in Integers, or whole numbers, adfirmative; in parts or fractions negative." Thus, $\overline{1}$ stood for negative 1, and, assuming base 10 as Oughtred did here, $\overline{1}$ was the exponent giving 1/10, whereas 1 was the exponent giving 10. Negative numbers, then, entered *The Key* not as roots of equations but as exponents or vital parts of the useful theory of logarithms.[35]

34. For a brief history of logarithms, see Wolfgang Kaunzer, "Logarithms," in *Companion Encyclopedia of the History and Philosophy of the Mathematical Sciences*, ed. I. Grattan-Guinness, 2 vols. (London: Routledge, 1994), 1:210–228. On Napier, see Yannis Thomaidis, "Aspects of Negative Numbers in the Early 17th Century: An Approach for Didactic Reasons," *Science and Education* 2 (1993): 69–86, esp. 73–76.

35. Oughtred, *The Key*, 1. Oughtred did not employ modern exponential notation. Moreover, the *Clavis* of 1631, unlike subsequent editions, did not open with a reference to logarithms.

Once in his algebra, the "negatives" had to be defined, and so on page 3 Oughtred assigned two basic meanings to the signs + and −: as standing for operations and for qualities of numbers. "The signe of adding or of adfirmation," he defined, "is + plus, or pl: as 34 or +34. . . . The signe of diminishing or of Negation, is − minus, or mi: as −34, are denied to be at all." The last phrase was quite telling, for, despite his introductory table giving negative exponents, Oughtred (like Viète) seemed more comfortable with negatives as terms of polynomials than as isolated algebraic entities. In illustration, he wrote shortly after his definition of the sign − that: "I use the signes + and − when one single magnitude is adfirmed or denied of another single one; but I use the signes of pl, and mi, when a compound magnitude is adfirmed or denied of one single; or contrariwise a single of a compound."[36] Similarly, although he stated the rule of signs for multiplication, his examples referred only to those cases where the sign − appeared between two terms of a binomial, that is, where it denoted subtraction and not an isolated negative quantity.[37]

Still, isolated negatives appeared in Oughtred's early chapters on addition and subtraction. Significantly, the examples of isolated negatives were all drawn from specious (not numeral) arithmetic or from indices. That is, he wrote A and $-A$ separately and then added them; he similarly added $5A$ and $-3A$; and he wrote $3A$ and $5A$ separately and subtracted the latter from the former to get $-2A$. But he did not in general add or subtract negative numbers per se or take a greater number from a lesser. That is, there were no corresponding examples of the addition of 5 and −3 or of the subtraction of 5 from 3.[38]

In these examples as in his definitions of addition and subtraction, he implied that isolated negative quantities and unlimited subtractions were legitimate for specious but not for numeral arithmetic. He explicitly distinguished between (numeral) "addition" and "specious addition" and between (numeral) "subduction" and "specious subduction." He gave a definition or at least a rule for each. Whereas addition led to "the particular Summes of the severall rowes (of numbers)," specious addition "joyneth together all the magnitudes given, keeping their figures."

36.   Ibid., 3. In the posthumous edition of 1694, + and − were defined exclusively as signs of operations (Oughtred, *Key of the Mathematicks: Newly Translated from the Best Edition* [London: John Salusbury, 1694], 4). Cajori argues that "the 1694 edition marks a recrudescence" (Cajori, *William Oughtred*, 25).

37.   Oughtred, *The Key*, 10–11. He took the rule without proof.

38.   Ibid., 4–6.

Or, as he hinted, addition involved a real sum; specious addition, the mechanical joining of magnitudes. Similarly, whereas subduction led to "the particular differences of the severall rowes [of numbers]," specious subduction "joyneth together both the magnitudes, changing all the signes of the magnitude to be subducted."[39] It is perhaps not a stretch to summarize Oughtred's message in the following example: one could in specious subduction take the greater $5A$ from the lesser $3A$, getting $-2A$, although in numeral arithmetic such subtraction was impossible and, in fact, -2, like -34, could be "denied to be at all."

The realities of early modern algebra, however, fought against such an ideal distinction and made for strange compromises. Apparently pushed by the necessity of and precedent for adding and subtracting exponents associated with logarithms, Oughtred left exponents to straddle the dividing line between the numeral and the specious. Although exponents were numbers, the examples of their addition and subtraction appeared under specious addition and subduction, but in special corners of their respective pages. Thus at the tail end of the chapter on subduction, under the category of "Specious Subduction," readers found: "So in the Subduction of Indices,

$$
\begin{array}{c|c}
3 & \overline{3} \\
\hline
2 & 2 \\
\hline
5 & \overline{5}.
\end{array}
$$
"

That is, Oughtred wrote isolated negative exponents, added and subtracted them, even taking a greater number 2 from the lesser $\overline{3}$ to get the negative, $\overline{5}$.[40]

Even as it exemplified the best of the new symbolical style, Oughtred's *Key* thus also captured the tensions inherent in the subject as it had come from Viète. *The Key* taught numeral along with "specious" arithmetic; each of the chapters on the basic operations of addition, subtraction, multiplication, and division was divided into two sections: one dealing with the numeral and the other, the specious. The relationship between the two, however, was never directly addressed. Nor was there any explicit discussion of arithmetic situations that made sense in species, but perhaps not in numbers – for example, the subtraction of a greater from a lesser. Nor was the issue of negative and imaginary roots directly engaged. Possibly because *The Key* was designed for beginning

39. Ibid.   40. Ibid., 5–6.

students, Oughtred concentrated on solutions of quadratic equations and typically recognized only their positive roots. Serious students of mathematics had to read between the lines and, indeed, brave paradoxes such as amphibian-like exponents that seemed to be both numeral and specious.

5

Oughtred's avoidance of negative and imaginary roots is understandable. Not only was he a product of a period in which the negatives and imaginaries lay on the very frontier of the mathematical sciences, but his intent in *The Key* was not to offer a complete theory of equations. He gave only those results of early modern equation theory that were, in his opinion, essential to beginning students of analysis, including a nearly general solution of the quadratic equation;[41] the coefficients for binomial expansions through the tenth degree;[42] and rules for simplifying equations, such as transposition, division by the unknown, and division by a constant.[43] In addition, he offered many examples of the application of the analytic art to geometry. Indeed, the latter aspect of *The Key*, along with its striking defense of the symbolical style, accounted for its profound influence on the direction of English mathematics in the seventeenth century.

The final chapters of *The Key* dealt with analysis applied to geometry. Chapter 18, which (in his homey way) Oughtred entitled "The Analytical Store," explained that there were many – in fact, innumerably many – equations that mathematicians could write. This "Analyticall furniture," he declared, "is no lesse precious than plentious." He promised to offer examples of equations beyond those appearing in earlier chapters and challenged "the studious in Analytice . . . [to] devise more for their exercise." Such equations could come from geometry as well as arithmetic,

> wheresoever, whether in Arithmetique, Geometrie, or in any other Art, he [the student of mathematics] shall light upon some magnitude, to which another is understood to bee equall; he shall turne, wrest and vary, that Æquality by whatsoever meanes and comparings he can; that from thence he may find out a new Instrument of

41. Ibid., 56–59. Cajori has analyzed Oughtred's qualified solution (Cajori, *William Oughtred*, 29–32).
42. Oughtred, *The Key*, 40.
43. Ibid., 54–56. Although he here followed Viète rather closely, Oughtred gave five rules in all.

Art; which afterwards he shall keep in Store; and wheresoever occasion serves, bring it forth for the helpe and advancement of the Art.[44]

Not content to set his readers loose on their own just yet, he gave multiple examples of reducing geometric situations to symbolical equations and then solving the equations. In a strategy that seemed to be extremely effective with his English audience, he used as examples specific Euclidean propositions. He developed his case for applying the analytic art to these propositions in a series of stages. In chapter 18 he stated propositions 5 through 10 of book II of Euclid's *Elements* in prose and simply rewrote them as equations. Next, turning temporarily to pure geometry, he abruptly proclaimed that those intending to be "analysts" needed to know certain geometric theorems and problems, and so he listed key Euclidean results (in most cases, giving no analytic equivalents).[45]

The point of all this became obvious as Oughtred opened his next and final chapter, which bore a lengthy, descriptive heading: "Examples of Analytical Æquations, for inventing of Theoremes, and resolving of Problemes: At which marke (as it were) the Precepts hitherto delivered, do principally aime."[46] Chapter 18 had established the essential preliminaries for chapter 19. And here in chapter 19 the power of analysis shone, as Oughtred took specific Euclidean propositions, translated them into symbolical equations, and proceeded thereby to offer eminently concise demonstrations of them.

In the first problem of chapter 19 he sought "the invention" of Euclid's proposition 11 of book II: to cut a given straight line so that the rectangle contained by the whole and one of the segments is equal to the square on the remaining segment. Euclid's solution ran a good many lines (certainly for more than a page), and, although it produced the desired effect, its path to that effect was not at all obvious. That is, as Oughtred had claimed of so many of the results of classical geometry, Euclid's solution did not tell how he had actually come to his proof. Oughtred, touting the inventiveness of the analytic art, translated proposition 11 into a quadratic equation and then promptly solved the equation. B stood for the given straight line. Then if A stood for the greater segment, the lesser was $B - A$, and the rectangle contained by the whole and the lesser segment was $Bq - BA$, which was required to be equal to Aq. Thus the equation of the problem was $Aq + BA = Bq$, to which Oughtred merely applied his quadratic formula.[47] The latter solution

44. Ibid., 68–69.   45. Ibid., 69–70, 72–80.
46. Ibid., 81.   47. Ibid.

was much faster and, in Oughtred's opinion, much tidier and revealing than Euclid's.

Oughtred tried hard to convince his readers of the benefits of analysis applied to geometry. For some of the geometric propositions studied in chapter 19, he gave both algebraic and corresponding geometric treatments. In his discussion of proposition 11 of book II of the *Elements*, for example, he followed the pattern of: (1) Euclid's statement of the proposition, (2) translation of the proposition into a symbolical equation and the algebraic solution of the equation, (3) the prose statement of the equation's solution, and (4) the corresponding geometric demonstration of the proposition (with this last section beginning "Geometrically thus . . .").[48]

In one cramped textbook, then, Oughtred not only had introduced the symbolical style and the analytic art to England but had mounted a mighty defense of early modern algebra based primarily on its clarity and inventiveness. No timid reformer, he had engaged Euclid head-on in a battle for the control of mathematics. From 1631 on, there would no longer be an exclusively geometric tradition in English mathematics; there would be a parallel analytic or algebraic one as well. Furthermore, one strand of the analytic tradition, that developed by Wallis, would come more and more to assert the existence of "pure" algebra, that is, one independent of geometry.

6

The *Artis analyticae praxis* (usually referred to as the *Praxis*) of Thomas Harriot (ca. 1560–1621) was published a full decade after its author's death, and thus the same year as *The Key*. Harriot and Oughtred shared more than the year of the publication of their textbooks. Both had succumbed to the lure of Viète's symbolical style, and used and coined symbols freely; both had established around themselves circles of mathematical thinkers whom they infected with the analytic way. Oughtred's disciples promoted his *Key* and supervised new editions of the work.[49] Following Harriot's death, his disciples collected some of his unpublished algebraic manuscripts in the *Praxis,* which proved to be an important work in British algebra but not as influential as *The Key*.

48. Ibid. Oughtred's *Key* and, especially, his application of algebra to geometry deserve further historical study.
49. Robert Wood translated the greater part of the first English edition, and John Wallis saw the editions of 1652 and 1667 through the press (Cajori, *William Oughtred*, 18–19).

Harriot spent the period 1577–1580 as a student at St. Mary's Hall, Oxford University. According to the statutes of Oxford, dating from 1564/1565, the program for the bachelor of arts degree, which was Harriot's terminal degree, required some grammar, rhetoric, dialectics, arithmetic (specifically three terms of Boethius or Gemma Frisius), and music.[50] Despite the slim mathematical component of this program, Harriot, like Oughtred, managed to acquire a basic grounding in science and mathematics while an undergraduate. It is unclear whether Harriot studied alone or – as Mordechai Feingold has implied – took advantage of opportunities for mathematical instruction that, despite their official statutes, Cambridge and Oxford offered at this point.[51] What is clear, however, is that at the end of his undergraduate career Harriot's knowledge and reputation supported an offer of a position as a mathematical tutor to Sir Walter Ralegh.

Unlike Oughtred, Harriot supported himself through science and mathematics. In late 1583 or early 1584 he began his long association with Ralegh, who employed him as an expert in cartography and the theory of oceanic navigation, and in 1585 he accompanied Ralegh's colonizing expedition to Virginia. Returning to England in 1586, he continued to serve Ralegh through the late 1590s, when Henry Percy, ninth earl of Northumberland, became his scientific patron.[52] A combination of talent and favorable patronage – Northumberland was a man of science as well as a patron[53] – led to Harriot's significant work in optics, astronomy, ballistics, and especially mathematics.

Although a polymathic scientist and a prolific writer, Harriot published only one book, *A Briefe and True Report on the New Found Land of Virginia,*[54] before his death from cancer in 1621. His major mathematical work, the *Praxis,* was published posthumously from notes among the voluminous manuscripts that he willed to his mathematical executor, Nathaniel Torporley (1564–1632), "to pervse and order and to separate the Chiefe of them from my waste papers, to the end that after hee doth vnderstande them hee may make vse in penninge such doctrine that belonges vnto them for publique vses as it shall be thought Convenient."[55] Like Harriot whom he recognized as his teacher, Tor-

50. On the Oxford that Harriot knew, see John W. Shirley, *Thomas Harriot: A Biography* (Oxford: Clarendon Press, 1983), 38–69.
51. Feingold, *The Mathematicians' Apprenticeship,* 86, 104.
52. For the details of Harriot's association with Ralegh and Northumberland, see Shirley, *Thomas Harriot.*
53. Feingold, *The Mathematicians' Apprenticeship,* 211.
54. (London, 1588).   55. Quoted in Shirley, *Thomas Harriot,* 2.

porley was an Oxford man and a mathematical practitioner.[56] After receiving his B.A. from Christ Church in 1584, he had served as Viète's amanuensis at the time of the publication of *The Analytic Art* "and may well have been the link between" Viète and Harriot.[57] Torporley, however, did not fulfill the charge to publish Harriot's mathematical writings. He worked through the mathematics, and by 1630 had transcribed most of Harriot's algebraic manuscripts. But he apparently became disenchanted with Harriot, possibly when he uncovered his support for the atomic theory. An ordained minister, he saw the latter theory as inconsistent with the biblical account of creation.[58] Finally, an anonymous editor – Walter Warner (1550?–1636), who was also an Oxford man, a mathematical practitioner, and an associate of Harriot in the Northumberland circle – collected Harriot's more elementary writings on arithmetic and algebra in the *Praxis*.[59] This textbook, intended for mathematical amateurs, captured Harriot's enthusiasm for the symbolical style and his skill in equation theory but not his full commitment to the expanding universe of algebra.

7

Like Oughtred's *Key*, Harriot's algebra textbook took its lead from Viète's. Its title referred to the "analytic art," and implied a complementary relationship between it and Viète's work: whereas Viète's was an

56. E. G. R. Taylor, *The Mathematical Practitioners of Tudor and Stuart England* (Cambridge: Cambridge University Press, 1967), 187.

57. Edward Rosen, "Harriot's Science: The Intellectual Background," in *Thomas Harriot: Renaissance Scientist*, ed. John W. Shirley (Oxford: Clarendon Press, 1974), 1–15, on 10. Harriot knew Torporley before his service to Viète and during the service Torporley sent Harriot at least one of Viète's mathematical problems (Shirley, *Thomas Harriot*, 3).

58. Rosen, "Harriot's Science," 5–6. Shirley still regards the reasons for Torporley's abandonment of the algebra publication an open question (Shirley, *Thomas Harriot*, 4).

59. On Warner, see Taylor, *Mathematical Practitioners*, 178. Torporley disapproved of Warner's edition of the *Praxis* because it omitted some of Harriot's major algebraic accomplishments and distorted his notation (Shirley, *Thomas Harriot*, 4–5). He condemned its "editors," indicating that Warner had at least one assistant, probably Sir Thomas Aylesbury. See Rosalind C. H. Tanner, "Thomas Harriot as Mathematician: A Legacy of Hearsay," parts 1 and 2, *Physis: Revista Internazionale di Storia della Scienza* 9 (1967): 235–247, 257–292, esp. 268–269.

Introduction to the Analytic Art, Harriot's concerned *The Practice of the Analytic Art.*[60] "[A] demonstration of what was first accomplished by Viète in carrying through his undertaking," the editor(s) of the *Praxis* wrote, "may permit a better understanding of the later contributions by our very highly informed author Thomas Harriot, who followed Viète in this pursuit of analytics."[61] Significantly, Harriot's "application of the analytic art" referred not to any geometric use of the art but rather to the advancement of equation theory. In this way as well as in some of his new notation, Harriot, even more so than Oughtred, promoted an independent algebra.

In adopting Viète's analytic program, Harriot seemed most taken by the symbolical style. As Florian Cajori has remarked, his "algebra is less rhetorical and more symbolic than perhaps any other algebra that has ever been written. . . . Some pages . . . have hardly a word of explanation. Often the ideas are conveyed to the eye by mathematical symbols alone."[62] There is evidence, moreover, that Harriot's interest in symbolism stretched beyond mathematics. Prior to his year in Virginia, which was to be spent among American Indians who spoke a different language, he developed "a scientific kind of writing" that permitted him to record the exact sounds of foreign words as they were spoken to him. In detail, Harriot's phonetic alphabet was related to the symbolical style of algebra. In designing it, he did not employ the characters of the English language, but rather invented "a wholly new cursive script" that was modeled on the algebraic symbols of Stifel and Clavius.[63]

So comfortable with the symbolical style that he adapted it innovatively to linguistics, Harriot did not hesitate to modify the algebraic

60. The title of Harriot's work, published in Latin in London in 1631, was *Artis analyticae praxis, ad aequationes algebraicas novâ, expeditâ, et generali methodo, resolvendas* (The Practice of the Analytic Art for the Purpose of Solving Algebraic Equations by a New, Convenient, and General Method).

61. Quoted from the *Praxis* in Rosen, "Harriot's Science," 7–8. This assignment of priority to Viète is significant, especially since in 1610/1611 Sir William Lower wrote Harriot: "Do you not here startle, to see every day some of your inventions taken from you? . . . Vieta prevented [anticipated] you of the gharland of the greate Invention of Algebra" (quoted in ibid., 7).

62. Florian Cajori, "A Revaluation of Harriot's Artis Analyticae Praxis," *Isis* 11 (1928): 316–324, on 317–318. On this point, see also R. C. H. Tanner, "The Ordered Regiment of the Minus Sign: Off-beat Mathematics in Harriot's Manuscripts," *Annals of Science* 37 (1980): 127–158, esp. 136.

63. Shirley, *Thomas Harriot,* 109–112.

symbols of his predecessors or to coin new ones. He was familiar with the symbols used by earlier algebraists, including not only Stifel and Clavius but also Diophantus, Viète, and Simon Stevin.[64] He modified Viète's symbols for unknowns and knowns, continuing to denote them by vowels and consonants, respectively, but using lowercase letters throughout.[65] Through a change in notation he discarded the remnants of the dependence of algebraic powers on geometric analogues, found in Viète's symbols. Thus, in what is generally taken as an important contribution to the emergence of pure algebra, he indicated powers by iteration of symbols; for example, in place of Viète's *A quadratus* and *A cubus,* he wrote *aa* and *aaa.* Through his posthumous *Praxis,* he also popularized the sign = for equality, which Robert Recorde had suggested in 1557.[66] In addition, Harriot deserves some credit for helping to establish the notion of inequalities in early modern algebra. In his unpublished manuscripts, he used the concepts of "greater than" and "less than" and expressed both in symbols. His symbols, however, were not exactly the modern ones of > and <, even though the modern symbols appeared in the posthumous *Praxis.*[67]

In its ignoring of the negative and imaginary roots of equations, just as in its embracing of the analytic program with the symbolical style, Harriot's *Praxis* fell squarely in the tradition of Viète's *Analytic Art.* As represented in the *Praxis,* Harriot accepted specious negatives. For example, he wrote the isolated $-d$ and added it to $a + b$ to get $a + b - d$. Still, he did not use the terms "plus" and "minus," and, as Cajori has pointed out, only once and incidentally in the *Praxis* did he employ the terms "affirmative" and "negative" in connection with such numbers.[68]

Furthermore, in the theory of equations, which formed the major part

64. Ibid., 110.
65. Tanner sees this modification as "somewhat accidental" since Harriot "tended to write all letters small" (Tanner, "Minus Sign," 136).
66. Rosen notes that Harriot's manner of denoting powers came from a suggestion of Stifel (Rosen, "Harriot's Science," 9).
67. Historians have seen Harriot as the inventor of the modern signs of inequality (e.g., Rosen, "Harriot's Science," 9). But: "There is good reason to believe that Harriot himself never knew the modern signs of inequality to which his name is attached" (Tanner, "Thomas Harriot as Mathematician," 240–241; quotation from p. 240).
68. Cajori, "Revaluation," 317. My analysis here follows Cajori who worked through the *Praxis* to determine: "What number-system did Harriot employ? What kind of roots of equations did he admit?" (316). Based on Cajori and Tanner's observations, we may conclude that Harriot's failure to name symbols such as $-d$ was not exceptional for him, since he often expressed himself almost exclusively in symbols, with little or no rhetorical commentary.

of the *Praxis,* he generally recognized only positive roots. As one of his major contributions to equation theory, he wrote equations in the "canonical form," that is, as products of binomial factors. Even for those equations where the canonical form seems (at least to modern eyes) to highlight negative roots, he either overlooked the root(s) or totally dismissed those equations with negative roots alone. When he expressed an equation through its factors

$$(a - b)(a - c)(a + d),$$

where a is the unknown, he gave its roots as b and c.[69] When discussing a class of equations constructed of such factors as:

$$(a + b)(a + c)$$

and

$$(a + b)(a + c)(a + d),$$

he wrote that, "since they cannot be derived except on the supposition of negative roots, they will be disregarded as quite useless." Moreover, the *Praxis* alluded only once to an imaginary number. The number, which went unnamed, appeared in the cubic equation,

$$e\,e\,e = c\,c\,c + \sqrt{-d\,d\,d\,d\,d\,d},$$

where e was the unknown. Harriot dismissed the equation as "impossible to reduce," because of its "inexplicable" last term.[70]

8

As scholars of the nineteenth and twentieth centuries have shown, the conservatism of the *Praxis* was not representative of Harriot's deepest reflections on the expanding algebraic universe. In manuscripts other than those published in the *Praxis,* Harriot used negative and imaginary roots, calling the latter "noetical." He completely solved a quartic equation with a positive, a negative, and two complex roots.[71]

69. Ibid., 319, 323–324. On the latter pages Cajori summarizes Harriot's other (rather minor) contributions to equation theory.

70. My discussion of the want of negative and imaginary roots in the *Praxis* (along with the translated quotations) follows Cajori, "Revaluation," 317–320.

71. In the 1830s S. P. Rigaud discovered this solution in Harriot's manuscripts (Tanner, "Thomas Harriot as Mathematician," 243, 262). On the term "noetical," see J. A. Lohne, "Thomas Harriot," *Dictionary of Scientific*

He also privately experimented with deviationary versions of the rule of signs for multiplication.[72] His experiments, which were similar to, and perhaps inspired by, those of Cardano's *De aliza*, were the products of an inquisitive, daring mind (not unlike his Italian predecessor's), which was able in the early seventeenth century to see that some mathematical rules are arbitrary. More specifically, the experiments were a response to the problem of the complex numbers. As Tanner has suggested, Harriot's speculations on the rule of signs probably originated in exchanges on the negative and imaginary numbers with his close associates Warner and Sir William Lower. In these exchanges, Lower, Harriot's pupil and friend who had been a resident of Exeter College from 1586 to 1593, perhaps "reflect[ed] the prevailing discomfort regarding negative numbers, and the hope of a device to elude imaginaries in resolving algebraic equations that involve them."[73]

Lower's handwriting appears on a manuscript sheet on which Harriot explored the consequences of various deviations from the rule of signs, including (1) minus times minus gives minus and (2) minus times plus, or plus times minus, gives plus. On the sheet, which is devoid of explanatory prose, Harriot squared 5 − 2, first according to the standard rule of signs and second according to the standard rule modified by deviationary rule (1). He arrived at the same result both ways, since, in the deviationary case, he added the results from right to left:

$$
\begin{array}{r}
5 \ - \ 2 \\
5 \ - \ 2 \\
\hline
- \ 10 \ - \ 4 \\
25 \ - \ 10 \\
\hline
25 \ - \ 20 \ - \ 4 \\
\hline
25 \ - \ 16 \\
\hline
9
\end{array}
$$

The same sheet preserves an aborted attempt to square 5 − 2 using the standard rule modified by deviationary rule (2).[74]

Biography, ed. Charles C. Gillispie, 18 vols. (New York: Scribner's Sons, 1970–1990), 6:124–129, at 125. Aristotle distinguished between the "noetical" and the sensible.

72. My discussion of Harriot's deviationary mathematics is based largely on Tanner's research (Tanner, "Minus Sign").

73. Tanner, "Minus Sign," 128.

74. British Library Additional Manuscripts 6785, fol. 386; reproduced, with a long explication, in Tanner, "Minus Sign," 129–133. On another sheet,

That Harriot accepted isolated negative numbers and that his deviationary multiplication was at least sometimes connected with reflections on complex numbers are evidenced by another of his surviving manuscripts. Here he solved the quadratic equation:

$$25 = 6{,}a - aa.$$

He proceeded as follows,

$$aa - 6{,}a = -25,$$
$$aa - 6{,}a + 9 = -16,$$

arriving at the two roots:

$$a = 3 + \sqrt{-16}$$
$$3 - \sqrt{-16} = a.$$

After checking that each root satisfied the given equation, he drew a double line across the page and stated:

$$aa - 6a = -25$$
$$a = 3 + {-4}$$
$$a = 3 - {-4}.$$

He checked the solution $a = 3 + -4$, using the standard rule of signs modified by deviationary rule (1). For aa, he calculated

$$
\begin{array}{r|l}
3 + - 4 & a \\
3 + - 4 & a \\
\hline
-16 & \\
+ -12 & \\
+9 + -12 & \\
\hline
-7 + -24 & = aa.
\end{array}
$$

For $6a$, he calculated:

$$
\begin{array}{r|l}
3 + \ -4 & a \\
6 & \\
\hline
18 + -24 & = 6,a.
\end{array}
$$

Harriot multiplied specious numerals using the standard rule modified by deviationary rule (1) (Tanner, "Minus Sign," 133).

He concluded:

$$- 7 + \ -24, \ - 18 \ - \ -24 \quad = aa - 6a.^{75}$$

$$-25$$

The objective of this exercise, according to Tanner, was "the resolution of an otherwise 'impossible' equation by means of positive and negative numbers alone, kept at arm's length from each other."[76]

Interpreted even more liberally, Harriot's solution can be seen as a step toward the treatment of complex numbers as pairs of real numbers. Harriot's root $3 + \ -4$ is written in terms of two real numbers, 3 and 4. The positioning of 4 behind two operational signs, $+$ and then $-$, however, designates that 4 is a special kind of number – we might say an "imaginary" number. Any number in this position bears the immediate sign $-$, and its multiplication obeys deviationary rule (1).

Incomplete and captured on seemingly scrap paper, Harriot's excursions into deviationary multiplication were probably not intended for publication.[77] These episodes of mathematical experimentation, where almost anything went,[78] were heavily dependent on his symbolical style. Indeed, his extremely symbolical style freed him to experiment with problematic numbers. Relieved of the need to express himself in prose, of the need to capture in words the ideas behind such numbers and their manipulations, he recognized roots of all kinds, modified the rule of signs, and quickly pursued the consequences.

These were also episodes of mathematical instruction, witnessed by Harriot's brief note referring Warner to Cardano's *De aliza* as support for deviationary rule (1). Crossing out and inserting words as he wrote, Harriot directed his disciple to read chapter 32 of *De aliza,* where he

75. British Library Additional Manuscripts 6784, fol. 401 v; transcribed and discussed in Tanner, "Minus Sign," 137–139. (I substituted the sign " $=$ " for Harriot's sign "   ", used throughout these calculations.) Harriot did not complete the example by checking that the root $3 - \ -4$ satisfies the given equation.

76. Tanner, "Minus Sign," 139. In the roots, $3 + \ -4$ and $3 - \ -4$, " -4 stands as a negative number, not to be combined with the positive 3, by the operational $+$ or $-$ positive and negative forming a binomial like real and imaginary above" (ibid.).

77. Ibid.

78. On a sheet with a reference to "W.W." (Walter Warner), Harriot wrote the equation $xx = 2x - 8$. Although it has complex roots, he used deviationary rule (1) to verify that -4 is a root of the given equation (ibid.).

would "find him [Cardano] iust of my opinion." He recommended that Warner also see Frederico Commandino "where he blameth those that . . . thinke that minus per minus shal produce plus."[79] As Warner would find, in his version of Euclid's *Elements* Commandino (1509–1575) had referred to, and seconded, Cardano's objections to the Euclidean-based proof of the rule of signs.[80]

Harriot shared not only Cardano's conclusion that the standard proof of minus times minus gives plus was invalid but also his early recognition that some mathematical rules are arbitrary. In a good example of poetry used to express metamathematical ideas that are perhaps too daring for mathematical prose, he recorded his views that:

Yet lesse of lesse makes lesse or more	(1)
Use which is best keep both in store	(2)
If lesse of lesse you will make lesse	(3)
Then bate the same from that is lesse.	(4)
But if the same you will make more	(5)
Then adde to it the signe of more.	(6)
The rule of more is best to use	(7)
Yet for some cause the other choose	(8)
So both are one, for both are true	(9)
Of this inough and so adeu.[81]	(10)

In line 1, Harriot clearly posed the two alternatives: that $- \times - = -$ or $- \times - = +$. Although in line 7 he endorsed the standard rule, he never excluded the deviationary rule. Already in the seventeenth century envisioning mathematicians as agents of mathematics and mathematical rules as their somewhat arbitrary creations, he explicitly referred to the mathematician's right to "choose" which rule to use (line 8) and to the legitimacy of both rules as mathematical axioms (line 9).

As Cardano's deviationary rule of signs was buried in *De aliza,* which was probably seldom read and certainly seldom cited, Harriot's reflections on the rule of signs, imaginary roots, and mathematical arbitrariness did not find their way onto a printed page until two centuries after his death. To his immediate successors, his *Praxis* offered, on the one hand, a striking endorsement of the symbolical style but, on the other,

79. British Library Additional Manuscripts 6783, fol. 121; quoted in Tanner, "Minus Sign," 144.

80. For Commandino's remarks on Cardano's *De aliza,* see Tanner, "Minus Sign," 145–146.

81. British Library Additional Manuscripts 6785, fol. 321 v. The verses are reproduced (following Torporley's transcription) and interpreted in Tanner, "Minus Sign," 148–150.

little more than a basic introduction to an equation theory that recognized only positive, real roots.

The strengths and omissions of Harriot's *Praxis*, then, help explain the nature of early modern English algebra. Harriot and Oughtred conveyed to their disciples, above all, an almost unbounded enthusiasm for the symbolical style. Thus, from its very inception in England, the new algebra was fundamentally symbolical. Only a few serious English thinkers, most prominently Thomas Hobbes, would later debate its symbolical nature. But English and later Scottish and Irish mathematicians would continue for two centuries to struggle with the role of the negative and imaginary numbers, an aspect of algebra which neither Oughtred's *Key* nor Harriot's *Praxis* had resolved. Furthermore, in one of the many curious twists in the history of algebra, the struggle would proceed without the benefit of Harriot's private reflections on these numbers or on the arbitrariness of mathematical rules.

9

Through their writings and disciples Harriot and, even more so, Oughtred not only influenced the style and substance of mathematics but also bolstered the subject's place among the scholarly pursuits of seventeenth-century England. Neither held an academic post beyond the fellowship that Oughtred enjoyed in his early career; a curacy and enlightened patronage enabled Oughtred and Harriot, respectively, to concentrate on mathematics. There were in fact no mathematical professorships at England's universities until Sir Henry Savile endowed the Savilian professorship of geometry at Oxford in 1619 and Henry Lucas, the Lucasian professorship at Cambridge in 1662.[82] Through the early seventeenth century neither university could boast of a formal mathematical program. In the late sixteenth and early seventeenth centuries there were "scientifically oriented members" of the Cambridge and Oxford communities who offered motivated students some mathematical instruction. But mathematics remained a "semi-liberated" subject, subordinate to philosophy,[83] and one into which no student was obliged to delve deeply.

During this period mathematics was evolving from a subject for tradesmen and other applied practitioners to a profession suitable for scholars. When Wallis enrolled in Emmanuel College, Cambridge, in the

---

82. Ball, *Study of Mathematics at Cambridge*, 29, 47.
83. Feingold, *The Mathematicians' Apprenticeship*, 86 (quotation), 17.

early 1630s, he knew but a little arithmetic, which his brother had taught him. As he reminisced late in life, he found at the college "two (perhaps not any) who had more of *Mathematicks* than [he] . . . And but very few, in that whole University." According to him, "Mathematicks . . . were scarce looked upon as *Accademical* studies, but rather *Mechanical;* as the business of *Traders, Merchants, Seamen, Carpenters, Surveyors of Lands,* or the like."[84] Into the early seventeenth century, many continued to value mathematical applications above theory. Students from the upper classes were typically advised to "confine their studies only to those elements of the mathematical sciences directly relevant to practical affairs."[85]

Harriot and Oughtred, then, worked with and wrote for a diverse band of mathematical practitioners: university-trained men, some with academic affiliations but many without, as well as the nonuniversity men who swelled the ranks of what E. G. R. Taylor has described as the "lesser men [of mathematics] – teachers, text-book writers, technicians, craftsmen."[86] Harriot spent his career in small circles of practitioners surrounding Ralegh and Northumberland. Both circles had connections with Oxford University and, just three years after the publication of the *Praxis,* Archbishop Laud purchased a copy of the textbook, presumably with the intent of donating it to the library of St. John's College, Oxford.[87] Oughtred reached a wide audience of nonuniversity and university men. His mathematical inventions, including the horizontal instrument, and some practical publications entitled him to good standing among mathematical craftsmen.[88] On the other hand, his *Key* was used

84. Christoph J. Scriba, "The Autobiography of John Wallis, F.R.S.," *Notes and Records of the Royal Society of London* 25 (1970): 17–46 (quotation from p. 27). Scriba gives both the draft and final versions of the autobiography; all my quotations are from the final version. For a warning against taking Wallis's account of his undergraduate experience at face value, see Feingold, *The Mathematicians' Apprenticeship,* 86–88.

85. Feingold, *The Mathematicians' Apprenticeship,* 191.

86. Taylor, *Mathematical Practitioners,* xi. This classic study offers 582 biographies of English mathematical practitioners.

87. Nicholas Tyacke, "Science and Religion at Oxford before the Civil War," in *Puritans and Revolutionaries: Essays in Seventeenth-Century History Presented to Christopher Hill,* ed. Donald Pennington and Keith Thomas (Oxford: Clarendon Press, 1978), 73–93, on 86.

88. Oughtred's works included *The Circles of Proportion* (1632), *The Double Horizontal Dial* (1636), *Sun-Dials by Geometry* (1647), and *The Horological Ring* (1653). On his ties with instrument makers, see Taylor, *Mathematical Practitioners,* 192–193.

at Sidney Sussex College, Cambridge, by 1642,[89] and among his students and correspondents was a select band of university men, including Wallis and Ward, who in the late 1640s were called to the Savilian professorships of mathematics and astronomy, respectively.

There was justice in the appointment of his students to England's early mathematical professorships, for symbolical algebra and Oughtred's persuasive writings did much to transform mathematics into an academic discipline. It was early modern algebra that set Oughtred, Wallis, Ward, and John Pell afire with mathematical zeal, and made them reluctant to abandon mathematics for established academic pursuits. The new algebra, especially in its most advanced symbolical forms, was far removed from the everyday applications of the mathematical practitioners and from the practical mathematics thought appropriate for study by the upper-class students of Cambridge and Oxford. Viète, moreover, had presented the analytic art not as a mechanical one but as a scientific and humanistic discipline. Zeal for algebra thus seemed to go hand in hand with zeal for mathematical reform at the university level.[90]

Oughtred spent most of his career as a harbinger of scholarly mathematics. His ambiguous position – between the practitioners and the emerging academic mathematicians – sometimes put him on the defensive to justify the kind of mathematics he did and why he did mathematics at all. The occasion for his most famous statement on these subjects was a pamphlet in which a former student, Richard Delamain, charged him with abusing his clerical calling by studying mathematics, denigrating mathematical practitioners, and stealing an invention. Oughtred's reply, entitled *To the English Gentrie,* was a carefully reasoned defense of mathematics as a scholarly discipline, which left room for mathematical practitioners who combined practice with theory. This defense proved useful to later generations of English thinkers who would argue not so much that practitioners needed to know theory but that all educated men needed to study mathematics. When elaborated in the late seventeenth century by Isaac Barrow and throughout the eighteenth century, the defense would elevate mathematics to the paradigm of human knowledge. Mathematics would become a solidly academic discipline and the very core of undergraduate studies at Cambridge.

But the intellectual temper of the early seventeenth century was quite

---

89. Feingold, *The Mathematicians' Apprenticeship,* 89.
90. Algebraic enthusiasm was not however a prerequisite for support for academic mathematics. Isaac Barrow (Chapter 6) disliked algebra but argued eloquently for a central role for mathematics, especially geometry, in the Cambridge curriculum.

different. Aubrey reported that Oughtred's "neighbour ministers say that he was a pittiful preacher; the reason was because he never studied it, but bent all his thoughts on the mathematiques."[91] Delamain publicly charged him with "neglecting . . . [his] calling" by studying mathematics,[92] a charge that Oughtred countered by defending mathematics as not only a pursuit worthy of a divine but also an academic discipline appropriate for every scholar. Oughtred was not alone in perceiving the urgency of formulating a public defense for mathematics. A few years earlier, Pell, another Cambridge man smitten with the analytic way, called for preparation of a "*Catalogue* of Mathematicians and their works" and a "consiliarum" answering the question: "What *fruit* or *profit* ariseth from the studie of *Mathematics?*" This question needed addressing, Pell had written, since so many "men . . . want *will, wit, means* or *leisure* to attend . . . [mathematical] studies." Pell's defense of the subject – that it was "*profitable . . .* to the student, and to mankinde" and that there was "refined *pleasure* . . . [in] hunting out hidden truths, wrastling with difficult Problemes, and getting the victorie"[93] – was tame in comparison with Oughtred's.

With a keen sense of the intellectual and moral concerns of his time, Oughtred wrote forcefully of mathematics as "good literature," and elaborated two reasons why ministers and, in fact, all educated men were almost duty bound to study the subject. The subject was, first and foremost, a path to the understanding of God: "[I]n no other thing, after his sacred word, Almighty God (who creating all things in number, weight, and measure, doth most exactly Geometrize), hath left more expresse prints of his heavenly & infallible truth, then in these Sciences." The study of mathematics, moreover, improved men's reasoning powers. Whereas "all other knowledges . . . [were] involved with a thicke mist of ignorance and obscurity," the

> exercise of these [mathematical] Arts accustomed to the certainty of demonstration, quickeneth the understanding, rousing it up from a

91. Aubrey, *Brief Lives,* 2:111.   92. Oughtred, *To the English Gentrie,* 8.
93. John Pell, *An Idea of Mathematics* (London: William Du-Gard, 1650), 34, 33, 36. The *Idea* appeared here as an appendage to John Durie, *The Reformed Librarie-Keeper with a Supplement to the Reformed School* . . . (London: William Du-Gard, 1650). The *Idea,* however, circulated "in manuscript in an early version before 1630" and was originally published in 1638 (P. J. Wallis, "John Pell," *Dictionary of Scientific Biography,* 10:495–496 [quotation from p. 495]). For an earlier defense of mathematics, see John Dee, *The Mathematicall Praeface to the Elements of Geometrie of Euclid of Megara* (London, 1570; reprint, with an introduction by Allen G. Debus, New York: Science History Publications, 1975).

lasie and drowsie indormition and servile assent to dialecticall and
conjecturall probabilities and spurring it forward, and supplying it
with meanes, unto the accurate investigation of true and undeceiv-
able principles.

There was at this point in Oughtred's rejoinder but one question to ask:
"Now tell me *R.D.* are these studies worthy of a Divine, or no?" The
answer underscored mathematics as an essential academic discipline,
and Oughtred declared: "Indeed to know no more thereof then you
know . . . is unworthy of a Divine, yea of a rationall man: worthy onely
of some rude and reasonlesse dulman."[94] And thus were some of the
seeds of the English (and later Scottish) academic glorification of mathematics sown.

As he promoted mathematics as an academic discipline, Oughtred
argued that the practice of mathematics ought to be combined with and
guided by theory – a position that Delamain exaggerated in an attempt
to drive a wedge between Oughtred and the English upper classes who
patronized mathematical practitioners. Delamain seized on remarks that
Oughtred had made concerning the education of William Forster, a
nonuniversity man who had trained under Oughtred for a career as a
teacher of mathematics. Oughtred, determined to make a scholar of
Forster, had admonished him "That the true way of Art is not by
Instruments, but by demonstration: and that it is a preposterous course
of vulgar Teachers, to beginne with Instruments, and not with the Sciences, and so in stead of Artists, to make their Schollers onely doers of
tricks, and as it were jugglers." Delamain twisted this admonition into
Oughtred's supposed accusation that "noble Personages" with practical
mathematical interests were no more than "doers of tricks" and "jugglers."[95]

In response, Oughtred lavishly praised the English gentry and then
blamed any mathematical deficiency on the gentry's part on inept mathematical teachers, like Delamain. "[N]o Land under the cope of heaven,"
he wrote of England, "is more happy with a gallant, and glorious flower
of Gentry, . . . which is more liberally enriched by nature with ingenuity,
and all excellent endowments both of wit, courage, and abilities of mind
and body." But, he stressed, the development of mental talents depends
on "the wise . . . choice of Teachers," and it was Delamain, not he, who
had associated the nobility with "doers of tricks" and "jugglers." Putting
these points together, he concluded, "having so long ordered your studies, disposed of your times, and received your money, [Delamain] hath
even in his owne conscience done you so little good, that there being but

94. Oughtred, *To the English Gentrie*, 8.   95. Ibid., 27–28.

the very name of *losse of time, jugling, and ignorance,* . . . is himself first of all ready . . . [to] pinne it upon you." Labeling Delamain a "*vulgar Teacher,*" Oughtred demanded his credentials: "I may . . . aske you, how you obtayned *that calling and profession.* . . . What University, what degree, what court of faculties, what other lawfull way, conferred it upon you?"[96]

Oughtred's *Letter to the Gentrie* thus provided the broad rationale for the emergence of academic mathematics in England. In his defense of the study of mathematics he was clearly speaking for a small band of English thinkers, mostly analytic mathematicians, who hoped to find a formal place for the exciting mathematical developments of the late sixteenth and early seventeenth centuries at Cambridge and Oxford. His arguments for mathematics as an academic subject were to be elaborated as the seventeenth century progressed, and eventually accepted to the point where mathematics formed the core of the undergraduate program at Cambridge University.

96. Ibid., 28–29, 11.

3

John Collins's Campaign for a Current English Algebra Textbook

The 1660s and 1670s

Although in the second third of the seventeenth century Oughtred's *Key* and Harriot's *Praxis* converted key English thinkers to the analytic way, neither work seemed to satisfy the needs of less talented students wanting to pursue algebra. *The Key* was too brief, and both it and the *Praxis* were too dependent on symbols for the general student. The *Praxis*, designed as a supplement to Viète's *Analytic Art,* was not a general algebra textbook. *The Key* – with its focus on quadratic equations and positive real roots – had been, to a certain extent, obsolete as an introductory algebra textbook even at the time of its original publication. Furthermore, subsequent editions of *The Key,* which continued to be reissued through 1702, took no explicit account of René Descartes's *Géométrie* of 1637 or other later algebraic developments.[1]

Of all the analytic mathematicians in England during the second half of the seventeenth century, John Collins (1625–1683) most persistently pushed for the preparation of an English algebra textbook that both elaborated the principles of the subject in a fashion appropriate for university men and sophisticated practitioners, and covered the main algebraic developments of the century. Collins coaxed Isaac Newton and, at times, John Wallis to produce the desired book. He encouraged John Pell to complete his English edition of a German algebra as well as John Kersey to finish his textbook on the subject.

Collins died in 1683 without seeing the best fruits of his campaign for

---

1. The relationship between Descartes's *Géométrie* and *The Key* merits further study. On Oughtred's possible influence on Descartes, see H. Bosmans, S.J., "La première édition de la *Clavis Mathematica* d'Oughtred. Son influence sur la *Géométrie* de Descartes," *Annales de la Société scientifique de Bruxelles* 35, pt. 2 (1910–1911): 24–78. Bosmans argued for at least an indirect influence, which Cajori contested (Florian Cajori, *William Oughtred: A Great Seventeenth-Century Teacher of Mathematics* [Chicago: Open Court, 1916], 71–73).

a current English algebra textbook. The two major English algebras he helped inspire, Wallis's *Treatise of Algebra* and Newton's *Universal Arithmetick,* were published in 1685 and 1707, respectively. Still, by the end of his life Collins could have taken some satisfaction in the role he had played in the evolution of algebra into an academic subject in England. He had repeatedly brought algebraic developments from the continent to the attention of Wallis, Newton, and other English mathematicians; he had kept Wallis abreast of the burgeoning interest in algebra among Cambridge University scholars and London mathematical practitioners. Moreover, he had fostered the interest of Wallis and Newton in (what the three men saw as) a major technical problem of the new algebra – the extraction of the cube roots of binomials, including cases where the binomials were complex numbers. Finally, in 1668 he had assisted Pell in the printing of the translation of J. H. Rahn's *Algebra* and in 1673 he had seen Kersey's *Algebra* through the press. Neither an algebraist of the caliber of Wallis and Newton nor a textbook author like Pell and Kersey, Collins was nevertheless a major facilitator of early modern algebra in England, one of those influential but "second-rate" mathematical thinkers who are too often rendered invisible by historians of mathematics.

I

Collins[2] was the son of a nonconformist minister who had died young. He was basically a mathematical autodidact who in his early years had moved through an apprenticeship to a bookseller, a clerkship in the kitchen of Prince Charles, and a stint as a seaman in the service of the Venetians. If as a young man Collins had a mathematical tutor, he was William Marr, then clerk to the kitchen of the prince and later a maker of sundials and a surveyor, who seems to have taught Collins the arithmetic essential to the position of assistant clerk.[3]

From midcentury, Collins supported himself through occupations tied

2. See A. M. C., "John Collins," *Dictionary of National Biography,* ed. Leslie Stephen and Sidney Lee, 22 vols. (Oxford: Oxford University Press, 1921–1922), 4:824–825; E. G. R. Taylor, *The Mathematical Practitioners of Tudor and Stuart England* (Cambridge: Cambridge University Press, 1967), 228; and Derek T. Whiteside, "John Collins," *Dictionary of Scientific Biography,* ed. Charles C. Gillispie, 18 vols. (New York: Scribner's Sons, 1970–1990), 3:348–349.

3. Lesley Murdin, *Under Newton's Shadow: Astronomical Practices in the Seventeenth Century* (Bristol: Adam Hilger, 1985), 48. There is a note on Marr (fl. 1640–1684) in Taylor, *Mathematical Practitioners,* 222.

to some extent to his mathematical knowledge: through teaching and later through a series of minor posts as a government accountant. He published a few books on practical mathematics, including *An Introduction to Merchants Accompts* of 1653, and a few papers, principally on arithmetic and algebra; but his greatest service to English mathematics was his extensive correspondence with the period's major mathematical figures. Isaac Barrow, John Flamsteed, James Gregory, Newton, and Wallis willingly exchanged letters with Collins, who by personality was the perfect correspondent for proud men of attainment – "a man of good arts, and yet greater simplicity; able, but no ways forward."[4] Collins filled his letters with news of the latest mathematical developments from all parts of western Europe,[5] and with pleas for increasing England's mathematical reputation through original research and publication. A trusted friend of his mathematical compatriots, he also used his knowledge of the London book trade to see some of their publications through the press.[6]

His most persistent and at times frustrating efforts were directed toward the publication of an English algebra textbook suitable for university students and advanced mathematical practitioners. Whereas Oughtred's *Key* proved a stimulating introduction to analytic mathematics for the likes of Wallis and Newton, lesser students had complained primarily about its terseness and called for an expanded version of *The Key*. In 1633 William Robinson, who believed that the "analytical way is indeed the only way," had written Oughtred, "I shall long exceedingly till I see your Clavis turned into a picklock; and I beseech you enlarge it, and explain it what you can."[7] A few years later, Robinson had made a second appeal for an expanded *Key*, this time based on a stereotype of the general student as too "lazy or dull" to take to "abstruse" mathematics quickly. "Brevity may well argue a learned author, that without any

4. Sir Phillip Warwick, quoted in A. M. C., "John Collins," 825. The franking privileges attached to Collins's government posts also made correspondence with and through him attractive (Taylor, *Mathematical Practitioners*, 228).

5. Collins corresponded "not only with his compatriots but with Bertet, Borelli, and (through Oldenburg) with Huygens, Sluse, Leibniz, and Tschirnhausen" (Whiteside, "John Collins," 348).

6. He assisted with many of Barrow's publications.

7. W. Robinson to Oughtred, 11 June [1633], *Correspondence of Scientific Men of the Seventeenth Century*, ed. Stephen Jordan Rigaud, 2 vols. (Oxford, 1841; reprint, Hildesheim: Georg Olms, 1965), 1:16. Rigaud tentatively suggests that the author was William Robinson, the archdeacon of Nottingham from 1635 to 1660 (1:20).

excess or redundance, either of matter or words, can give the very substance and essence of the thing treated of," he had complimented Oughtred, "but it seldom makes a learned scholar; and if one be cap[able, twenty are not."[8]

Oughtred, however, had not proved sympathetic to such pleas. He had written to Robinson that even students with "mathematical genius" needed to "carefully study" *The Key* if they were to apply its results.[9] Much later when Richard Stokes, after returning to Cambridge from studies at the Albury rectory, had asked him to clarify a few mathematical points (including some from his *Key*), Oughtred had answered Stokes's questions but reminded him that "whosoever will rightly study my book, that it may be a Clavis to him, ... must be attentus, operans, constanterque per ipsa vestigia insequens."[10] In short, serious students of *The Key* had urged its author to make it more accessible through lengthening, but Oughtred had simply advised that the work – available from 1647 with some appended tracts – be studied more diligently.

After Oughtred's death in 1660, Collins and Wallis discussed the prospect of expanding *The Key*. Wallis knew Margaret Lichfield, who had an impression of *The Key* that she was willing to sell to the publisher Moses Pitts. Wallis apparently asked Collins, because of his connections in the publishing industry, to pursue the matter. Collins reported that Pitts was ready to negotiate with Lichfield provided that the publisher would have the right to add to *The Key* any "comments or explications" that his friends or Wallis advised.[11]

Whereas Pitts and seemingly Collins saw merit in elaborating on Oughtred's book, even if only by attaching a commentary, Wallis thought that *The Key* ought to stand untouched as a mathematical classic. Here Wallis seems to have been in the minority, for, as Collins observed, other commentaries were already in the works. A Mr. Bunning, described as an aged minister from Warwickshire, "hath commented on the Clavis, which he left with Mr. Leybourne to be printed." Too, Collins had heard of a commentary prepared by Gilbert Clerke, a graduate of Sidney Sussex College, Cambridge, who taught mathematics privately. Collins intimated, furthermore, that there was by this time not only a growing consensus that *The Key* needed explication but also disagreement on the book's basic merits. John Kersey was reported to

8. William Robinson to Oughtred, 2 July 1636, ibid., 1:26–27.
9. "Oughtred's Answer to Robinson," ibid., 1:10.
10. Oughtred to Stokes, [1654–1655], ibid., 1:83–84; Stokes to Oughtred, 6 Feb. 1654–1655, ibid., 1:81–82.
11. Collins to Wallis, 2 Feb. 1666–1667, ibid., 2:471.

admire nothing in it except its treatment of Euclid's *Elements*. Robert Anderson, a weaver, had declared that *The Key* and Bunning's commentary on it "were immethodical, and the precepts for educing the roots of an adfected equation maim and insufficient."[12] Even the late Samuel Foster, who held a M.A. from Emmanuel College, Cambridge, and served as the professor of astronomy at Gresham College in London, had "utter[ed] his dislike of" *The Key*.[13]

Wallis, who saw two Latin editions of *The Key* through the press,[14] defended it as a mathematical classic and, as such, a book worthy of reprinting in its original form. In 1666/1667 he told Collins that he valued Oughtred's work "as a very good book, . . . which doth in as little room deliver as much of the fundamental and useful part of geometry (as well as of arithmetic and algebra) as any book I know." Mathematical "fashions will daily alter." No mathematical classic, he implied, is perfect; each uses notation, methods, and so on, that are subject to later improvement. In *The Key*, for example, Oughtred had "delivered . . . in a more advantageous way" some important propositions that Euclid and Archimedes had stated earlier. Nevertheless, Euclid and Archimedes "did not cease to be classic authors and in request." "And the like I judge of Mr. Oughtred's Clavis," Wallis concluded, "which I look upon (as those pieces of Vieta who first went in that way) as lasting books and classic authors in this kind; to which, notwithstanding, every day may make new additions." Precisely because of its standing as a classic and because Oughtred himself had steadfastly refused to write anything but "a small epitome," Wallis argued against expanding *The Key* posthumously. His final opinion on the matter was that "comments, by enlarging . . . [*The Key*], do rather destroy it" and therefore "those, who may publish any thing that way, . . . [should] do it as a work of their own, than as a comment on this."[15]

Such arguments seem to have turned Collins from promoting a revision of *The Key* to calling for a new English algebra textbook. In 1667

12. Ibid. There is a brief biography of Gilbert Clerke in Taylor, *Mathematical Practitioners*, 230. Robert Anderson wrote on stereometry, gunnery, and the value of mathematical studies (Taylor, *Mathematical Practitioners*, 249).

13. Collins to Wallis, n.d., *Correspondence of Scientific Men*, 2:483. On Foster, see Mordechai Feingold, *The Mathematicians' Apprenticeship: Science, Universities and Society in England, 1560–1640* (Cambridge: Cambridge University Press, 1984), 80, 114.

14. Cajori, *William Oughtred*, 19. These were the editions of 1652 and 1667.

15. Wallis to Collins, 5 Feb. 1666/1667, *Correspondence of Scientific Men*, 2:474–476.

he told John Pell that there were Cambridge tutors and students who needed a textbook of arithmetic and algebra similar to the "small anonymous Jesuit's Euclid."[16] Four years later he called for an advanced algebra textbook covering equations of higher order, which would appeal not only to academic mathematicians but also to "divers ingenious mechanics, gaugers, carpenters, shipwrights, some seamen, lightermen, &c., whose whole discourse is about equations."[17] In short, according to Collins, by the 1660s English university men and mathematical practitioners needed an algebra textbook that explained the foundations of the subject, was up-to-date and accessible to general students, and perhaps employed more words than either *The Key* or the *Praxis*.[18]

2

Oughtred and Harriot were not totally responsible for the analytic enthusiasm that lay behind the increasing calls for an up-to-date English algebra textbook. The work of René Descartes (1596–1650) on analytic geometry profoundly affected English as well as continental mathematics. In 1664 Newton studied Frans van Schooten's second Latin edition of Descartes's *Géométrie,* a work that subsequently determined the direction of his mathematical research.[19] Prior to Newton's Cartesian conversion, other English mathematical thinkers had turned to *La géométrie* and the major commentaries on it for insights into analytic geometry as well as early modern equation theory. Wallis read at least part of the work in the original French as early as 1648,[20] and, like Newton, he, Collins, James Gregory, Pell, and Kersey were all in one way or another indebted to it.

Descartes's *Géométrie,* one of three appendixes to his *Discours de la méthode,* had been published in 1637. The work achieved its greatest

16. John Collins to Dr. Pell, 9 Apr. 1667, ibid., 1:125. The Jesuit work was *Euclidis Elementa Geometrica, novo ordine ac methodo fere, demonstrata* (London, 1666).
17. Collins to Wallis, 21 Mar. 1671, *Correspondence of Scientific Men,* 2:526.
18. Collins believed that students could learn specious division, e.g., more easily when the rules were given in words rather than symbols (Collins to Dr. Pell, 9 Apr. 1667, 126).
19. D. T. Whiteside, "Sources and Strengths of Newton's Early Mathematical Thought," in *The "Annus Mirabilis" of Sir Isaac Newton, 1666–1966,* ed. Robert Palter (Cambridge, Mass.: MIT Press, 1970), 69–85, on 73–75.
20. Wallis to Collins, 29 Mar. 1673, *Correspondence of Scientific Men,* 2:559.

popularity not in its original French edition but rather in the Latin editions prepared by Frans van Schooten Jr. Schooten, a professor of mathematics at Leiden University and a leader of the strong seventeenth-century Dutch mathematical school, had been among the first mathematicians to recognize the power of Descartes's new geometry. Nevertheless, he had found *La géométrie* unsatisfactory, since Descartes seemed therein neither to aim at nor to achieve clarity. As later commentators have agreed, the book had "many gaps in . . . [its] arguments," "esoteric elements which were difficult to grasp," and a structure that impeded its use as a textbook.[21]

Schooten translated *La géométrie* from French to Latin, which remained the universal language of educated western Europeans, and commented on and extended the work.[22] His first annotated edition of the work appeared in 1649; his second edition, in two volumes of 1659 and 1661. This second edition, which included papers by his students Johann Hudde and Hendrick van Heuraet, combined direct commentary on *La géométrie* with an overview of the early modern results on normals, tangents, and extreme values, all accompanied by applications. In so extending Descartes's work, Schooten and his students "created a link between Descartes' *Géométrie* . . . and the invention of the infinitesimal calculus by Newton and Leibniz."[23]

Moreover, *La géométrie* – although first and foremost a geometric work – affected algebra in fundamental ways. The treatise introduced analytic geometry, for which Descartes and Pierre de Fermat (ca. 1608–1665) are often cited as co-inventors. Developed in the treatise, but almost as a "side issue," was the fundamental insight of analytic geometry that curves can be studied through their equations and equations through their curves. The treatise thus made a compelling case for the usefulness of algebra applied to geometry and, generally, the interconnections between these two mathematical subdisciplines.[24] In addition,

21. Jan A. van Maanen, "Facets of Seventeenth Century Mathematics in the Netherlands" (Ph.D. diss., Utrecht University, 1987), 20–31 (quotations from p. 24).
22. Ibid., 24. He "explained and completed points where proofs or steps were missing, he laid foundations, he referred to other publications, he simplified and corrected . . . he widened the scope of the text (i.e. he generalized, answered questions that Descartes had left open, worked out similar cases, studied new problems and extended the theory)."
23. Ibid., 1 (quotation), 23, 30.
24. For an exceptionally coherent analysis of *La géométrie,* which informs my discussion, see Henk J. M. Bos, "The Structure of Descartes's *Géométrie,*" in *Lectures in the History of Mathematics* (Washington, D.C.: American

book III of *La géométrie* formalized the elementary equation theory of the early modern period as well as establishing many of the symbols and terms of modern algebra.

The intermingling of algebra with geometry reflected Descartes's dissatisfaction with the prevailing approaches to the two mathematical subdisciplines. In the *Discours de la méthode,* he explained how in his search for true knowledge he had reexamined his earlier philosophical studies, including logic, geometric analysis, and algebra, and how these studies had all come up short. "With regard to logic," he declared, ". . . syllogisms and most of its other techniques are of less use for learning things than for explaining to others the things one already knows." Eschewing the traditional emphasis on synthetic, Euclidean geometry, Descartes turned to the evaluation of the two main branches of analytic mathematics, namely, classical geometric analysis and the new algebra. He found neither totally satisfactory. Both dealt with

> highly abstract matters, which seem to have no use. Moreover the former is so closely tied to the examination of figures that it cannot exercise the intellect without greatly tiring the imagination; and the latter is so confined to certain rules and symbols that the end result is a confused and obscure art which encumbers the mind, rather than a science which cultivates it.[25]

Continuing nevertheless to pursue mathematics as the only science offering "any demonstrations – that is to say, certain and evident reasoning," he claimed to go to the core of the subject; to find at that core "nothing but the various relations or proportions that hold between . . . objects"; and to construct a new mathematical approach to studying

Mathematical Society, 1991), 37–57 ("side issue" on p. 37). Historians have debated Descartes's contributions to analytic geometry, some claiming to find significant forerunners of his approach and others stressing the differences between his work and modern analytic geometry (e.g., he did not restrict himself to coordinate axes at right angles to one another). A recent synopsis of Descartes's contributions is given by J. J. Gray, "Algebraic and Analytic Geometry," in *Companion Encyclopedia of the History and Philosophy of the Mathematical Sciences,* ed. I. Grattan-Guinness, 2 vols. (London: Routledge, 1994), 2:847–859, esp. 849–853. For references to important literature on the topic, see Bos, "Structure," 54 n. 2. My discussion centers on Descartes's analytic geometry, since Fermat, by failing to publish during his lifetime, did not exert as powerful an influence on seventeenth-century mathematics.

25. *Discourse on the Method,* in *Descartes: Selected Philosophical Writings,* ed. John Cottingham, Robert Stoothoff, and Dugald Murdoch (Cambridge: Cambridge University Press, 1988), 28–30.

these relations. He supposed the relations "to hold between lines, because . . . [he] did not find anything simpler, nor anything that . . . [he] could represent more distinctly to . . . [his] imagination and senses." Thus his approach had the advantages of geometric analysis in that it dealt with objects capable of sensible representation. But what of the possible straining of the human imagination, which Descartes had portrayed as a major side effect of geometric analysis? To "keep the lines in mind or understand several together," he prescribed that they be "designate[d] . . . by the briefest possible symbols." "I would," he gloated, "take over all that is best in geometric analysis and in algebra, using the one to correct all the defects of the other."[26] Hence, according to Descartes, was born analytic geometry – in which the imagination could assist the intellect but, because of algebraic symbolism, was not called upon at every step.

La géométrie, then, was an elaboration of Descartes's new blend of algebra with geometric analysis. As Viète had claimed that the analytic art could solve all problems, Descartes pursued here nothing less than the developing of "a universal method of finding the constructions for any problem that could occur within the tradition of geometrical problem solving."[27] Book I of *La géométrie,* which focused on problems that could be solved by straightedge and compass, opened: "Any problem in geometry can easily be reduced to such terms that a knowledge of the lengths of certain straight lines is sufficient for its construction." As he had hinted in the *Discours,* Descartes now reduced geometric problems to problems of straight lines; he denoted the lines by algebraic symbols; and he subjected the lines to geometric operations involving constructions with straightedge and compass and corresponding to the five basic operations of arithmetic – addition, subtraction, multiplication, division, and extraction of square roots.[28]

His very first definition, that of the multiplication of two line segments, set the new geometry apart from the geometric demonstrations of algebraic results that Arabic mathematicians and Cardano had offered. In Descartes's geometry, multiplication of two line segments did not produce a plane figure. Rather, as the product of two numbers is a number, so he defined, the product of two line segments is a line segment: "Let AB be taken as unity, and let it be required to multiply BD by BC. I have only to join the points A and C, and draw DE parallel to

26. Ibid.   27. Bos, "Structure," 43.
28. *The Geometry of René Descartes with a Facsimile of the First Edition,* trans. David Eugene Smith and Marcia L. Latham (New York: Dover, 1954), 2–5.

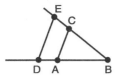

Figure 2. Descartes's multiplication of line segments (*The Geometry of René Descartes,* trans. David Eugene Smith and Marcia L. Latham [New York: Dover, 1954], 4)

CA; then BE is the product of BD and BC" (see figure 2). To assure that his readers caught this new definition, Descartes (who wasted no words in *La géométrie*) added: "by a^2, b^3, and similar expressions, I ordinarily mean only simple lines, which, however, I name squares, cubes, etc., so that I may make use of the terms employed in algebra."[29] Thus he stripped terms like "square" and "cube" of traditional geometric connotations but assigned them a new meaning that exploited a fundamental correspondence between arithmetic and geometry.

After covering his preliminary definitions in a few pages, he outlined his method of solving geometric problems: he began with a problem stated geometrically; translated the problem into an equation; if necessary, reduced that equation to an irreducible one; and then geometrically constructed the root(s) of the equation. Since this work began and ended in geometry, algebra was, in a very basic way, a means to an end for Descartes.

But it was not simply a means to an end, for algebra determined some key methodological aspects of Descartes's geometry. Algebra proved a good guide to constructions for many geometric problems – those translating into equations of the first and second degrees, whose roots are constructible by straightedge and compass. But, as Henk Bos has stressed, in his quest for a method to find constructions for all geometric problems Descartes had to consider as well problems for which algebra does not lead to constructions based on the straightedge and compass. This second group of problems is hardly insignificant since Cardano's solution of cubic and quartic equations involves cube roots, which cannot be constructed with a straightedge and compass, and there are no general algebraic solutions to equations of the fifth and higher degrees.

29. Ibid., 5. As Boyer has pointed out, however, Descartes seemed sometimes to think of such terms as having geometric dimension. See ibid., 6, and Carl B. Boyer, *History of Analytic Geometry* (New York: Scripta Mathematica, 1956), 84.

Curves other than the straight line and circle are clearly needed for some constructions, and Descartes had to decide which other curves to use and why.[30] In making key methodological decisions about acceptable and unacceptable ways of constructing the roots of equations produced by the more intractable second group of problems, he relied heavily on algebra. In two respects he allowed algebra to determine his geometric constructions. First, in book II he implicitly recognized as legitimate curves of construction only those that are algebraic, that is, expressible in equations. Secondly, discussing in book III problems for which different constructions are possible, he mandated that the simplest curves be used, and he defined those curves as those with equations of the lowest algebraic dimension.[31] This was algebra dictating the methodology of geometry.

3

Further muddying distinctions between the two mathematical subdisciplines, *La géométrie* was not only an avant-garde geometric textbook but an algebraic one as well. Book III offered a compendium of equation theory, intended to aid in such essential parts of Descartes's universal method as the simplifying of equations into irreducible forms.[32] Regardless of whether readers embraced, or even understood, the universal method, they found in book III perhaps the best compendium of equation theory available on the continent – and, Wallis's later objections to it notwithstanding, in the whole of Europe.

Throughout *La géométrie*, and especially in book III, Descartes brought algebraic symbolism to new heights of order and simplicity, and in doing so he became a major arbiter of the language of early modern algebra. He used the first letters of the alphabet to denote known quantities and the last letters to denote the unknowns; he offered a compact notation for powers, writing x, x^2 (sometimes, xx), x^3, and so on. Although not entirely original, his compendium of equation theory collected many of the major sixteenth- and seventeenth-century results of

30. These important points are made in Bos, "Structure," 46–47.
31. On Descartes's conclusion that acceptable curves are those with algebraic equations, see Bos, "Structure," 47–49; on his considerations of simplicity, see ibid., 49–51.
32. This is Bos's interpretation, although he admits: "At first sight, . . . this [equation] theory seems totally unrelated to geometry" (Bos, "Structure," 51).

this rapidly developing subject, including solutions of cubic and quartic equations. He defined the concepts of equation and root, and cast the theory in a form and with rules that soon became standard for European mathematicians. "An equation," he wrote, "consists of several terms, some known and some unknown, some of which are together equal to the rest; or rather, all of which taken together are equal to nothing; for this is often the best form to consider." Roots are "values of the unknown quantity."[33] From this point he consistently assumed the "best" form for equations involving a single unknown – placing terms in descending order of dimension of the unknown and setting the result equal to zero.

He then stated general rules of equation theory, including "the sum of an equation having several roots is always divisible by a binomial consisting of the unknown quantity diminished by the value of one of the true [positive] roots, or plus the value of one of the false [negative] roots. In this way, the degree of an equation can be lowered." An example was the equation

$$x^2 - 5x + 6 = 0,$$

with roots of 2 and 3 and thus the factors $x - 2$ and $x - 3$.[34] Assuming the standard form of an equation, he proposed what came to be known as "Descartes's rule of signs": "An equation can have [*peut y avoir*] as many true roots as it contains changes of sign, from $+$ to $-$ or from $-$ to $+$; and as many false roots as the number of times two $+$ signs or two $-$ signs are found in succession." In illustration, he determined the number of positive and negative roots of the equation:

$$x^4 - 4x^3 - 19x^2 + 106x - 120 = 0.$$

Here the signs of the coefficients change first from $+$ to $-$ (evidencing one positive root), then another $-$ follows (a negative root), then $-$ changes to $+$ and $+$ back to $-$ (two more positive roots). And indeed he gave three positive roots (2, 3, and 4) and one negative root (-5) for the equation.[35]

He also tentatively enunciated the fundamental theorem of algebra, a limited statement of which Albert Girard had given in 1629. Toward the

---

33. Descartes, *The Geometry,* 156–159.   34. Ibid., 159.

35. Ibid., 160. Note that Descartes does not indicate that his rule applies only to equations with real roots. Also, Descartes "merely sets a limit to the number of positive and negative roots." On these points, see J. F. Scott, *The Scientific Work of René Descartes (1596–1650)* (London: Taylor & Francis, 1976), 140–141.

beginning of book III, Descartes wrote: "Every equation can have as many distinct roots (values of the unknown quantity) as the number of dimensions of the unknown quantity in the equation."[36] Later in the same book he proposed a stronger version of the theorem, which came close to the statement that every polynomial of order *n* actually has *n* roots.[37]

Acceptance of the strong version of the fundamental theorem involved, first of all, acceptance of negative numbers. In illustration of his weak version, Descartes considered the roots $x = 2$, $x = 3$, and $x = 4$, and showed that multiplication of the equations $x - 2 = 0$, $x - 3 = 0$, and $x - 4 = 0$ produced

$$x^3 - 9x^2 + 26x - 24 = 0, (*)$$

an "equation, in which *x*, having three dimensions, has also three values, namely, 2, 3, and 4." In the paragraph following this example, he admitted that: "It often happens . . . that some of the roots are false or less than nothing." "Suppos[ing] . . . *x* to represent the defect of a quantity 5," he then multiplied equation * by $x + 5 = 0$ to get:

$$x^4 - 4x^3 - 19x^2 + 106x - 120 = 0.$$

The new equation, in which *x* has four dimensions, he observed, has also "four roots, namely three true roots, 2, 3, and 4, and one false root, 5."[38]

Although Descartes regularly listed negative roots of equations, and they were crucial to the fundamental theorem, he saw negative numbers quite differently from their positive counterparts. As Cardano had called negative roots "fictitious," he called them "false," and he described them as being "less than nothing." Furthermore, he stopped short of formulating the concept of an isolated negative. As in the previous example, he often indicated a "false root" not by a number preceded by a minus sign, but rather by a number preceded by the expression "the false root." In some cases where the context made it clear that the root was "false," he simply wrote the absolute value of the root. For example, in a later reference to the previous equation, he gave its roots as 2, 3, 4, and 5.[39] Nevertheless, he regularly symbolized and worked with isolated specious negatives, for example $-q$. But even then he conceived of $-q$ not as a single entity in itself but rather as the quantity *q* marked by the sign $-$.[40] This latter view seems in fact to have been at the core of his

---

36. Descartes, *The Geometry,* 159. As Smith and Latham note, Descartes here wrote "can have" *(peut-il y avoir)* and not "must have."
37. Ibid., 175.   38. Ibid., 159.
39. Ibid., 160.   40. "[L]a quantité *q* est marquée du signe - -" (ibid., 201).

ontological legitimation of what are now called negative numbers. If a negative number was "real" – and Descartes so classified the negatives – it was real because there was some quantity that corresponded to the number. The negative number itself however did not stand for that quantity but for a "defect" of that quantity. This was, in fact, the way that Descartes first introduced negative roots: "[Q]uelques unes de ces racines sont fausses, ou moindres que rien. comme si on suppose que *x* designe aussy le defaut d'une quantité."[41]

Difficulty with negative numbers limited also his analytic geometry. Indeed, "he knew in a general sort of way that negative lines are directed in a sense opposite to that taken as positive." When dealing with the intersection of a circle and a parabola, for example, he explained that the positive and negative roots lie on opposite sides of E, the center of the circle. Still, "he did not realize the applicability of the idea as a general principle in connection with coordinate systems. He occasionally made use of negative ordinates but not of negative abscissae."[42]

The strong version of the fundamental theorem entails acceptance of roots involving $\sqrt{-1}$ as well as negative ones. "Neither the true nor the false roots," Descartes observed,

> are always real; sometimes they are [only] imaginary [*seulement imaginaires*]; that is, while we can always conceive of as many roots for each equation as I have already assigned, yet there is not always a definite quantity corresponding to each root so conceived of. Thus, while we may conceive of the equation $x^3 - 6x^2 + 13x - 10 = 0$ as having three roots, yet there is only one real root, 2, while the other two, however we may increase, diminish, or multiply them in accordance with the rules laid down, remain always imaginary.[43]

In short, the stumbling block to Descartes's unqualified enunciation of the fundamental theorem was the "imaginary" number, which defied geometric construction,[44] corresponded to no "definite quantity," and

41. Ibid., 175, 158. Compare with Napier's remarks: "But the Logarithmes which are lesse then [*sic*] nothing, we call Defective, or wanting" (quoted in Yannis Thomaidis, "Aspects of Negative Numbers in the Early 17th Century," *Science and Education* 2 [1993]: 74).

42. Boyer, *History of Analytic Geometry,* 86 (quotations); Descartes, *The Geometry,* 200.

43. Descartes, *The Geometry,* 174–175.

44. According to Descartes, complex roots were not without value in analytic geometry. If the equation for a geometric problem led to complex roots only, then the solution to the original problem was not constructible (Descartes, *The Geometry,* 187).

therefore existed only to the extent that it was "conceived of" or imagined.

Descartes's final statement of the fundamental theorem and accompanying endorsement of complex roots were highly qualified. Mathematicians could "conceive of" n roots for a polynomial of degree n. Some of these were true, standing for specific quantities; others, false, standing for defects of quantities. But still other roots corresponded in no way to quantities. These roots were mere products of the human imagination, or only "imaginary"[45] – a term he seems to have coined. Thus, although Descartes rather consistently recognized complex roots and thereby promoted their use by algebraists across Europe, his very name for these roots and his explanation of them took the legitimation of the complex numbers little beyond Cardano's *Great Art,* where such roots had already appeared as entities to be "imagined" rather than realized in arithmetic or geometry.

4

Not in one way, but in many, did the ideas of *La géométrie* – including analytic geometry, the fundamental theorem of algebra, false roots, and imaginary roots – shape English mathematics. Pell read the book and translated large sections by 1640.[46] Wallis proved himself one of the finest of the new breed of analytic geometers, so fine that Carl Boyer has observed that "whereas Descartes made . . . arithmetization [of geometry] a possibility, Wallis made it a fact."[47] Nevertheless, in his later publications, Wallis questioned Descartes's originality and significance. As an undergraduate Newton read Schooten's second Latin edition of *La*

45. On the evolution of Descartes's views on the imagination, see Dennis L. Sepper, "Descartes and the Eclipse of Imagination, 1618–1630," *Journal of the History of Philosophy* 27 (1989): 379–403. "Through December 1629," Sepper writes, "we find him still using the term imagination to refer to a power that is at the foundation of science. . . . But then, though he continues to use the terms *imagination, imaginer,* and *fantaisie,* they lose the cognitive connotation and are employed just in the narrow sense of image-making without any intrinsic truthfulness and even with overtones of unreliability, even deceptiveness" (ibid., 398).

46. Jean Jacquot, "Sir Charles Cavendish and His Learned Friends," *Annals of Science* 8 (1952): 13–27, on 27.

47. Boyer, *History of Analytic Geometry,* 109.

géométrie and began immediately to develop the mathematics contained therein.[48] But Newton expressed regrets late in life for the analytic turn in his mathematics that *La géométrie* had occasioned.

Although extremely influential, *La géométrie* was thus to enjoy no lasting acclaim among English mathematicians. The latter discounted the treatise's significance at least partially because of national chauvinism. More fundamentally, in spirit and in content *La géométrie* ran against the early English analytic tradition that had so quickly formed around Oughtred and Harriot. Whereas these English mathematicians cared for algebra in and of itself, Descartes treated the subject largely as a means to geometric ends. He translated geometric problems into equations, but – because his aims were geometric and perhaps also because he did not "regard . . . an equation as an adequate definition of the curve"[49] – he refused to accept roots coming from algebraic formulas as final solutions. Oughtred and Harriot did algebra; Descartes did geometry. But this comparison is too simple. Harriot, especially, was a forerunner of pure algebra. Descartes, whose essential work was analytic geometry, practiced a different type of algebra, algebra mixed with geometry, which Boyer has described as "Cartesian geometrical algebra"[50] but which in its "purest" form of book III is perhaps better described as algebra in the service of geometry.

On the other hand, even as *La géométrie* emphasized geometry at the expense of algebra, it subverted traditional geometry by emphasizing the analytic method over the synthetic and by permitting algebra to dictate geometric methodology. It was this subversion that the older Newton found objectionable. Later, when Wallis and Newton seemed to struggle for control of the direction of English mathematics, the former saw Descartes's work as an unoriginal treatise derivative of Harriot's *Praxis* and, to some extent, as an example of the geometric contamination of algebra, whereas the latter saw the work as a subversion of classical geometry. At the very minimum, *La géométrie* assured that, in England as well as on the continent, the relationship between algebra and geometry would be debated into the modern period.

48. Whiteside, "Newton's Early Mathematical Thought," 74–75.
49. Boyer, *History of Analytic Geometry,* 88.
50. See Carl B. Boyer, "Cartesian and Newtonian Algebra in the Mid-Eighteenth Century," *Actes du XIᵉ congrès internationales d'histoire des sciences* (Warsaw, 1968), 3, 195–202, on 198.

5

But in the 1660s Wallis had not yet published his charges of near plagiarism against Descartes, and Newton was quite taken with the new analysis. Collins, Newton, and Wallis agreed at this point on the necessity of exposing English students to the recent continental advances in mathematics, including analytic geometry and Descartes's equation theory. As an alternative to an expanded version of *The Key*, Collins searched the foreign literature for an up-to-date textbook that was suitable for use by English students, either in its original form or in translation.

By 1666 he was recommending Gerard Kinckhuysen's *Algebra Ofte Stel-konst* of 1661 to his Cambridge and London friends.[51] Written in Low Dutch, Kinckhuysen's book was an elementary algebra that was intended to serve as a companion to the author's earlier work on the elements of Cartesian analytic geometry. In Collins's opinion, the book met the particular needs of English students of the 1660s. It opened with introductory material on algebra and the basic algebraic operations as well as a rather lengthy treatment of irrational numbers. The latter especially caught the eye of Collins, who wrote Pell that Kinckhuysen's *Algebra* "had in it the doctrine of surd numbers, binomials and residuals and their roots, and more than either the Principia or yours."[52] In addition, the *Algebra* provided "a useful if somewhat pedantic summary of recently published work in algebra," mainly the third book of Descartes's *Géométrie*, with additions from Schooten's *Commentarii*, Hudde's *De reductione æquationum*, and Kinckhuysen's own researches.[53] Collins felt so strongly about the book that he recommended it in its original Dutch to the (unidentified) Cambridge scholar who had alerted him to the university's need for an algebraic work comparable

51. Collins to Pell, 28 Aug. 1666, *Correspondence of Scientific Men*, 1:118. In an undated letter to Wallis, Collins described the work as "an excellent introduction" (ibid., 2:484). My brief account of Kinckhuysen's *Algebra* and Collins's efforts to have it translated into English and published, with commentary, and my later discussion of Newton's role in the project follow Whiteside's reconstruction. See *The Mathematical Papers of Isaac Newton*, ed. D. T. Whiteside, 8 vols. (Cambridge: Cambridge University Press, 1967–1981), 2:277–294.

52. Collins to Pell, 9 Apr. 1667, *Correspondence of Scientific Men*, 1:126. The *Principia* is Schooten's *Principia matheseos*.

53. Whiteside, "Introduction" to part 3, *Mathematical Papers of Newton*, 2:278.

with the Jesuit's *Elements*,[54] and then arranged to have the book translated into Latin by Nicolaus Mercator. In late 1669 he recruited Newton
to write a commentary on the translation, which Newton finished a year
later.

Collins's timing had been poor. Assuredly, in the early 1670s mathematical practitioners, such as Richard Towneley, looked forward to
Newton's "Observations" on Kinckhuysen's *Algebra*.[55] By the time
Newton and Collins were ready to release the work for publication,
however, the London and Cambridge booksellers were no longer interested in it. London publishers were still reeling from the effects of
the Great Fire of 1666, a few recent mathematical treatises had been
commercial failures, and the supply of algebra textbooks already seemed
to exceed the demand:[56] the English version of Rahn's *Algebra* had
appeared in 1668 and Kersey's *Algebra* was far advanced in its press
proofs. By 1676, then, Newton and Collins abandoned the project.
There was a touch of irony as well as scientific selflessness and coopera-
tion here – the latter speaking well of the small band of analytic enthusi-
asts who, spurred on by Collins, responded to the need for a late-
seventeenth-century English algebra textbook. Collins had assisted in the
publishing of the translation of Rahn's *Algebra*, with Pell's additions,
and seen Kersey's *Algebra* through the press; and both he and Newton
solicited subscriptions to the latter work.[57]

54.   Collins to Pell, 9 Apr. 1667, 1:126.
55.   Richard Towneley to John Collins, 4 Jan. 1671/1672, *The Correspondence
      of Isaac Newton*, ed. H. W. Turnbull, J. F. Scott, A. R. Hall, and Laura
      Tilling, 7 vols. (Cambridge: Cambridge University Press, 1959–1977),
      1:78. A nephew of Christopher Towneley of Towneley Hall, Richard
      Towneley was an amateur instrument maker (Taylor, *Mathematical Prac-
      titioners*, 248–249).
56.   See Whiteside, "Introduction" to part 3, *Mathematical Papers of Newton*,
      2:287–291, and "Introduction" to part 1, ibid., 3:5–7.
57.   For a more detailed account of Newton's work on Kinckhuysen's *Algebra*,
      and the fate of his "Observations," see Chapter 7.

4

John Pell's English Edition of Rahn's Algebra *and John Kersey's* Algebra

John Pell's English edition of Rahn's *Algebra* and John Kersey's *Algebra* were, then, the immediate fruits of the late-seventeenth-century call for a current English algebra. Although these textbooks helped forestall the publication of other English algebras, including those by Wallis and Newton, neither totally satisfied the period's algebraic devotees. The revision of Rahn's *Algebra* for the English edition failed to live up to contemporary expectations of Pell. Despite the substantial merits of Kersey's *Algebra,* some complained that it was too long in length and too spare in the theory of higher equations.

Yet the influence of these early algebras was substantial, for, with them, English algebra took on a distinctive cast. Owing a clear debt to Descartes's *Géométrie,* they laid the foundations for English acceptance of the negative numbers while at the same time perhaps obstructing appreciation of the imaginary numbers. Rahn's *Algebra,* to a limited degree, and Kersey's *Algebra,* to a greater degree, were responsible for importing into England the definition of negative numbers as "less than nothing." Moreover, both algebras referred to, and categorized, numbers of the form $a\sqrt{-1}$ as "impossible." And, offering a glimmer of the direction in which English algebra would move from the mid-eighteenth century on, Kersey argued that algebra was a demonstrative art, leading to certain conclusions.

I

John Pell (1610/1611–1685) – who not only revised Rahn's *Algebra* for its English edition but also probably guided Rahn as he wrote the original work – was born in Southwick, Sussex, into an old English family. At the age of thirteen he enrolled at Trinity College, Cambridge; by his twenties he had held teaching positions at Collyer's School in Horsham and Samuel Hartlib's short-lived Chichester academy. Despite his resolve to dedicate himself to the pursuit of mathematics, Pell was unable to find within England a suitable academic position. In 1643 the right connections helped him procure the professorship of mathematics at Amsterdam, which he held for three years before accepting the offer of the prince of Orange to become the professor of mathematics at the newly founded College of Orange in Breda. His life then touched by the Civil War, he returned to England in 1652 at the request of Oliver Cromwell, who first appointed him a mathematical lecturer and subsequently sent him abroad as his resident in Zurich. He next returned to England in 1658, just weeks before Cromwell's death. He subsequently took orders, made his peace with the royalists, and obtained two preferments, which provided his main support through the end of his life.[1]

The foreign professorships, as well as the familiarity with continental mathematics that he had acquired while residing abroad, helped Pell to earn a reputation as "a very learned man, more knowing in algebra, in some respects . . . than any other."[2] Too, Pell was connected to Harriot's mathematical legacy through Walter Warner. Pell's remark, recorded by Collins, that Harriot "was so learned . . . that had he published all he knew in algebra, he would have left little of the chief mysteries of that art unhandled," suggests that he knew more of Harriot's algebra than that which made its way into the *Praxis*.[3] But Pell never fulfilled his contemporaries' expectations. Perhaps his talents were confined to limited aspects of algebra, for, according to Jan van Maanen, in 1646 he

1. Biographies of Pell include "John Pell," in John Aubrey, *"Brief Lives,"
    Chiefly of Contemporaries, Set Down by John Aubrey, between the Years
    1669 & 1696,* ed. Andrew Clark, 2 vols. (Oxford: Clarendon Press, 1898),
    2:121–131; A. M. C., "John Pell," *Dictionary of National Biography,* ed.
    Leslie Stephen and Sidney Lee, 22 vols. (Oxford: Oxford University Press,
    1921–1922), 15:706–708; and P. J. Wallis, "John Pell," *Dictionary of
    Scientific Biography,* ed. Charles C. Gillispie, 18 vols. (New York: Scrib-
    ner's Sons, 1970–1990), 10:495–496. See also Jan A. van Maanen, "The
    Refutation of Logomontanus' Quadrature by John Pell," *Annals of Science*
    43 (1986): 315–352, and Christoph J. Scriba, "John Pell's English Edition
    of J. H. Rahn's *Teutsche Algebra,*" in *For Dirk Struik: Scientific, Historical
    and Political Essays in Honor of Dirk J. Struik,* ed. R. S. Cohen, J. J.
    Stachel, and M. W. Wartofsky, Boston Studies in the Philosophy of Science,
    vol. 15 (Dordrecht: D. Reidel, 1974), 261–274, esp. 262–264.
2. Collins to Dr. Beale, 20 Aug. 1672, *Correspondence of Scientific Men of
    the Seventeenth Century,* ed. Stephen Jordan Rigaud, 2 vols. (Oxford,
    1841; reprint, Hildesheim: Georg Olms, 1965), 1:196–197.
3. John Collins to [Francis] Vernon, n.d., ibid., 1:152–153. Between 1630
    and 1640 Pell collaborated on an antilogarithm table with Warner
    (A. M. C., "John Pell," 707).

evidenced little interest in "Cartesian mathematics" even though he had access to a copy of *La géométrie*.[4] Moreover, clerical duties, lingering poor health, and financial worries distracted him from mathematical research and writing.[5] Whatever the main causes, he published so little that Collins eventually compared inciting him to publish anything to "grasping the Italian Alps, in order to their removal."[6]

2

Although on his own Pell published nothing of significance on the algebra in which he supposedly excelled, his initials were attached to the English edition of Rahn's *Algebra*, to which he added extensive commentary.[7] The original work, *Teutsche Algebra*, had drawn on Diophantus, Viète, Descartes, Schooten, and others to provide the first introduction to the subject written in German; it had been published by Johann Heinrich Rahn (1622–1676) in Zurich in 1659. In his "Vorberricht" to the book, Rahn had acknowledged a considerable debt "to a high and very learned person," whom scholars have subsequently identified as Pell. Already in the late seventeenth century, John Aubrey related that Rahn had studied algebra under Pell while he was the English resident in Zurich, and "Rhonius's Algebra, in High-Dutch, was (indeed) Dr. Pell's."[8] Modern scholars have unearthed additional evidence of a pupil–teacher relationship between Rahn and Pell and of the writing of the *Teutsche Algebra* while Pell resided in Zurich. Thus, some influence by Pell on the *Teutsche Algebra* is definite, although there seems to be insufficient evidence to determine its exact extent.[9]

This earlier relationship explains Pell's involvement in the publication

---

4. Maanen, "Longomontanus' Quadrature," 342.
5. On Pell's health, see Lesley Murdin, *Under Newton's Shadow: Astronomical Practices in the Seventeenth Century* (Bristol: Hilger, 1985), 58; on Pell's finances, see Aubrey, *Brief Lives*, 2:124, 126–127.
6. Collins to Dr. Beale, 20 Aug. 1672, *Correspondence of Scientific Men*, 1:196.
7. For a detailed account of Pell's role in the English edition, see Scriba, "John Pell's English Edition," 264–272.
8. Aubrey, *Brief Lives*, 2:125.
9. Scriba, "John Pell's English Edition," 264–266. "Without further evidence, it is best to assume that there was joint responsibility for these innovations [in Rahn's *Algebra*] and that Pell's contemporary reputation as a mathematician, and particularly as an algebraist, was not unearned" (Wallis, "John Pell," 495).

of the English translation of the work. Not Pell, but Thomas Brancker, another correspondent of Collins and an Oxford man with mathematical interests, took the lead by translating the book into English. As Brancker related, when he sent his manuscript to the press in 1665 he was referred to "a Person of Note very worthy to be made acquainted with my design, before I made any farther progress in the Impression."[10] The person was Pell, who subsequently took an active role in Brancker's project and made additions to the original, which doubled the book's length. With Pell writing as the textbook was being printed, the printing stretched through 1668. It was delayed by the Great Fire, Brancker's move from London to Lancashire, and some bickering about its contents.[11]

Seeing its early pages in 1667, Collins had written Brancker that many thought the introduction "might have been more plain, and ought to have been more large than it is."[12] For example, John Twysden,[13] described here by Collins as a student of algebra for two decades, was supposed to have been unable to understand the introduction. Collins suggested directly to Pell "a Latin preface to explain the symbols, and to signify that the greatest part of the book may be understood by others, ignorant of our tongue." Some scholars, he added, believed that students learned division and other algebraic operations more easily when they were presented in prose than merely symbolically.[14] But Pell was apparently ignoring his advice, and so Collins had diplomatically urged Brancker "to incline the Doctor to admit the first seven sheets of the Introduction, enlarged out of Kinckhuysen . . . to come out as your translation, as soon as may be; the Doctor taking what time he pleaseth to supply the defect at the beginning, and to enlarge and complete the book."[15]

His call for a supplementary introduction coming to nought, Collins still saw the final sheets of the English edition of Rahn's *Algebra* through the press. Presented with the book (the title page of which, apparently

10. *An Introduction to Algebra, Translated out of the High-Dutch into English,* by Thomas Brancker, M.A., Much Altered and Augmented by D.P. (London: Moses Pitt, 1668), "The Translator's Preface."
11. Scriba, "John Pell's English Edition," 269–271.
12. Collins to Brancker, June [?], *Correspondence of Scientific Men,* 1:134.
13. Twysden (1607–1688), a graduate of University College, Oxford, author of several tracts for instrument makers, and a member of the Inner Temple (E. G. R. Taylor, *The Mathematical Practitioners of Tudor and Stuart England* [Cambridge: Cambridge University Press, 1967], 212), had met Oughtred, whose *Key* he praised.
14. Collins to Dr. Pell, 9 Apr. 1667, *Correspondence of Scientific Men,* 1:126.
15. Collins to Brancker, June [?], 1:136.

according to his wishes, gave only his initials),[16] Pell complained of numerous "pressfaults" that he would have to correct by hand in copies intended for his friends.[17] His compatriots, like Collins, seem to have judged the work more harshly since it failed to undergo any reprintings.

3

"According to *Vieta's* Way," the book's introductory paragraph began, "we represent the concerned Quantities by Letters. As [a, b, c, d,] stand for four several Quantities; whereof we call one [a,] another [b,] &c." The paragraph also stated that Viète used the term "coefficients" for numbers prefixed to quantities, and that 2a, for example, "stand[s] for (a) taken twice," whereas a quantity without a coefficient "stands for it self once taken."[18] Following this and a few other terse paragraphs, which explained nothing of the methods of algebra or of its general symbolical language, the work went on to cover basic arithmetic and algebra in about two hundred pages.

Not the compendium of early modern algebra that Collins hoped for, the English edition of Rahn's *Algebra* yet brought important glimpses of Cartesian algebra to the English student. The work adopted some aspects of Cartesian notation, but not all. For example, it used Descartes's notation for exponents. "So $a^7b^5c^8 = aaaaaaabbbbbcccccccc$," it noted, "And therefore [is] used by Modern *Analysts* for *Compendium*." On the other hand, it denoted unknown quantities by lowercase letters of the alphabet, and known quantities by capitals (A, B, C, and so on).[19] The second part of the *Algebra* specifically invoked Descartes's name as it restated major parts of his equation theory. It was here that, in tackling negative and imaginary roots, the work departed substantially from existing English texts.

This section opened with a statement of the fundamental theorem – "As great as is the *Number* of the highest *Power* in any *Equation,* so many Roots doth it contain . . ." – and the observation that the roots of equations were of three kinds. Although inspired by Descartes, the section challenged his typology of roots in curious and perhaps signifi-

16. Pell had wavered about how credit for the work should be assigned him (Scriba, "John Pell's English Edition," 268).
17. John Pell to Moses Pitt, 3 June 1668, *A Collection of Letters Illustrative of the Progress of Science in England,* ed. James Orchard Halliwell (1841; reprint, London: Dawsons of Pall Mall, 1965), 103–104.
18. *Introduction to Algebra,* by Thomas Brancker, 1.   19. Ibid., 7, 56.

cant ways. Students were told that the roots are "either *Affirmative* or *Negative*, or *Impossible*. These Negative Roots *Des Cartes* cals [*sic*] *False Roots*." In defense of its change of names from "true" to "affirmative" and "false" to "negative," the work argued that negative roots "express Quantities of a Denomination opposite to those that are counted Affirmative, and in *Geometrical Projection* they require a *situation* contrary to the Affirmative Roots; therefore they are no less *true* than the Affirmative ones." The *Algebra* thus used geometric representation of quantities in opposite directions to defend the "truth" of negative and positive roots. As roots of the equation,

$$x^4 - 4x^3 - 19xx + 106x - 120 = 0,$$

it gave $+4$, $+3$, $+2$, and -5, and not simply (Descartes's) 5.[20]

Although there was precedent in English algebra for the designations "affirmative" versus "negative" – Oughtred had after all written of "adfirmation" versus "negation" – the English version of Rahn's *Algebra* contained perhaps the first clear-cut defense of negative roots to appear in the English language. Even this *Algebra*, however, did not consistently refer to such numbers as "negative." It brought Descartes's description of negative numbers as "numbers less than nothing" into the English algebraic vocabulary. Specifically, in a discussion of Schooten's *Sectiones miscellaneae* of 1657, Schooten's failure to produce a complete solution of a problem was attributed to his "being discouraged by meeting with *numbers under nothing*."[21]

Roots of the third kind, which Descartes had denominated "imaginary," were called "impossible" in the *Algebra* and described as "such as by their own nature or their Signs are so ordered and *entangled* as that they render the Probleme *unreasonable* and *impossible*." Focusing on these roots as essentially signs of impossibility seems to have been a defense against their being (mis)taken for quantities or numbers. Tellingly, the *Algebra* gave a new symbol, ƆI, by which all impossible roots were to be marked.[22] Use of a single symbol for all such roots was tantamount to a statement that these roots, unlike quantities, had no separate values, or, at the least, their separate values were largely insignificant. Thus the solution of the equation, $cc + 20c = -364$, proceeded as follows:

---

20. Ibid., 47–48.   21. Ibid., 121.
22. Ibid., 48. Compare: "These impossible roots, saith Dr. Pell, ought as well to be given in number as the negative and affirmative roots, their use being to shew how much the data must be mended to make the roots possible" (Collins to Wallis, n.d., *Correspondence of Scientific Men*, 2:481).

$$cc + 20c + 100 = -264,$$
$$c + 10 = \pm \sqrt{-264}$$
$$c = + \sqrt{-264} - 10 = \mathrm{OI}$$
$$c = - \sqrt{-264} - 10 = \mathrm{OI}.^{23}$$

From a modern perspective, the above solution is startling, for it denies the individuality of two different complex numbers. Still, two different complex expressions appear as roots of the given equation, and here and elsewhere Rahn's *Algebra* gave roots involving $\sqrt{-1}$ to an extent unprecedented in English algebra textbooks. In this way, thanks to the Cartesian influence, the imaginaries were off to a start in English algebra, but it was an inauspicious start that did not bode well for their speedy or full acceptance as quantities or numbers.

4

When the English version of Rahn's *Algebra* failed to live up to Pell's reputation, and publishers expressed no interest in the translation of Kinckhuysen's *Algebra,* Collins pinned his hopes for a successful English algebra textbook on John Kersey (1616–1701). A nonuniversity man, Kersey was a London surveyor and a well-respected teacher of mathematics.[24] He not only read widely in mathematics but took a special interest in the preparation of mathematical textbooks, as evidenced by the popular editions of Edmund Wingate's *Arithmetic* that he published from 1650 through 1683.[25] By 1667, according to Collins, Kersey's algebra textbook had already "been . . . ten or twelve years on the anvil" and was now "ready for the press." "What Mr. Kersey hath hitherto published," Collins added, "hath been well esteemed; and this is his masterpiece."[26]

In a modest way, Kersey's textbook, which was not published until 1673–1674, lived up to Collins's expectations. As he wrote its final sections, Kersey showed the earlier ones to friends and circulated a synopsis in the hopes of obtaining subscribers, needed to induce the printer to continue with the work.[27] From this early stage through its

23. *Introduction to Algebra,* by Thomas Brancker, 75.
24. "John Kersey," *Dictionary of National Biography,* 11:69–70, and Taylor, *Mathematical Practitioners,* 219.
25. Taylor, *Mathematical Practitioners,* 205.
26. Collins to Brancker, June [?], 1:135.
27. Collins to Newton, 30 Apr. 1672, *Correspondence of Scientific Men,* 2:320.

publication, the work enjoyed quite a bit of popular support. In his letters urging Brancker and Pell to expand the introduction to Rahn's *Algebra,* Collins reported that "Mr. Leake, Mr. Gunton, and others account it [Kersey's unpublished algebra] an excellent tract, his Introduction copious and easy."[28] In the early 1670s Richard Towneley – who remarked that "the want of such books . . . is the cause we have so few that understand any thing of algebra" – and Newton subscribed to Kersey's algebra. With Newton's aid, at least four copies of the work were sold to Cambridge men in advance of publication.[29]

When finally completed, Kersey's *Elements of That Mathematical Art Commonly Called Algebra, Expounded in Four Books* met many of the criteria that Collins and his circle had set for an English algebra. With "the younger Students of Symbolical *Arithmetick* and *Analytical Doctrine*" in mind, Kersey explained in his preface, he had "earnestly endeavoured to render the Fundamentals, and most important Rules of the *Algebraical* Art in both kinds, to wit, Numeral and Literal, very clear and easie."[30] He offered adequate introductory materials: a three-page preface on mathematical methods, an introductory chapter on algebra and its notation, and later chapters with substantial prose and symbolical descriptions of addition and the other basic algebraic operations.

In his preface, Kersey made a special effort to promote algebra by highlighting arithmetic and geometry as exemplary demonstrative sciences and then explaining why algebra deserved to rank alongside, if not above, these sciences in the hierarchy of knowledge. "It is an undoubted truth," the preface opened, "that among all Humane Arts and Sciences, *Arithmetick* and *Geometry* have obtain'd the greatest evidence of Certainty." He associated this certainty with the "verity and perspicuity" of the first principles – definitions, postulates, and axioms – of arithmetic and geometry. "Hence it is, that all Propositions which are proved by those certain Principles are likewise certain, and called Demonstrative Truths, by which are meant strictly and properly, infallible Conse-

28. Collins to Brancker, June [?], 1:135. Collins was probably referring to John Leake (fl. 1650–1686), a London mathematical practitioner, who was appointed master of Christ's Hospital in 1673 (Taylor, *Mathematical Practitioners,* 236–237).

29. Towneley to Collins, 13 May 1672, *Correspondence of Scientific Men,* 1:190; Newton to Collins, 25 May 1672, ibid., 2:323; and Newton to Collins, 13 Jul. 1672, ibid., 2:332.

30. John Kersey, *The Elements of That Mathematical Art Commonly Called Algebra, Expounded in Four Books,* 2 vols. (London: William Godbid, 1673–1674), 1: preface (iii). Pagination of the preface is mine.

quences, or Conclusions, deduced from clear and undeniable Premisses."[31]

Kersey then measured algebra against arithmetic and geometry. Based on his reckoning, algebra's merits seemed to exceed even those of the traditionally exemplary mathematical sciences. He argued that algebra was based on the analytic method; the method could be used to solve the difficult problems of mathematics; it provided insights into mathematical invention; and it, like synthesis, produced certain conclusions. He stressed especially the latter point, writing that algebra employed "a more easie, and not less sure Method than that called *Synthetick*." Also, algebra "first assumes the Quantity sought . . . as if it were known, and then, with the help of one or more Quantities given, proceeds by undeniable Consequences, until that Quantity which at first was but assumed or supposed to be known, is found equal to some Quantity certainly known, and is therefore known also." Furthermore, the "Analytical way of Reasoning" not only produces theorems and canons but also "discovers Demonstrations of the certainty of the resulting *Theorem* or *Canon,* in the *Synthetical* Method, or way of Composition, by the Steps of the *Analysis,* or *Resolution.*"[32] In short, he argued forcefully that certainty was built into the analytic method, which involved a form of mathematical demonstration (proceeding "by undeniable Consequences"); in addition, analytic results could be demonstrated synthetically. If certainty and basic usefulness recommended the study of arithmetic and geometry to "all ingenious Minds," certainty and a higher-level usefulness made algebra a subject for "Lovers of Art . . . [who] have hours to spare" and are desirous of "the choicest pieces in the Common-wealth of Learning."[33]

Algebra's special usefulness in solving difficult problems extended to both arithmetic and geometry. But, seeming to indicate algebra's independence, Kersey turned immediately in his textbook proper to algebra, not arithmetic or geometry. However important algebraic applications were, they were confined to later chapters. Thus Kersey treated arithmetic questions per se only in chapter 14 of the first book and, like Oughtred, he segregated geometric questions to his final book. More by structure than by any specific declaration, he thus supported an independent algebra.

5

Chapter 1 of the *Algebra* offered definitions of synthesis and analysis, a discussion of the kinds of truth to which algebra led, explanations of

31. Ibid., 1: preface (i).   32. Ibid., 1: preface (ii); see also ibid., 1:2.
33. Ibid., 1: preface (ii).

"numeral algebra" and "literal, or specious algebra," a lengthy treatment of algebraic powers (their meaning and notation), and a long section on "the Signification of Characters used in the First Book." The organization of the chapter and its clear explanations spoke well of Kersey's pedagogical abilities. Carefully laying out all the basics, he noted, for example, that the unknown of an equation "must be design'd or represented by some Symbol or Character, at the will of the Artist." He explained why he followed Harriot in using vowels to stand for unknowns and consonants for knowns. Acknowledging that, on the question of symbols for powers of the unknown, there was "much diversity among *Algebraical* Writers, every one pleasing his fancy in the choice of Characters," he recommended Harriot's notation for powers.[34]

In particular, he developed an algebraic concept of powers, one explicitly stripped of geometric signification and one thus aptly denoted by repetitive symbols. "From the premises it is evident," he summarized,

> that upon an Arithmetical foundation, a Scale or Rank of Algebraick Powers may be raised and continued as far as you please; the three first of which have an affinity with, and may be expounded by Geometrical dimensions. . . . But none of the rest of the Algebraick powers can properly be explain'd by any Geometrical quantity, in regard there are but three dimensions in Geometry, to wit, Length, Breadth, and Depth.

This said, there was no place in Kersey's algebra for the *A quadrato-quadratum* of Viète or the *Aqq* of Oughtred. Instead he gave a "Table shewing the two wayes (now most in use) to express simple Powers by Alphabetical Letters" – namely, the notations of Harriot and Descartes (*aaa* vs. a^3, for example). He recommended Harriot's longer notation as "the plainest for Learners."[35]

The chapter ended with a long list, accompanied by explanations, of the major algebraic characters. No explanations so well illustrate Kersey's pains in making the signs and concepts of algebra reasonable to beginning students as those he offered for the symbols + and −. He tried a barrage of approaches, drawn directly or indirectly from Cardano, Oughtred, Descartes, and even the English edition of Rahn's *Algebra*.[36] He first distinguished between the two uses of each of the

34. Ibid., 1:4.
35. Ibid., 1:4–5. Substituting *a* for Descartes's *x*, Kersey offered notation "after the manner of *Renates des Cartes*" (*a*, a^2, a^3, and so on).
36. There is a need for a detailed study of Kersey's *Algebra*, including its background. In his preface Kersey mentioned the analytic work of not only Diophantus, Cardano, Viète, Oughtred, Harriot, Descartes, Pell, Seth Ward, Wallis, and Isaac Barrow but also Clavius, Stevin, Schooten, Hudde, Fermat, and still other continental mathematicians (*Algebra,* 1: preface [ii–iii]).

signs: "This Character + is a sign of *Affirmation,* as also of *Addition.*"
"This Character − is a sign of *Negation,* as also of *Subtraction.*"
Although these preliminary definitions were close to those offered in *The Key,* Kersey did not give Oughtred's example of "−34, [which] are denied to be at all." Rather, in a strange explication, he further defined negative numbers and even offered analogies to help legitimate them while, at the same time, he, like Cardano, called them "fictitious" and, like Descartes, described them as "less than nothing." In his words, " +*a* affirms the quantity denoted by *a* to be real, or greater than nothing"; "−5 is a fictitious number less than nothing by 5; *viz.* as +5*l.* may represent five pounds in money, or the Estate of some person who is clearly worth five pounds; so −5*l.* may represent a Debt of five pounds; owing by some person who is worse than nothing by five pounds." Just a few pages later, in his chapter on "Addition of Algebraical Integers," he similarly taught students to add +*c* to −*c* through analogy:

> [T]he Summ [of +*c* and −*c*] will be 0, or nothing; for the Affirmative quantity will destroy or extinguish the Negative: ... For supposing −*c*, or −1*c* to be a Debt of one Crown that I owe; and +*c*, or +1*c* to be one Crown in my purse, it is evident that one Crown in ready money will discharge or strike off a Debt of one Crown; and so that Debt and Credit being added or compared together, the Summ amounts to 0.[37]

On the whole, Kersey thus accepted negative numbers as entities of algebraic discourse, while admitting implicitly that a tidy, coherent explanation of their nature was beyond him. Affirming his acceptance of negative numbers, the frontispiece of his *Algebra* featured a book opened to pages dealing with the addition and subtraction of negative numbers;[38] Kersey argued that the negatives were analogous to debts; he called them "quantities"; and he consistently expressed isolated negatives as roots of specific equations. For example, in a later chapter, he gave $a = +5$ and $a = -11$ as the two roots of $aa + 6a = 55$ and demonstrated that each root actually satisfied the given equation.[39] Yet, as we have seen, he wrote of negative numbers as "fictitious numbers" and, to a greater degree than Rahn and Pell, he was responsible for importing Descartes's definition of a negative number as "less than nothing" into the genre of the English algebra textbook, where it would survive and trouble English mathematicians into the early nineteenth century.

37. Kersey, *Algebra,* 1:5, 3, 9. The *Algebra* includes a few misnumbered pages (e.g., p. 5 immediately precedes p. 3).
38. The examples given were $+5a + -3a = +2a$ and $+12 - -18 = +30$.
39. Kersey, *Algebra,* 1:83–84.

Besides including extensive introductory material, Kersey's *Algebra* proved appropriate for its time and place in other ways. He covered some of the recent algebraic developments on the continent, particularly in chapter 11 of book II, "Extractions out of the Algebraical Treatises of Vieta and Renates des Cartes, concerning the Constitution and Resolution of Compound Equations in Numbers." At the beginning of the chapter, he directed students to set their equations equal to zero – the form, he observed, that Descartes prescribed. He went on to discuss other aspects of Descartes's equation theory, including constructing equations from binomial factors and the rule of signs, which he noted did not apply to equations with roots involving $\sqrt{-1}$.[40]

Also in the chapter on the algebraic work of Viète and Descartes, Kersey described the three different "kinds" of roots that an equation may have, "*viz.* either Affirmative, or Negative, or Impossible." In naming these roots as well as in developing his ideas of them, he actually followed Rahn's *Algebra* more closely than Descartes's *Géométrie*. Thus Kersey's main definition of a negative root matched almost word for word the definition offered in the English edition of Rahn's *Algebra*: "a negative Root (which *Cartesius* calls a false Root) expresseth a quantity whose Denomination is opposite to an affirmative."[41]

If Kersey contributed anything new to the discussion of roots involving $\sqrt{-1}$, it was a careful explanation of their impossibility. "[I]mpossible Roots are such whose values cannot be conceived or comprehended either Arithmetically or Geometrically: As in this Equation, $a = 2 - \sqrt{-1}$, where $\sqrt{-1}$, that is, the square Root of -1 is no manner of way intelligible, for no number can be imagined, which being multiplied by it self according to any Rule of Multiplication, will produce -1."[42] Since mathematicians could not imagine such a number, he implied, so they should not even write it. Unlike the English edition of Pell's *Algebra*, Kersey's *Algebra* stated no roots involving $\sqrt{-1}$ (except in the previously mentioned definition). In an earlier section on the general solution of a quadratic equation of the form, $ca - aa = n$, where a is the unknown, he dismissed roots involving $\sqrt{-1}$ without giving them. He arrived at

$$a = \frac{1}{2}c + \sqrt{\frac{1}{4}cc - n:,}$$

and then made a "determination" that "The Absolute number given [n] must not exceed the Square of half the Coefficient [$\frac{1}{4}cc$]." Justifying the determination, he noted that "when in any Equation of the . . . form,

40. Ibid., 1:268–272.   41. Ibid., 1:269.
42. Ibid.

the given Absolute number exceeds the Square of half the Coefficient that Equation is impossible, and likewise the Question that produced it."[43] In short, although compelled to mention the "imaginary" roots of Descartes, Kersey seemed to believe them to be no roots at all. Whereas "imaginary roots" had permitted Descartes to generalize to the strong version of the fundamental theorem of algebra, Kersey's "impossible roots" restricted him to the weaker version.[44]

The inclusion of Kersey's sole discussion of roots involving $\sqrt{-1}$ under "Extractions out of . . . des Cartes" stands as impressive evidence of the role of *La géométrie* in helping such roots gain a foothold within the universe of English algebraic discourse. It certainly seems to have been *La géométrie* that led Kersey and Pell (as attested in Rahn's *Algebra*) to mention a third kind of root. There was after all no significant precedent in English mathematical textbooks for discussing these roots. Kersey and Pell, moreover, still seemed unable to shake the view that such roots were fundamentally impossible. Perhaps Kersey spoke for his countrymen when, in defining "impossible roots," he seemed to say that, whereas Descartes could pretend to imagine such roots, he knew that these roots were, in the last analysis, impossible.

6

Of additional appeal to the increasingly chauvinistic English mathematicians of the late seventeenth century, Kersey's *Algebra* was an English work in origin as well as spirit. Although derivative as most textbooks are, it was algebra composed afresh in the English language by an English practitioner. Nevertheless, substantial introductory material, coverage of some continental mathematics, a perhaps English twist on complex roots, and English origins were insufficient to win full approval for the textbook, even from Collins. Although Collins saw the work through the press and received a special acknowledgment in Kersey's preface,[45] he criticized the work's length and its emphasis on equations of degree three or less.

43. Ibid., 1:91–92.
44. Ibid., 1:271. "[S]ometimes an Equation may have as many Roots as there be unities in the Index of the highest unknown Term; I say sometimes, not alwayes." His counterexample to the strong fundamental theorem was a cubic equation with two roots involving $\sqrt{-1}$, neither of which he stated (ibid.).
45. Ibid., 1: preface (iii). Kersey here described Collins as a friend responsible for "animating" him "to Compile" the work.

Kersey's prose explanations, coverage of continental algebra, numerous examples, and attempt at completeness swelled the first two books of the *Algebra* to over three hundred pages. In taking into account the needs of beginning students, Kersey made some chapters tedious as well as comprehensive. In chapter 11 of book I, which dealt with the standard topic of reduction of equations, he not only stated the "axiom" upon which "reduction by addition" was based ("If equal quantities, or one and the same quantity, be added to equal quantities, the wholes shall be equal"); he also offered seven examples of such reduction.[46] Other well-thought-out pedagogical devices, perhaps applied too often, added to the textbook's length. In one example after another of chapter 14 of book I, dealing with the algebraic solution of arithmetic problems, Kersey posed a problem and then gave simultaneous numeral and literal solutions, with a prose description of the whole process running alongside and a prose statement of the conclusion following.[47] That he thereby failed to strike that happy medium of the textbook that is fairly comprehensive, accessible, and yet compendious, had become evident already in the late 1660s when Collins had described the forthcoming work as "a laborious treatise."[48] When book III (on the work of Diophantus and Fermat) and book IV (on algebra applied to geometry)[49] finally appeared in 1674, they spread over an additional 416 pages, sealing the standing of the *Algebra* as an elementary but ponderous tome.

Kersey's decision to concentrate on elementary algebra generated additional complaints. "Nor doth Mr. Kersey treat of the roots or limits of equations of any high degree," Collins wrote Wallis in 1671, "and therefore [it] might be another treatise apart to be sold with his, . . . collecting what is scattered in Hudden, Bartholinus, Dulaurens, and the Dutch writers, Kinkhuysen, Ferguson, &c. well digested in Latin, [or] especially in English."[50] Indeed, only quadratic and cubic equations fell within the main focus of Kersey's *Algebra*, and only such equations as involved no complex numbers. As Kersey had dismissed quadratic equations leading to complex roots, so, when presenting Cardano's solu-

---

46. Ibid., 1:52.
47. See, e.g., ibid., 1:64.
48. Collins to Wallis, 2 Feb. 1666–1667, *Correspondence of Scientific Men*, 2:471.
49. In book IV, which merits further study, Kersey stressed that algebraic solutions of geometric problems could be turned around to provide synthetic demonstrations (Kersey, *Algebra*, 2:223).
50. Collins to Wallis, 21 Mar. 1671, *Correspondence of Scientific Men*, 2:526.

tion of the cubic equation, he noted that it covered all cases but the irreducible one.[51]

As early as 1672 Wallis, in turn, seems to have thought of preparing a sequel to the *Algebra*: "I am glad Mr. Kersey's book is so forward," he told Collins, "and shall be willing enough to see the whole of it abroad. As to what you mention to follow it, I shall not be wanting, as to my part, when there is occasion."[52] But Collins had already secured a promise from Lord William Brouncker to pay Michael Dary, a self-taught mathematician, to prepare such a sequel. He explained to Wallis that Dary was "well versed in those authors and matters, and at present very poor and void of employment," but that Wallis's "frequent advice and assistance" with the project would be welcomed.[53] Dary seems never to have prepared the sequel, and Wallis published his own *Treatise of Algebra* in 1685.

In another twist in the history of English algebra, however, posterity would vindicate Kersey's *Algebra* as a useful textbook, for, unlike Wallis's *Treatise* which was published separately just once, Kersey's work (in its entirety or volume 1 alone) was published a half-dozen times and last in 1741. In 1707 it appeared on a list of algebras recommended for Cambridge students.[54]

51. Kersey, *Algebra*, 1:280–283. Kersey, however, mentioned the work of Florimond de Beaume on finding the limits of roots of equations of degree four and less, and gave an example of his routine for a cubic equation (ibid., 1:268).

52. Wallis to Collins, 14 Nov. 1672, *Correspondence of Scientific Men*, 2:552.

53. Collins to Wallis, 27 Mar. 1673, ibid., 2:556. Dary (1613–1679) worked as a tobacco cutter and later as an exciseman and gunner. Collins tried in various ways to gain Dary mathematical employment (Taylor, *Mathematical Practitioners*, 217). Lord Brouncker, a founding member of the Royal Society of London, was a patron of the mathematical arts and a member of the Mathematical Club in Moorfields (ibid., 220–221).

54. W. W. Rouse Ball, *A History of the Study of Mathematics at Cambridge* (Cambridge, 1889), 95. Also included on the list were the books of Oughtred, Harriot, Pell, Wallis, and Newton.

5

The Arithmetic Formulation of Algebra in John Wallis's Treatise of Algebra

As Oughtred and Harriot fostered the symbolical style, and Pell and Kersey assured negative roots a place within English algebra, John Wallis (1616–1703) endorsed imaginary numbers and, it can be argued, gave algebra its first arithmetic formulation. Indeed, Wallis was responsible for assuring that by the late seventeenth century English algebra was "early modern" in every respect, not only in its language but in its generality, universe of objects, and methods as well. Wallis as a mathematician superseded both Oughtred and Harriot at least partially because he built on their ideas and those of continental thinkers, and because an Oxford professorship permitted him to devote much of his life to mathematics. More fundamentally, he surpassed his algebraic predecessors because of his stubborn commitment to arithmetic as the foundation of algebra and hence to "pure algebra" – his own term, which proclaimed an algebra that was independent of and perhaps superior to geometry.

I

Wallis was the eldest son of John Wallis, a graduate of Trinity College, Cambridge, who served as the minister at Ashford until his death in the early 1620s. Raised by his mother, the young Wallis studied at the grammar school at Tenterden, Kent, the school of Martin Holbeach at Felsted, Essex, and finally Emmanuel College, Cambridge, which was at that point the "Puritan College."[1] According to Wallis's autobiography,

1. See Christoph J. Scriba, "John Wallis," *Dictionary of Scientific Biography,* ed. Charles C. Gillispie, 18 vols. (New York: Scribner's Sons, 1970–1990), 14:146–155, and G. Udny Yule, "John Wallis, D.D., F.R.S.," *Notes and Records of the Royal Society of London* 2 (1939): 74–82. See also the draft and final versions of Wallis's autobiography in Christoph J. Scriba, "The Autobiography of John Wallis, F.R.S.," *Notes and Records of the Royal Society of London* 25 (1970): 17–46. All references are to the final version.

the grammar schools of his period concentrated on Latin and Greek, and his first introduction to mathematics came through a younger brother. This brother had learned the subject as preparation for a trade and during the Christmas break of 1631 taught Wallis to "Cipher, or Cast account." The brother's lessons covered the four basic operations of arithmetic, "The Rule of Three (Direct and Inverse,) the Rule of Fellowship (with and without, Time,) the Rule of False-Position, Rules of Practise and Reduction of Coins and some other little things." Although Wallis was "pleased" with his new subject, his studies at Emmanuel, which began the next year and concentrated on divinity, left him little time for serious mathematics. He read mathematical books in his spare time and, if his autobiography is accurate, only those books that he happened upon. He claimed: "I had none to direct me, what [mathematical] books to read, or what to seek, or in what method to proceed."[2]

Even if Wallis missed some opportunities for mathematical instruction that were available at early-seventeenth-century Cambridge,[3] he studied other subjects that helped to prepare him for a major role in the English Scientific Revolution. One of these, logic, related closely to mathematics. His "first business" at the university was "the study of Logick," through which – he later claimed – he had become "Master of a *Syllogism*, as to its true Structure, and the Reason of its Consequences, however Cryptically proposed."[4] Armed with this strong foundation in logic, he would later publicly defend the certainty of mathematics against claims coming from the Renaissance that mathematics did not conform to the highest type of syllogistic reasoning. He would argue in his *Institutio logicae* of 1687 that all mathematical demonstrations were certain although some were more evident than others,[5] and in his *Treatise of Algebra* of 1685 that symbolical algebra "doth *(without altering the manner of demonstration, as to the substance,)* furnish us with a short and convenient way of Notation."[6] Meanwhile, at Cambridge he also "imbib'd the

2. Scriba, "Autobiography of John Wallis," 26–27.
3. For criticism of Wallis's account of the state of mathematics at early-seventeenth-century Cambridge, see Mordechai Feingold, *The Mathematicians' Apprenticeship: Science, Universities and Society in England, 1560–1640* (Cambridge: Cambridge University Press, 1984), 86–88.
4. Scriba, "Autobiography of John Wallis," 28.
5. See Paolo Mancosu, "Aristotelian Logic and Euclidean Mathematics: Seventeenth-Century Developments of the *Quaestio de Certitudine Mathematicarum*," *Studies in History and Philosophy of Science* 23 (1992): 241–265, esp. 251–255.
6. John Wallis, *A Treatise of Algebra, Both Historical and Practical, Showing the Original, Progress, and Advancement Thereof, From Time to Time,*

Principles of . . . the *New Philosophy.*" He seems to have been exposed to the philosophy of the Scientific Revolution, with its emphasis on observation and experimentation, through the subjects of astronomy and medicine. Studying under Francis Glisson, then Regius professor of medicine at Cambridge, he became the university's first student-disputant to defend William Harvey's theory of circulation of the blood.[7]

Having also pursued some ethics, metaphysics, and theology, he received his undergraduate degree in 1637 and his master's in 1640, and then turned to the ministry. Ordained in 1640, he served for a while as a private chaplain to Sir Richard Darley and then Mary, Lady Vere, who sympathized with the parliamentarians. In 1642, while in the service of Lady Vere, he discovered his special talent for deciphering coded messages – a talent that would make him the premier English cryptographer of his period. Coming to the attention of the parliamentarians, he began to use his cryptographic skills in the campaign against Charles I.[8] Possibly from this curious, and later controversial, service to the parliamentarians Wallis gained professional and even mathematical preferments. He received the living of St. Gabriel's, Fenchurch Street, London, and was named a secretary to the Assembly of Divines at Westminster, which was called during the Civil War to discuss new forms of church government. In addition, a parliamentary ordinance of 1644 "require[d]" Queens College, Cambridge, "to receive the said Mr John Wallis and . . . [eight others] as Fellows."[9] Wallis, however, married in 1645 and thus relieved Queens College of his fellowship. He then lived off money inherited from his mother and a ministry at St. Martin's, Ironmonger Lane. During these fallow years he published little but cultivated contacts with England's other scientific men, a select group of whom, including Wallis, began around 1645 the weekly scientific meetings that led ultimately to the founding of the Royal Society of London.

In 1647–1649, when Wallis was already in his thirties, two events set him on a career as an academic mathematician. The first was a personal experience: in 1647 or 1648 he happened upon a copy of Oughtred's *Key,* which seems to have cast a "symbolic spell" on the young divine and made him a partisan of the new analytic mathematics through his

    and by What Steps It Hath Attained to the Height at Which Now It Is (London, 1685), preface. Emphasis is mine.
7. Scriba, "Autobiography of John Wallis," 29.
8. Ibid., 37–38, 45 n. 23. On Wallis's cryptographic talent and his contributions to the intelligence work of the regime of Charles II, see Alan Marshall, *Intelligence and Espionage in the Reign of Charles II, 1660–1685* (Cambridge: Cambridge University Press, 1994), 23, 91–95.
9. Quoted in Yule, "John Wallis," 75.

dying days. According to his later account, under the influence of *The Key* and its powerful notation, he soon rediscovered a version of Cardano's solution of the cubic equation (which *The Key* had not covered).[10]

There was, however, a big gap between mathematical enthusiasm and a mathematical professorship, which few in England had any hope of bridging. But in 1648–1649 parliamentary connections eased Wallis's way to a professorship at Oxford. Royalist troops had occupied the city of Oxford from 1641 to 1646, but parliament controlled the city from summer of the latter year until the Restoration of 1660. Among the major changes instituted at Oxford University by the parliamentarians was the appointment of a new warden of Wadham College in 1648. The warden was John Wilkins, author of *Mercury, or the Secret and Swift Messenger* of 1641, a work dealing with cryptography and containing a sketch of a general theory of language and communication. Shared interest in the latter kind of subjects, as well as political, religious, and larger scientific ties, brought Wilkins and Wallis together. When in late 1648 the parliamentarians stripped Peter Turner, a royalist, of the Savilian professorship of geometry, Wilkins recommended Wallis for the professorship.[11] In June of the next year Wallis was named Turner's successor, probably at least as much because of his parliamentary connections and service as because of his demonstrated mathematical prowess.

Wallis seized the opportunity to pursue mathematics as a full-time career. In his own words, "*Mathematicks* which had before been a pleasing Diversion, was now to be my serious Study. And (herein as in other Studies) I made it my business to examine things to the bottom; and reduce effects to their first principles and original causes."[12] The sheer number of Wallis's purely mathematical publications pointed to the striking difference in the level of professionalism separating him and his immediate English predecessors Oughtred and Harriot. But Wallis excelled in quality and depth, as well as quantity. In his writings he

10. For Wallis's account of the influence of Oughtred's *Key* (he worked with the *Clavis* of 1631) on his early mathematics, see Wallis, *Treatise of Algebra*, 68, 175–177.

11. On Wilkins, see Barbara J. Shapiro, *John Wilkins, 1614–1672: An Intellectual Biography* (Berkeley: University of California Press, 1969), esp. 46–48, 82–87. On Turner, see Feingold, *The Mathematicians' Apprenticeship*, 69–70.

12. Scriba, "Autobiography of John Wallis," 40–41.

roamed over the areas of traditional and analytic geometry, the germs of the calculus, arithmetic, and early modern algebra.

2

His major algebraic work, *Treatise of Algebra, Both Historical and Practical,* was published in 1685, as Wallis neared his seventieth birthday. However, as we have seen, he had thought about an English algebra textbook much earlier in his career. In 1666/1667 he had advised algebraic authors to "do . . . a work of their own, [rather] than . . . a comment on" Oughtred's *Key,*[13] and in 1672 he had expressed his willingness to write a sequel to Kersey's *Algebra.* The *Treatise of Algebra,* the body of which Wallis claimed to have finished by 1676,[14] was no mere sequel. An entirely new textbook, it covered the basics of algebra while attempting to legitimate the negative and imaginary numbers, to establish English priority in early modern algebra, and to formulate the arithmetic foundations of algebra.

Already by the time of the publication of Kersey's *Algebra,* Wallis had outlined ideas of the negative and imaginary numbers that would make his *Treatise* different from all earlier algebra texts. He had accepted the negatives and imaginaries as "quantities" and had decided on a new strategy for their legitimation. As he related in a letter to Collins of May 1673, he had been "of opinion from the first" that algebraists should use both negative and imaginary numbers. But, on the use of the imaginaries, he related: "I durst not without a precedent, when I was so young an algebraist . . . , take upon me to introduce a new way of notation, which I did not know of any to have used before me." And: "I was then too young an algebraist to innovate without example." In his caution about bucking mathematical tradition, he resembled other major figures of the Scientific Revolution, including Galileo who through middle age had worried about being "hissed off the stage" should he publicly endorse the heliocentric theory.[15] But, eager to use the imaginaries,

13. Wallis to Collins, 5 Feb. 1666/1667, *Correspondence of Scientific Men of the Seventeenth Century,* ed. Stephen Jordan Rigaud, 2 vols. (Oxford, 1841; reprint, Hildesheim: Georg Olms, 1965), 2:476.

14. Wallis, *Treatise of Algebra,* preface.

15. Wallis to Collins, 6 May 1673, *Correspondence of Scientific Men,* 2:577–578. On Galileo, see Arthur Koestler, *The Watershed: A Biography of Johannes Kepler* (Garden City, N.Y.: Doubleday, 1960), 177.

Wallis eventually found an easy way out: he discovered $\sqrt{-dddddd}$ in Harriot's *Praxis* and elevated it to a key example of precedent for the imaginary numbers, and an English precedent at that.

In the letter of May 1673 as well as three letters of the preceding two months – letters providing an exceptional window to the trial-and-error world of early modern algebra – Wallis described his path to the argument from precedent. These letters highlight the pivotal role of the irreducible case of the cubic equation in forcing early modern algebraists to deal seriously with the imaginaries. As Pell's edition of Rahn's *Algebra* had shown, algebraists could give complex roots while still viewing such roots as mere signs of impossibility. On the other hand, as the work of Bombelli and now Wallis showed, algebraists could not generally solve the irreducible case of the cubic without accepting imaginaries as some sort of quantities or numbers to be manipulated according to definite rules.

Comments by Collins provided the occasion for Wallis's account of his assault on the irreducible cubic. In 1669 Collins referred Wallis to John Jacob Ferguson's *Labyrinthus algebrae,* claiming that this Dutch work solved "cubic and biquadratic equations by such new methods as render the roots in their proper species."[16] When Wallis finally received extracts from Ferguson's algebra with Collins's supposed clarifications, he concluded that the solution of the irreducible cubic had eluded Ferguson. He then tried to set Collins straight in his four letters of March through May 1673, in which he related his personal struggle with the irreducible case and his coming to terms with imaginary quantities. Assuming the role of teacher, he first carefully explained that Cardano's formula sometimes called for extraction of the cube roots of complex binomials, $a \pm \sqrt{-b}$. "The main difficulty," he stressed, "is extracting that cubic root, which all binomials do not admit."[17] When Collins, following Ferguson, implied that some sort of multiplication of the unknown root would facilitate the extraction process, Wallis responded that "barely changing the value of the unknown root, whether by addition, subtraction, multiplication, or division, will not at all help the matter, to make the designation by Cardan's rules any whit more possible, if before it were not."[18]

16. Collins to Wallis, 17 June 1669, *Correspondence of Scientific Men,* 2:515. Newton's reaction to Ferguson's work, and to Collins's claims for the latter, is discussed in Chapter 7.

17. Wallis to Collins, 29 Mar. 1673, *Correspondence of Scientific Men,* 2:557. Later mathematicians would conclude that the cube roots of complex binomials cannot always (generally) be extracted arithmetically but they can be found trigonometrically.

18. Wallis to Collins, 6 May 1673, 2:576–577.

Wallis claimed to have derived the solution of the cubic equation independently in 1648, before he learned of Cardano's or Descartes's work on the cubic.[19] As he reported to Collins, in his original assault on the cubic he had been "troubled" that the formula led sometimes to cube roots of complex binomials, mentioned earlier, and other times to cube roots of surd binomials. Thus application of the formula to the equation $x^3 + 12x = 112$ gave

$$x = \sqrt[3]{\sqrt{3200} + 56} - \sqrt[3]{\sqrt{3200} - 56},$$

where $\sqrt{3200}$ cannot be reduced to a rational number.[20]

Wallis had resolved the problem of the surd binomials before the complex binomials. An essential first step was deciding how to manipulate surds. It was one thing to accept surds, he implied, but quite another to decide exactly how to operate on them. "It was not without some diffidence," he confessed, "that I ventured on $2\sqrt{3}$, instead of $\sqrt{12}$, not having then met with any example of a number so prefixed to a surd root." Thus for the young Wallis the reduction of a surd was a "venture," a risky or daring undertaking. The mathematical usefulness of the reduction and the existence of a precedent, however, helped him conquer his diffidence. He explained:

> I found it [the reduction] so expedient, not only for the discovering the root of a binomial ... but for the adding and subducting of commensurable surds, that I resolved to use it for my own occasions, before I knew whether others would approve of it or no; especially having found in the first edition of Oughtred's *Clavis* ... one instance or two for it, to justify myself if it should be questioned. But since that time it is grown more common, and I perhaps have somewhat contributed thereunto.[21]

Using this process as he solved $x^3 + 12x = 112$, he reduced $\sqrt{3200}$ to $40\sqrt{2}$. Thus the remaining problem was to find the cube roots of $40\sqrt{2} + 56$ and $40\sqrt{2} - 56$. He now assumed the cube roots to be $a\sqrt{2} \pm e$. Cubing the latter, he arrived at:

$$2a^3\sqrt{2} \pm 6a^2 e + 3ae^2\sqrt{2} \pm e^3.$$

Having decided on a rule for the equality of binomials involving surds, he set

$$2a^3 + 3ae^2 = 40 \text{ and}$$

19. Wallis to Collins, 29 Mar. 1673, 2:558.
20. Wallis to Collins, 12 Apr. 1673, *Correspondence of Scientific Men*, 2:566–567.
21. Wallis to Collins, 6 May 1673, 2:578.

$$6a^2e + e^3 = 56, \text{ and}$$

worked from there to $x = 4$.[22]

Wallis's solution of cubic equations involving surd binomials went hand in hand with his elaborating rules for manipulating surds. Seeing no compelling mathematical argument upon which to base the manipulation, including the reduction, of surds, he turned to precedent. Such a reduction was extremely useful or (as he often described algebraic novelties for which there was no compelling justification) "expedient," and at least one other major mathematician had already employed it – and that was all that could and needed to be said, according to Wallis.

As he now explained to Collins, one could extract cube roots of complex binomials by a process similar to that used for surd binomials. But he admitted that as a young algebraist he had not dared to extend the process to complex binomials, let alone to manipulate imaginary quantities freely. As he put it: "I durst not then be so venturous without a precedent" and "I had not then confidence enough to introduce, without example, this notation, $\sqrt{-N}$" for what appeared to be "an impossible quantity."[23] Indeed, an argument from precedent for treating the imaginaries as quantities was more difficult to make than that for the surds, given Oughtred's avoidance of the imaginaries and (later) Pell's and Kersey's insistence that they were signs of impossibility. "And had I but known of any precedent . . . ," Wallis wrote Collins of these earlier days, "I should not have scrupled to follow it." With the cards of the English mathematical tradition so stacked against the imaginaries, Wallis nevertheless did his best to discover – or, perhaps we should say more correctly, construct – the precedent he felt he needed to use imaginary quantities. He finally found a precedent: "one and I think but one $\sqrt{-dddddd}$" in Harriot's *Praxis*.[24]

Wallis would later make the most of Harriot's symbol. In his *Treatise*, he would argue for the imaginaries partially on the basis of precedent and elevate Harriot to one of the early promoters of negative and imaginary roots. This would be at best a half-truth: as we have seen,

22. Wallis to Collins, 12 Apr. 1673, 2:567–568.
23. Ibid. Here Wallis took the equation $x^3 - 63x = -162$, the solution of which required the extraction of the cube roots of $-81 \pm \sqrt{-2700}$. Reducing $\sqrt{-2700}$ to $30\sqrt{-3}$, he wrote the cube roots of $-81 \pm 30\sqrt{-3}$ as $-a \pm e\sqrt{-3}$, etc., and reasoned to $x = -9$.
24. Wallis to Collins, 6 May 1673, 2:578. Note that Wallis had already related to Collins that geometric construction had helped convince him of the solvability of irreducible cubic equations (Wallis to Collins, 12 Apr. 1673, 2:568–572).

although in the *Praxis* Harriot mentioned negative roots and cited one imaginary root, he dismissed the equations involving the former as "useless" and that involving the latter as "impossible."[25] But these qualifications mattered little to Wallis who, as attested by his remarks to Collins, simply wanted "any precedent" to bolster his case for negative and complex roots.

This argument from precedent for the use of the imaginaries both fueled and was fueled by Wallis's national chauvinism. In a period that witnessed the rise of a sense of English nationality, Wallis was eager to cite the exploits of his countrymen to bolster the English claim to prowess in modern science.[26] He, Collins, and other English scientists complained that continental scholars often ignored them or, when these scholars paid attention to their work, they frequently expropriated it. At the turn of the seventeenth century William Gilbert had lamented that "Frenchmen, Germans, and Spaniards of recent time ... misuse the teachings of others, and like furbishers send forth ancient things dressed with new names and tricked in an apparel of new words as in prostitutes' finery."[27] Wallis and Collins often reiterated this theme; both wished that English scholars would publish more and in a timely fashion, and thus "not let strangers reape ye glory of what those amongst us are ye Authors."[28] Harriot's *Praxis* – in which the supposed precedent for imaginary numbers appeared – had of course been published, even if posthumously. But, in the opinion of Wallis, continental mathematicians had still refused to give Harriot his due.

What a coup then for Wallis when he discovered the reference to an imaginary quantity in the *Praxis!* When suitably interpreted, the reference offered a partial solution to justification of the imaginary numbers as well as ammunition for a case for English priority in early modern algebra. In the *Treatise of Algebra* Wallis would write that the acceptance of imaginary roots was "a great discovery of *Harriot's* (...

25. For a discussion and refutation of "Wallis's extraordinary claims" that Harriot pioneered the use of negative and imaginary roots, see J. F. Scott, *The Mathematical Work of John Wallis, D.D., F.R.S. (1616–1703)* (London: Taylor and Francis, 1938), 141–145.

26. For a general discussion of science in the service of seventeenth-century England, see Liah Greenfeld, "Science and National Greatness in Seventeenth-Century England," *Minerva* 25 (1987): 107–122.

27. William Gilbert, *De Magnete,* trans. P. Fleury Mottelay (1600; reprint, New York: Dover, 1958), 15.

28. Wallis, 21 Mar. 1666–1667, letter 623, *The Correspondence of Henry Oldenburg,* ed. A. Rupert Hall and Marie Boas Hall, 9 vols. (Madison: University of Wisconsin Press, 1966–1973), 3:373.

wherein *Des Cartes* follows him.)" Generalizing from the imaginaries and other examples, as well as distorting the historical record in the service of England, he would also state that Harriot had "made very many advantagious improvements in this Art; and hath laid the foundation on which *Des Cartes* (though without naming him,) hath built the greatest part (if not the whole) of his *Algebra* or *Geometry*. Without which, that whole Superstructure of *Des Cartes* (I doubt) had never been."[29] In short, the causes of mathematical generalization and national pride would go hand in hand in Wallis's *Treatise*, perhaps nowhere more conspicuously than in the case for an English precedent for using imaginary quantities.

3

No matter what Wallis felt toward foreigners, his *Treatise* acknowledged English and foreign debts. Clearly inspired by his countrymen Oughtred and Harriot (and, to a lesser extent, Pell and Kersey) as well as the Frenchmen Viète and Descartes, the work was a first-rate synthesis of early modern algebra. It was a synthesis informed by its author's complete commitment to the symbolical style, deep reflections on the arithmetic foundations of the new algebra, daring use of the inductive method, and penchant for generalization. Furthermore, as its very title and preface stressed, the work was no "small epitome" of algebra – Wallis's description of Oughtred's *Key* – or mere manual of algebraic practice. Rather, as Wallis heeded Collins's call for an algebra suitable for English university students, he prepared a "treatise . . . , both historical and practical," which offered "an Account of the Original, Progress, and Advancement of (what we now call) Algebra."[30] The *Treatise* thus covered the history, results, language, universe, and methods of algebra, sometimes from the myopic perspective of an English chauvinist, but, more often and more importantly, from the vantage point of a first-rate mathematician, scholar, and teacher who had been actively involved in

---

29. Wallis, *Treatise of Algebra*, 174, 126. More than strictly nationalism – e.g., Wallis's strong antipapist feelings – explains his antagonism toward Descartes and his many other disputes with French mathematicians (see Scott, *Mathematical Work of John Wallis*, 135, 178–179). For a more general analysis of Wallis's slighting of French algebraists, see J. F. Scott, "John Wallis As Historian of Mathematics," *Annals of Science* 1 (1936): 335–357.
30. Wallis, *Treatise of Algebra*, preface.

the doing of early modern mathematics for nearly four decades and who had scoured much of the available literature on both the old and new algebra.

The *Treatise,* first of all, testified to Wallis's personal debt to and respect for Oughtred, whose *Key* had made a recruit of him for the new algebra. Personal and chauvinistic sympathies guiding him, Wallis extolled Oughtred as a major symbolical algebraist at the expense of Viète. "The method of *Vieta,*" he wrote, "is followed, and much improved, by Mr. *Oughtred* in his *Clavis* . . . and other Treatises of his." Wallis, moreover, left no doubts that many of the symbols he used in his everyday mathematics, as well as his penchant for coining new symbols, came from Oughtred.[31]

Oughtred's *Key* had, perhaps above all, impressed Wallis with the inventive power of the new symbolical language. As Wallis observed in the *Treatise,* symbols were: (1) abbreviations that permitted Oughtred to effect the characteristic brevity of the *Key,* and (2) instruments of discovery that enabled mathematicians to uncover "the true nature of divers intricate Operations . . . , which without such Ligatures, or Compendious Notes, would not be easily discovered or apprehended; but by the help thereof, appear obvious and conspicuous to the first view." For Wallis, then, symbols laid bare algebraic situations, which once dressed in impervious prose now appeared "obvious and conspicuous to the first view." And, indeed, as he wrote of "the first view," he, like Oughtred, seems to have had in mind the sense of sight. Symbols made algebra easier because they permitted mathematicians to paint, as it were, transparent thumbnail sketches of otherwise complicated algebraic situations. As he put it:

> Specious Arithmetick, which gives Notes or *Symbols* (which he [Viète] calls *Species*) to Quantities both known and unknown, doth (without altering the manner of demonstration, as to the substance,) furnish us with a short and convenient way of Notation; whereby the whole process of many Operations is at once exposed to the Eye in a short Synopsis.

Moreover, it was by means of the compendious symbolism that propositions became "intelligible, with much more ease than when involved in a multitude of Words, and long Periphrases [*sic*] of the several Quantities and Operations."[32]

The *Treatise of Algebra,* then, testified to great enthusiasm for symbolical reasoning and a corresponding mistrust of words (of which we have

31. Ibid., preface, 67–68.
32. Ibid., 67, preface, 68.

already spoken) that was shared by Oughtred, Wallis, Seth Ward, and other key members of the English scientific community of the seventeenth century. Wallis here and Ward in his *Vindiciae Academiarum* of 1654 not only reiterated the content of Oughtred's defense of the symbolical style but did so in words reminiscent of their symbolical mentor and words that fitted nicely with the contemporary English attack on the "Idols of the Market-place." Oughtred had exulted that symbolical algebra did not "racketh the memory with multiplicity of words . . . but plainly presenteth to the eye the whole course and processe of every operation and argumentation."[33] As we have just seen, Wallis complained of a "multitude of words" and argued that with symbols "the whole processe of many Operations is at once exposed to the Eye in short Synopsis."

Ward was even more explicit about the connections between the symbolical style and the contemporary attack on ornate prose. By the time he wrote *Vindiciae Academiarum*, Ward was a member of the select group of talented scientific and mathematical thinkers – parliamentarians and royalists alike – whom John Wilkins had helped to assemble at Oxford. Many within the group advocated algebra, cryptography, terse scientific prose, the theoretical study of language, or some combination of the latter. In addition to Wilkins, Wallis, and Ward, the group included Christopher Wren, Laurence Rooke (another of Oughtred's students), Thomas Sprat, and later Robert Boyle. Ward, an Anglican and royalist who had been ejected from Sidney Sussex College, Cambridge, early in the Civil War, had moved to Oxford and "made choice of *Wadham* Col. to reside in, invited thereto by the Fame of Dr. *Wilkins* . . . , with whom he soon contracted an intimate Acquaintance and Friendship."[34] In 1649 Ward had been named the Savilian professor of astronomy.

In *Vindiciae Academiarum* he was responding to John Webster's *Academiarum Examen, or the Examination of Academies* of 1654. Alarmed

33. William Oughtred, *The Key of the Mathematicks, New Forged and Filed* . . . (London: Thomas Harper, 1647), preface, 2.
34. Walter Pope, *The Life of Seth, Lord Bishop of Salisbury*, ed. J. B. Bamborough (Oxford: Basil Blackwell, 1961), 28. On Wilkins's role in attracting scientific thinkers to Oxford, see Shapiro, *John Wilkins*, esp. 118, 124. On the causes and many facets of the new study of language in the seventeenth century, see Vivian Salmon, *The Works of Francis Lodwick: A Study of His Writings in the Intellectual Context of the Seventeenth Century* (London: Longman, 1972). There is a wealth of information on the English scientists of this period and their networks in Charles Webster, *The Great Instauration: Science, Medicine and Reform, 1626–1660* (London: Duckworth, 1975).

by what he and his colleagues saw as Webster's ill-informed attack on the English universities, Ward had rushed to publish his *Vindiciae Academiarum* of the same year, for which Wilkins wrote an introduction.[35] Part of the Webster–Ward debate centered on the symbolical style. It was not that Ward praised the style and Webster rejected it, for both saw the style as a major innovation. Webster had even written of "our never sufficiently praised Countreyman Mr. *Oughtrede*." In discussing the symbolical style, however, Webster had linked it to hieroglyphics, the emblems of the alchemists, and the subject of grammar. Alarmed at the prospects of a revived magical aura for mathematics and the subordination of algebraic symbolism to the traditional subject of grammar, Ward went on the attack:

> [Webster] hath extreamely disobliged whosoever have been Authors of the Symbolicall way, either in Mathematicks, Philosophy, or Oratory, to bring them under the *ferula*, and make those who have exempted themselves from the encombrances of words to be brought *post liminio*, into the Grammar Schoole, it was little thought by *Vieta, M. Oughtred,* or *Herrigon,* that their designation of quantities by Species, or of the severall waies of managing them by *Symbols* (whereby we are enabled to behold, as it were, with our eyes, that long continued series of mixt and intricate *Ratiocination*, which would confound the strongest fancy to sustaine it . . .) should ever have met so slight a considerer of them, as should bring them under Grammar.[36]

Here Ward drew a sharp dividing line between the old prose and the new scientific language advocated by Bacon and his English followers. He clearly wanted there to be no mistake about where symbolical algebra fell or about the camp to which the symbolical algebraists belonged. Students at Oxford, he continued, knew "that the avoiding of confusion or perturbatio[n] of the fancy made by words . . . was the end and motive of inventing Mathematicall Symbols; so that it was a designe perfectly intended against Language and its servant Grammar."[37] Such

35. John Webster, *Academiarum Examen, or the Examination of Academies* (London, 1654); [Seth Ward], *Vindiciae Academiarum Containing, Some Briefe Animadversions upon Mr Websters Book, Stiled, The Examination of Academies* (Oxford, 1654). Both works are reprinted in Allen G. Debus, *Science and Education in the Seventeenth Century: The Webster–Ward Debate* (London: Macdonald, 1970).
36. Webster, *Academiarum Examen*, 24 (106); [Ward], *Vindiciae Academiarum*, 19 (213). The references are to pp. 24 and 19 of the original works and pp. 106 and 213 of Debus, *Science and Education*.
37. [Ward], *Vindiciae Academiarum*, 19 (213).

a dividing line having been drawn, Wallis's *Treatise of Algebra* was a work of linguistic as well as mathematical significance.

4

Wallis found an excellent example of the conceptual impact of the new symbolism in Harriot's use of iterated letters (*a, aa, aaa,* etc.) to enforce the algebraic concept of power. This concept helped to liberate algebra from geometry and also paved the way for a generalization of powers beyond whole-number exponents. Harriot, Wallis began, "waves the names of *Square, Cube,* &c. in the Designation of his Symbols or Species; (and the Letters *q, c,* &c. the Characters thereof)." Harriot's new symbolism had two major effects, the first of which was to end the confusion that had surrounded the notation for powers. *A quadrato-cubus* or *Aqc,* for example, had been interpreted by some algebraists (including Diophantus, Viète, and Oughtred) as the fifth power of *A,* and by others (Cardano and Stifel) as the sixth power. Harriot's notation, on the other hand, permitted the algebraist to write unambiguously either *aaaaa* or *aaaaaa.* And the change affected algebra in a second and more significant way: in writing *A quadrato-cubus,* Viète had implied a geometric interpretation for algebra; Harriot's symbols, on the other hand, carried no geometric connotation. Harriot, Wallis generalized, "doth thus, by degrees, disoblige this *Arithmetical* consideration of *Proportions,* from the Intanglements which the terms of *Square* and *Cube,* (borrowed from *Geometry,*) and those more *Ungeometrical* Terms of *Squared Square, Sursolids, Squared Cube,* &c. have imposed on our Fancies and Understandings."[38]

Perhaps going beyond the literal text of the *Praxis* and reiterating a point that Kersey had explicitly made, Wallis then moved toward proclaiming algebra's independence from geometry. Focusing on algebraic powers, he noted that geometric analogy ended with the third power, but that all powers, considered as numbers, made sense. "Nature, in propriety of Speech," he began, "doth not admit of more than *Three* (Local) Dimensions."

> A Line drawn into a Line, shall make a Plane or Surface; this drawn into a Line, shall make a Solid: But if this Solid be drawn into a Line, or this Plane into a Plane, what shall it make? a *Plano-plane?*

38.   Wallis, *Treatise of Algebra,* 126. In the *Treatise* Wallis used Harriot's notation for powers, avoiding Descartes's index notation, which he had employed in the earlier *Arithmetica infinitorum.*

This is a Monster in Nature, and less possible than a *Chimera* or *Centaure*. For Length, Breadth and Thickness, take up the whole of *Space*. Nor can our Fansie imagine how there should be a Fourth Local Dimension beyond these Three.

But if we consider a *Number* Multiplied by itself, and this again into the same Number, and so again and again as oft as you please; in this, there is nothing of impossibility or of Difficulty to apprehend.[39]

As Wallis emphasized, Harriot's new symbols for algebraic powers testified then not simply to a technical advance in the language of algebra but more fundamentally to an emerging arithmetic view of the foundations of the subject. Symbolism and conceptualization were intimately intertwined.

"Disentanglement" of algebraic powers from geometric underpinnings facilitated the extension of the concept of power beyond whole-number exponents. In his *Arithmetica infinitorum* of 1656, pushing for a general formula for the area under a curve of the form $y = x^n$, Wallis had extended the concept of power to fractional and negative exponents. While he did not use the modern symbols, he had in essence defined $a^{1/2}$ as $\sqrt{a}$ and a^{-n} as $1/a^n$.[40] Not surprisingly, then, early in the *Treatise*, proceeding "as Mr. *Oughtred* directs," he introduced negative exponents and a general rule for the multiplication of powers – "the Exponent of each particular Product, is the Aggregate of those which belong to the particular Factors." When he discussed the *Arithmetica infinitorum* later in the *Treatise*, he gave a table of powers and their corresponding exponents, moving from zero and whole-number exponents through fractional, negative, and irrational ones. Although he merely listed, for example, the "power" $1/a$ and its "exponent" -1, he noted that Newton had adopted the notation a^n, according to which, for example, a^{-1} stands for $1/a$.[41]

This move to a general concept of power evidenced Wallis's strong commitment to mathematical generality combined with a powerful faith in "interpolation" as a reliable path to generalized results.[42] But here, as so often in the *Treatise*, he was building on the hints and faltering steps of his predecessors. As he acknowledged (and as we have seen),

39. Ibid., 126–127. Compare with John Kersey, *The Elements of That Mathematical Art Commonly Called Algebra, Expounded in Four Books*, 2 vols. (London: William Godbid, 1673–1674), 1:4.
40. On the *Arithmetica infinitorum*, see Scott, *Mathematical Work of John Wallis*, 26–64, esp. 34–35.
41. Wallis, *Treatise of Algebra*, 288–289.
42. See, e.g., Scott, *Mathematical Work of John Wallis*, 34–35.

Oughtred had referred to negative exponents in his development of logarithms, and even earlier in the seventeenth century Henry Briggs had used negative or "defective" logarithms. Yet neither Briggs nor Oughtred had begun to reconstruct the foundations of algebra to legitimate negative powers. Only Wallis, with his penchant for generalizing and concern for foundations, did.

5

By claiming for Harriot the liberation of algebraic powers from geometric "intanglements," Wallis was establishing the priority of English mathematicians in pursuing the arithmetic foundations of algebra. The liberation of algebra from geometry lay at the heart of Wallis's definition of a new independent science of algebra and thus at the heart of his *Treatise*. Boldly, he divided algebra into two parts: (1) "pure algebra," which was based on arithmetic, and (2) its applications, including the geometric. As he was wont, he presented pure algebra as no new creation of his, but rather as an already recognized part of English mathematics – in this case, a part of Harriot's mathematics. He used Descartes as Harriot's foil in juxtaposing pure algebra with geometric algebra.

Harriot, he claimed, was responsible for the best of early modern algebra. Conceding some geometrically oriented contributions to Descartes, Wallis boasted that Harriot

> hath taught (in a manner) all that which hath since passed for the *Cartesian* method of *Algebra*; there being scarce any thing of (pure) *Algebra* in *Des Cartes,* which was not before in *Harriot*; from whom *Des Cartes* seems to have taken what he hath (that is purely *Algebra*) but without naming him.
>
> But the Application thereof to *Geometry,* or other particular Subjects, (which *Des Cartes* pursues,) is not the business of that Treatise of *Harriot*, . . . the design of this being purely *Algebra*, abstract for particular Subjects.

As if to assure that his readers did not miss Harriot's claim as a pioneer of pure algebra, Wallis returned to the point in his "Recapitulation of Particulars in Harriot's *Algebra.*" Harriot "meddles not at all with *Geometrical Effections,*" he summarized. "Nor doth he make particular application of his *Algebra,* to particular Subjects, whether Geometrical or others: His business being in this Tract, to treat of *Algebra* purely by itself, and from its own principles, without dependance on Geometry, or any connexion therewith."[43]

43. Wallis, *Treatise of Algebra,* preface, 198.

Even if Wallis was able to read into the lean *Praxis* a vision of pure algebra, it was Wallis himself who undertook the task of founding the subject on its "own principles, without dependance on Geometry." Early remarks in the *Treatise* – in which subtraction, division, and extraction of roots were presented as examples of analysis, but addition, multiplication, and composition of powers as examples of synthesis – made it clear that Wallis, unlike Viète and his followers, was not prepared to defend algebra under the rubric of the analytic method.[44] Instead, unlike both Oughtred and Harriot who had highlighted analysis as the special method of algebra, Wallis presented algebra essentially as an extension of arithmetic. In his *Defense of the Treatise of the Angle of Contact* of 1684, he had already observed that, whereas arithmetic had originally been

> confined to (what we now call) Whole Numbers, . . . now a days, we extend Arithmetick, not only to Fractions, but to Surds also; and even *Diophantus's* Treatise of (what we now call) *Algebra*, is intituled . . . *rerum Arithmeticarum libri*; and (what we call) *Specious Arithmetick*, pretends in a manner to whatsoever is capable of Proportion.[45]

Here he concluded that it was difficult to determine any boundaries for arithmetic. In the *Treatise of Algebra*, Wallis then developed this arithmetic view of algebra, continuing to describe the new algebra as "specious arithmetick" and portraying it as a mathematical subdiscipline whose abstractness endowed it with limitless possibilities for application.

At the beginning of chapter 16, for example, he observed that the rules of algebra followed from those of arithmetic, or that "The Algorism or Practical Operations in this Specious Arithmetick, are for substance the same as in the practise of Numeral Algebra [arithmetic], before in use, but the Notation somewhat altered." He then stated the basic rules of algebraic addition, subtraction, multiplication, and division, as well as composition and extraction of roots. Although in doing so he followed Oughtred's *Key* rather closely, there was a personal touch, for Wallis explicitly stated the commutative property of addition and multiplica-

---

44. Ibid., 2.
45. John Wallis, *A Defense of the Treatise of the Angle of Contact* (London, 1684), 92. Wallis here notes the importance of viewing the unit as divided into parts. For a fine discussion of this newer understanding of the unit, its late-sixteenth-century sources (mainly Simon Stevin), and its role in breaking down the rigid distinction between magnitude and number (geometry and arithmetic), see Charles V. Jones, "On the Concept of *One* As a Number" (Ph.D. diss., University of Toronto, 1978).

tion and the distributive property.[46] Chapter 17 concerned "the Grounds of the foregoing Operations" and opened with the claim that "The reason of these Rules above mentioned, most of them, is (to one who understands Ordinary Arithmetick, and doth a little consider them) very obvious, (and where it is not so, I shall briefly explain it)." Indeed, Wallis now gave explanations of how all the basic arithmetic operations extended to algebra. In arithmetic one added whole numbers according to certain rules; in algebra one could add positive and negative numbers according to the same rules, since

> In Addition it is manifest (from the common notion of Ordinary Arithmetick,) that if to 3 we add 2, it makes 5, whatever be the things so added, (provided they be all of the same kind, and such as are capable of such Addition.) If to 3 *Cows* we add 2 *Cows,* it makes 5 *Cows*; if to 3 *Sheep* we add 2 *Sheep*, it makes 5 *Sheep*: And by the same reason, if to 3 A's we add 2 A's, it makes 5 A's; (whatever quantity be meant by A, provided that A be in both places of like signification.) And therefore also, if to *the want of 3 A's,* (or 3 defects of A,) we add *the want of 2 A's* (or 2 defects of A,) it makes, *the want of 5 A's,* (or the defect of 5 A's,) that is, if to − 3A, we add − 2A, it makes − 5A.[47]

Even the rule of signs was but a consequence of "the true notion of [arithmetic] Multiplication [which] is . . . to put the Multiplicand, or thing Multiplied (whatever it be) so often, as are the Units in the Multiplier." From the latter definition Wallis argued consecutively that multiplication of a negative multiplicand and positive multiplier involved no more than taking the multiplicand the specified number of times, and thus getting a negative sum; multiplication of a positive multiplicand and a negative multiplier involved nothing more than taking the multiplicand away the specified number of times; and, finally, multiplication of a negative multiplicand and a negative multiplier involved "taking away a Defect or Negative," which "is the same as to supply it" − and thus getting a positive.[48]

Although the operations of arithmetic and algebra were thus "for substance the same," the two sciences differed markedly in the level of abstraction each permitted − a point that Wallis stressed. "*Pure Algebra,*" he defined, ". . . simply considers the Computation and Management of Proportion, (abstract from the consideration of any particular Subject;) . . . ." In the same vein: "I look upon this, as the great advantage of Algebra, that it manageth Proportions abstractly, and not as

46. Wallis, *Treatise of Algebra,* 69–72 (quotation from p. 69).
47. Ibid., 73.   48. Ibid., 74.

restrained to Lines, Figures, or any particular Subject; yet so as to be applicable to any of these particulars as there is occasion."[49]

In short, presenting his work as an elaboration on Oughtred's and Harriot's, Wallis took early modern algebra into the next stage of "pure algebra." Solidifying the hints of his predecessors, he tried to free algebra of any dependence on geometry, and carefully distinguished pure algebra from its applications. He provided an arithmetic basis for the operations of algebra, while refusing to restrict algebra to the study of numbers. According to this new view, pure algebra was a science of symbols or "notes" which "manageth Proportion" in an abstract fashion. Abstractness, the key to algebra's power, meant that its results were not only arithmetically valid but also "equally applicable to other Quantities as well as to Lines and Figures."

6

This new view entailed a redefinition of the relationship between algebra and geometry. Following Oughtred and Descartes, Wallis had used algebra extensively in the study of geometry. But, more important, in the *Treatise of Algebra,* he enunciated the general theme of algebra's superiority over geometry based on algebra's economy, intellectual transparency, and abstractness. On a more specific level, he recast the traditional concept of proportion as an algebraic concept.

Algebra was exceptionally economical and lucid because it was symbolical. As Bacon and his scientific followers juxtaposed scientific prose to ornate prose, Wallis juxtaposed symbolical algebra to traditional geometry. In the *Treatise* he discussed his earlier *Conic Sections: New Method Exposed,* in which he had defined the parabola, ellipse, and hyperbola by algebraic equations rather than as slices of a cone.[50] In a rather complete turning of the tables of mathematical tradition (in which he reinforced ideas already in Oughtred's *Key*), he displayed algebra as the preferred approach to conics. Prior to algebraic treatment, he claimed, the study of conic sections was viewed "as so perplex and intricate, as that most of those who pretended to Mathematicks, were deterred from medling with it." He even speculated that the complexity of Apollonius's *Conics* had deterred transcription of its last four books,

49. Ibid., 198, 298–299.
50. For a discussion of *De sectionibus conicis* of 1655, see Scott, *Mathematical Work of John Wallis,* 15–25.

thus ultimately resulting in their loss; and furthermore that Apollonius had used "some such *Art of Invention,* as what we now call *Algebra,*" in arriving at the results presented geometrically in the *Conics.*[51] Responding to Fermat's criticism of the use of algebraic rather than geometric demonstrations in his *Arithmetica infinitorum,* Wallis again emphasized algebra's economy and associated lucidity. "I chose the shorter way," he explained, "because by this means I might in a compendious continued discourse deliver that in brief, which in the other way must (with more pomp and solemnity,) be parcelled out into several *Lemma's,* and preparatory Propositions. Which, though it might look more August, would be less edifying than if I reduce the same to a brief Synopsis."[52]

Algebra's claim to superiority over geometry also rested on the abstractness and hence greater generality of algebraic arguments. Responding to Fermat's call for more classical geometry, Wallis criticized the "Pompous ostentation of Lines and Figures," and added:

> For though such Lines and Figures be necessary where the Truth of a Proposition depends on Local Position: And though they be otherwise of use, sometimes for assisting the Fansy or Imagination, (shewing that to the eye, by way of instance, in one particular case, as that of Lines; which is abstractly true in all kinds of Quantity whatever:) Yet where the truth of the Proposition depends merely on the nature of Number or Proportion; (and is equally applicable to other Quantities as well as to Lines and Figures:) It is much more natural to prove it abstractly from the nature of Number and Proportion; without such embarassing the Demonstration.[53]

In short, Wallis now argued that algebra ought to be employed wherever "Number or Proportion" was involved and that, in these instances, abstract, algebraic arguments were "more natural" than geometric ones.

This algebraic program depended heavily on Wallis's recasting of the mathematical concept of proportion in terms of algebraic equations.[54] The traditional definition of proportion had come from Euclid who had taken into account proportions involving commensurable numbers as well as those involving incommensurables or surds. "Numbers are proportional," according to Euclid's definition 20 of book VII, "when the first is the same multiple, or the same part, or the same parts, of the

51. Wallis, *Treatise of Algebra,* 290.   52. Ibid., 305.
53. Ibid., 298.
54. Some of the following discussion of proportion is dependent on a detailed article on the subject: Chikara Sasaki, "The Acceptance of the Theory of Proportion in the Sixteenth and Seventeenth Centuries," *Historia Scientiarum* 29 (1985): 83–116.

second that the third is of the fourth." Thus, 2 is to 4 as 4 is to 8 because $2 \times 2 = 4$ and $4 \times 2 = 8$. Euclid's definition 5 of book V addressed, moreover, the problem of ratios of magnitudes that are not whole numbers:

> Magnitudes are said to be in the same ratio, the first to the second and the third to the fourth, when, if any equimultiples whatever be taken of the first and the third, and any equimultiples whatever of the second and the fourth, the former equimultiples alike exceed, are alike equal to, or alike fall short of, the latter equimultiples respectively taken in corresponding order.[55]

Through the early seventeenth century mathematicians and physicists used the traditional concept of proportion extensively. For example, Johannes Kepler expressed his third law of astronomy in terms of proportions, writing essentially that, for any two planets, the ratio of the squares of their orbital periods is equal to the ratio of the cubes of their average distances from the sun. Nevertheless, by the late sixteenth century Viète had suggested an approach to proportion that eventually changed the very way mathematicians viewed relationships between quantities. He defined: "If there are three or four terms and the first is to the second as the second or third is to the last, the product of the extremes will be equal to the product of the means." Getting right to the import of the preceding definition, he had next written that "a proportion may be said to be that from which an equation is composed and an equation that into which a proportion resolves itself."[56] Viète had thereby begun the process by which mathematicians would move from relating quantities through proportions to relating quantities through algebraic equations.[57]

Continuing to bring proportions within algebra, Oughtred had found a ratio or (what he called) *habitudo* (habitude) of one number to another by "dividing the Antecedent by the Consequent, as the Ratio of 31 to 7, is $4\frac{3}{7}$." That is, for Oughtred, the ratio of two numbers was merely another number, the result of arithmetic division of the first number by the second. Correspondingly, four numbers *(a, b, c,* and *d)* were

---

55. *The Thirteen Books of Euclid's Elements,* trans. Thomas L. Heath, 2d ed., 3 vols. (New York: Dover, 1956), 2:278, 114.

56. François Viète, *The Analytic Art: Nine Studies in Algebra, Geometry and Trigonometry . . . ,* trans. T. Richard Witmer (Kent, Ohio: Kent State University Press, 1983), 15.

57. For a sketch of this process and that by which mathematicians next moved from equations to functions, see Carl B. Boyer, "Proportion, Equation, Function: Three Steps in the Development of a Concept," *Scripta Mathematica* 12 (1946): 5–13.

proportional if the product of the two extremes was equal to the product of the two means – or, $a{:}b{::}c{:}d$ if $a \times d = b \times c$.[58]

In his *Mathesis universalis* of 1657 and then again in the *Treatise,* Wallis synthesized the earlier work on the algebraic approach to proportion.[59] In the *Treatise* he assigned credit for it to Oughtred and minimized its revolutionary nature. He gave no indication that some "mathematicians hesitated to regard the ratio of two integers as a single number, . . . and to see that the proportion $a{:}b{::}c{:}d$ implies quite the same thing as the equation $a/b = c/d$."[60] Rather, he recounted matter-of-factly, "By *Proportion* he [Oughtred] understands that Habitude or Relation of two Numbers, (or other Homogeneous Quantities,) one to the other, which is found by Dividing the Antecedent to the Consequent." And, "universally, the Proportion of A to B, is that denominated by A/B, that is, by the Quotient of A divided by B."[61] Wallis thus quietly promoted a new algebraic definition of proportion covering all quantities. A concept that was applicable to arithmetic and geometry, and (Wallis maintained) a concept at the very heart of mathematics, proportion was now operationally reduced to an algebraic equation. Proportion and its management, for which Euclid's *Elements* had hitherto been the guide, were absorbed into algebra.

Given algebra's economy, lucidity, abstractness, and generality, Wallis comfortably proclaimed the thesis that arithmetic and algebra were superior to geometry. In his *Mathesis universalis* he had already observed that continuous quantity, the special subject of geometry, was measured by number, and thereby implied that number was antecedent to and in some respect governed continuous quantity. The key earlier passage, quoted by Wallis's critics, was: "Simply because a line of two feet added to a line of two feet makes a line of four feet, it does not follow that two and two make four; on the contrary the former follows from the latter."[62] In the *Treatise,* he used an example of multiplication to make the same general point. He outlined a geometric demonstration of the proposition that $3 \times 4 = 12$, involving calculation of the area of a

58. Oughtred, *The Key,* 16–17.
59. On Descartes's contribution to the algebraic approach as well as Wallis's development of the approach in *Mathesis universalis,* see Sasaki, "Acceptance of the Theory of Proportion," 90–95.
60. Boyer, "Proportion, Equation, Function," 11.
61. Wallis, *Treatise of Algebra,* 78–80.
62. John Wallis, *Mathesis universalis* (Oxford, 1657), 69. The translation is taken from Florian Cajori, "Controversies on Mathematics between Wallis, Hobbes, and Barrow," *Mathematics Teacher* 22 (1929): 147.

rectangle with sides of 3 and 4. "Though this [demonstration] be true," he summarized, "yet it is not to the purpose; nor doth it prove, that Three times Four Angles, are Twelve Angles . . . ; nor so much as that Three times four Miles are Twelve Miles, . . . It proves at most, but the truth of the Proposition as to one case." Rejecting geometry as a basis of arithmetic or algebraic propositions, he stressed rather the priority of algebra:

> It is much more natural to prove, that Three times Four makes Twelve, (whether of more Angles, or any thing else that is numerable,) from the nature of Number, and of Multiplication. . . . And I look upon this, as the great advantage of Algebra, that it manageth Proportions abstractly, and not as restrained to Lines, Figures, or any particular Subject; yet so as to be applicable to any of these particulars as there is occasion.[63]

7

In elaborating his arithmetic approach to algebra, Wallis admitted that arithmetic gave rise to "impossibilities," which, however, algebra resolved through the extension of number to cover negatives, fractions, surds, and complex numbers. When discussing the impossibility of the quadrature of the circle, he observed: "Nor is it at all strange, that such impossibility should arise; for the same happens in all the Resolutive (analytical) parts of Arithmetick." But, he continued, "each of these [arithmetic] impossibilities hath a peculiar expedient for itself, and is not to be salved by any of the other Expedients." As his first example, he cited impossible subtractions, where a greater is taken from a lesser. The expedient in these cases was a negative number, which expressed "that impossibility (and the measure of that impossibility) . . . imparting somewhat less than nothing." The next example was of impossible division "as if the number 2 be to be divided into 3 equal parts; which (in Integers) the nature of number doth not permit"; the expedient was a fraction. The third example concerned impossible extraction of roots, resulting in surd numbers, "as the Square Root of 8, or the Cubick Root of 9: For there is no (effable) number which being once Multiplied into it self will give 8; or Multiplied Cubically, will give 9." Here again there was "an expedient notation, which is the note of Radicality; as $\sqrt{q}$ 8, $\sqrt{c}$ 9: That is, a supposed (impossible) number, which being Multiplied Quadratically, shall give 8; or which Cubically Multiplied, shall give 9."

63. Wallis, *Treatise of Algebra*, 298–299.

He alluded finally to those equations "which are capable of no other Root (Positive or Negative,) but such only as are (commonly called) Imaginary."[64]

Not surprisingly, he devoted more attention to the newer "impossibilities" of negative and imaginary numbers than the more established fractions and surds. In the *Treatise* he discussed negative and imaginary numbers in three distinct stages, claiming to offer his readers synopses of what Oughtred, Harriot, and then he himself thought about these numbers. The negatives were first defined in chapters 16 and 17 of the *Treatise,* which promised "a short account" of the basic operations of specious arithmetic, "according to Mr. Oughtred's precepts" and with reasons. Following Oughtred rather closely here, Wallis introduced positive and negative quantities through a discussion of their signs. He explained that the sign + "is a Note of Position, Affirmation or Addition; the other [the sign −] of Defect, Negation, or Subduction." This meager introduction behind him, he quickly referred to "a Positive Quantity" and, in language somewhat reminiscent of Descartes, "a Defect, or Negative quantity."[65] At the end of chapter 17, he tackled the extraction of roots. Caught between his own predisposition toward imaginary numbers and Oughtred's ignoring of them, he seemed to compromise with a one-sentence statement that "if the Square be −, there can be no other Root but what they call *Imaginary,* for the Square being made by Multiplying the Root into itself, no Root, either Affirmative or Negative . . . can be so Multiplied into it self as to produce a Negative Square."[66]

Dealing specifically with Harriot's contributions to equation theory in chapters 31 and 32, Wallis made an implicit argument for the acceptance of negative and imaginary roots based on precedent. He claimed not merely that Harriot had referred to "Negative or Privative Roots" but that Harriot had found "every Quadratick Equation, if possible; (that is, if the case proposed be not an impossible case) to have *Two* (Real) Roots . . . That is, Two Affirmatives, or Two Negatives; or one of them Affirmative, and the other Negative."[67]

Arguing in a roundabout fashion (as if to minimize his stretching of

---

64. Ibid., 316. See also ibid., 2.
65. Ibid., 69, 74. Wallis had already referred to negative numbers in passing, when he had explained that Briggs had given as the logarithms of some numbers "Negative Numbers, or Numbers less than 0 . . ." (ibid., 59).
66. Ibid., 75.
67. Ibid., 128, 131. Wallis claimed that Descartes "followed" Harriot in recognizing negative roots (ibid., 128).

the truth), he then made the more tenuous case for the Harriot of the *Praxis* as a proponent of imaginary roots. He considered the equation,

$$aa + za + bc = 0,$$

with its roots,

$$-\frac{1}{2}z \pm \sqrt{\frac{1}{4}zz - bc:},$$

where bc is greater than $\frac{1}{4}zz$. The "Case is impossible," he declared, "and the Equation hath no *Real* Roots; but only (what they call) *Imaginary*." Promising to "further ... discourse [on the imaginaries] toward the end of ... [the] Treatise," he here gave but a brief defense of imaginary roots. Imaginary roots not only show that the "Case proposed" is impossible but also give the "Measure of that impossibility; how far it is impossible, and what alteration ... would make it possible." Then, without mentioning that in the *Praxis* Harriot had never bothered with the imaginary roots of quadratic equations, Wallis quickly observed that "of such imaginary Roots, we find Mr. *Harriot* particularly to take notice (in the Solution of Cubick Equations)."[68]

Ignoring the relevant discussions from Cardano's *Great Art* and Bombelli's *Algebra*, Wallis then crafted a careful argument that made of Harriot the first algebraist to recognize that cubic equations leading to complex binomials had real roots. In defense of this thesis, he cited Harriot's treatment of the cubic equation whose solution involved $\sqrt{-ddddd}$. He explained that Harriot had begun with the equation

$$aaa - 3bba = + 2ccc, (*)$$

where a is the unknown and c is less than b. Applying his own version of Cardano's formula and simplifying $ccccc - bbbbbb$ as $-dddddd$, he had found the root to be

$$a = \sqrt{C.ccc + \sqrt{-ddddd.}} + \sqrt{C.ccc - \sqrt{-ddddd.}}$$

Wallis admitted that Harriot had then described the equation

$$eee = ccc + \sqrt{-ddddd},$$

where e is the unknown, as "Impossible by reason of the Inexplicability of $\sqrt{-ddddd}$." By the latter remark, he continued, Harriot had meant only that "either member of the Root separately considered, $\sqrt{C:ccc} + \sqrt{-ddddd:}$ and $\sqrt{C:ccc} - \sqrt{-ddddd:}$ do imply what they call an

---

68. Ibid., 133–134.

Impossible Quadratick Equation, because of that Inexplicable Root of a Negative Square, $\sqrt{-dddddd}$." That Harriot viewed the original cubic equation (*) as possible was obvious, according to Wallis, for Harriot had already given a real root of this equation.[69]

Readers were thus to evaluate what Harriot did with the cubic equation in question (produced one of its real roots) rather than exactly what he said about it. Wallis boasted:

> [This] is also a great discovery of *Harriot's* (and wherein *Des Cartes* follows him.) Nor do I know, that any before him had shewed, that such a Root could not (in the received ways of Notation) be explicated in Species; otherwise than by those Imaginary Quantities. . . .
>
> . . . For it was before thought, (and so delivered by divers Algebrists,) that whenever (in pursuance of the Resolution) we are reduced to an impossible construction, (such as is the Square Root of a Negative Quantity;) the case proposed is to be judged impossible. Which is yet here discovered to be otherwise.[70]

Thus, the somewhat twisted and carefully worded arguments of the *Treatise* made of Harriot a pioneer of negative and imaginary roots, although the posthumous *Praxis* revealed no interest in negative roots and at best a muddled reference to imaginary expressions. Here, as throughout the *Treatise*, Wallis, the keeper of the archives at Oxford and a crystal-clear thinker, seemed to know exactly what he was doing. Motivated by the need for a convincing precedent for the use of these numbers and inspired furthermore by national chauvinism, he even rationalized his historical distortion. He wrote that, in describing the work of earlier algebraists (he should have written "English algebraists"), he had "been careful to put the best Construction on their Words and Meaning; . . . For it many times happens, that a man lights on a good notion; which he hath not the happiness to express so intelligibly, as perhaps another may do for him."[71]

Still, Harriot's supposed endorsement of negative and imaginary roots, combined with Oughtred's references to negative numbers, did not make a compelling case for the negatives and imaginaries. Indeed, Wallis seems to have concluded that behind the persistent avoidance and obfuscation of such numbers was the fact that no compelling argument for either kind of number could be formulated according to the prevailing canons of mathematics. He then went outside traditional formal mathematics and built a persuasive, rather than mathematically compelling, case for the negative and imaginary numbers. He would not define these numbers in any traditional way; rather he would persuade mathe-

69. Ibid., 172, 174, 179 (quotations from p. 179).
70. Ibid., 174.   71. Ibid., preface.

maticians to accept them on the basis of a series of cumulative arguments. In a bold and somewhat risky maneuver in chapter 66, he tied the cause of the imaginaries to that of the negatives: he stressed that negative numbers as well as numbers involving $\sqrt{-1}$ were imaginary, according to the canons of arithmetic; then he argued that the usefulness of these numbers as well as geometric analogy sanctioned the mathematician's "supposing" that there were negative *numbers* and (the traditionally) imaginary *numbers*.

In *Mathesis universalis* he had already described negative numbers as "imaginary quantities." He had approached the negatives as results of subtraction of greater quantities from lesser, and, using a specific example, he had warned that it was "impossible" to subtract 8 from 5.

> Yet, although this is impossible, mathematicians and especially algebraists, look upon it as though it were not impossible. For they suppose, besides real quantities, certain *imaginary* quantities which are less than nothing. . . .
>
> . . . Nor is this supposition absurd. For when they say, $5 - 8 = -3$, it is as though they said: He who supposes 8 to be subtracted from 5 supposes a certain third number less than 0.[72]

Descartes's influence shone here and indeed in the later *Treatise of Algebra*, as Wallis defended both the negative numbers and numbers involving $\sqrt{-1}$ as imaginary quantities. Unlike Pell and Kersey who referred to numbers involving $\sqrt{-1}$ as "impossible," Wallis generally used Descartes's term "imaginary." In certain ways, moreover, he seemed to conceptualize the negatives and imaginaries as Descartes had in *La géométrie*. Although Descartes had not actually described the negatives as "imaginary," he had observed that some roots of equations "are false, or less than nothing, as if one *supposes* that x designates also the defect of a quantity."[73] According to Wallis's version of the preceding (from the previously quoted passage from *Mathesis universalis*), algebraists "suppose, besides real quantities, certain *imaginary* quantities which are less than nothing."

This thesis that a negative number was imaginary – that mathematicians decided to suppose or imagine that a greater number might be subtracted from a lesser and also that the result was a "negative num-

---

72. Wallis, *Mathesis universalis*, 99–100. The translation is from Scott, *Mathematical Work of John Wallis*, 68–69.

73. ". . . comme si on *suppose* que x designe aussy le defaut d'une quantité" (*The Geometry of René Descartes with a Facsimile of the First Edition*, trans. David Eugene Smith and Marcia L. Latham [New York: Dover, 1954], 158–159). Emphasis is mine.

ber" – was a vital component of Wallis's argument for numbers involving $\sqrt{-1}$. The latter numbers, too, were imaginary numbers; and, as such, they were as legitimate as the negatives. That is, he implicitly built a "coattails argument": if mathematicians accepted the negatives, they ought to accept numbers involving $\sqrt{-1}$. His letter of 1673 to Collins, in which he had underlined the importance of precedent in mathematics, made it clear that he personally subscribed to the "coattails argument." Writing here specifically of the geometric representation of negatives, he noted that he

> was of opinion from the first, that a negative plane may as well be admitted in algebra as a negative length, both being in nature equally impossible; for there can no more be a line less than nothing than a plane less than nothing, both being but imaginable; and if we suppose such a negative square, we may as well suppose it to have a side, not indeed an affirmative, or a negative length, but a supposed mean proportional between a negative and positive.[74]

Thus, after chapters 16 and 17 of the *Treatise* had presumably lulled students into accepting the negative numbers, and chapters 31 and 32 had set Harriot up as a proponent of (the traditionally) imaginary as well as negative roots, chapter 66 argued forcefully for acceptance of numbers involving $\sqrt{-1}$ at least partially based on their analogy with negative numbers. Wallis first noted that roots of negative numbers were commonly called "imaginary quantities," and that such roots were "Impossible . . . as to the first and strict notion of what is proposed. For it is not possible, that any Number (Negative or Affirmative) Multiplied into itself, can produce (for instance) -4." Then he quickly drew a parallel between the impossibility of (the traditionally) imaginary numbers and that of the negative numbers: "But it is also Impossible that any Quantity (though not a Supposed Square) can be *Negative*. Since that it is not possible that any *Magnitude* can be *Less than Nothing*, or any *Number Fewer than None*." In short, if mathematicians were bound by "strict notions," which included traditional, meaningful definitions, they ought necessarily to reject negative numbers along with numbers involving $\sqrt{-1}$.[75]

Of course, negative numbers had already been embedded in the English algebraic tradition, by Pell and Kersey if not by Oughtred and Harriot. Wallis, as he well knew, was on safe ground as he argued that a strict interpretation of the canons of mathematics was not for him. Building on chapter 16 and stressing applicability (as had Pell and

74. Wallis to John Collins, 6 May 1673, 2:577–578.
75. Wallis, *Treatise of Algebra*, 264.

Kersey), he emphasized that the "Supposition (of Negative Quantities)" was neither "Unuseful" nor "Absurd." A negative number was, after all, susceptible of "Physical Application . . . [and] denotes as Real a Quantity as if the Sign were +; but to be interpreted in a contrary sense."[76]

Examples reinforced his point that the negatives were mathematically acceptable because, even if they defied definition, they could be interpreted or represented both geometrically and in the physical world. First he gave the example of movement 3 yards forward from a point *A* to a point *C* and 3 backwards from *A* to *D*. "And . . . − 3," he argued, "doth as truly design the Point *D*; as + 3 designed the Point *C*. Not Forward, as was supposed; but Backward, from A. . . . And each [− 3 and + 3] designs (at least in the same Infinite Line,) one Single Point: And but one." Similarly, he continued, if positive numbers are used to represent land gained from the sea, negatives represent that lost to the sea. Thus one might gain 10 acres, or 1,600 square perches, "Which if it lye in a Square Form, the Side of that Square will be 40 Perches in length." On the other hand, one might lose 10 acres. "That is to say," Wallis wrote, "The Gain is 10 Acres less than nothing. Which is the same as to say, there is a Loss of 10 Acres: or of 1600 Square Perches."[77]

He used the latter example as a bridge from his argument for the physical and geometric applicability of the negative numbers to his algebraic explanation and subsequent geometric "exemplification" of (the traditionally) imaginary numbers. Suppose, he continued, "this Negative Plain, − 1600 Perches, to be in the form of a Square." What is the side of the square? "We cannot say it is 40, nor that it is − 40." Rather, he concluded, "it is $\sqrt{-1600}$, (the Supposed Root of a Negative Square;) or (which is Equivalent thereunto) $10\sqrt{-16}$, or $20\sqrt{-4}$, or $40\sqrt{-1}$." But what was $\sqrt{-1}$? Wallis continued to elaborate the "true notion of . . . [an] imaginary root":

> $\sqrt{}$ implies a Mean Proportional between a Positive and a Negative Quantity. For like as $\sqrt{bc}$ signifies a Mean Proportional between + *b* and + *c*; or between − *b*, and − *c* . . . So doth $\sqrt{-bc}$ signify a Mean Proportional between + *b* and − *c*, or between − *b* and + *c*; either of which being Multiplied, makes − *bc*. And, this as to Algebraick consideration, is the true notion of such Imaginary Root, $\sqrt{-bc}$.[78]

76. Ibid., 265.
77. Ibid. As Wallis explained, the English acre equaled a plane of 40 perches in length and 4 in breadth.
78. Ibid., 265–266.

After ending chapter 66 with this algebraic explanation, Wallis devoted the next chapter to the exemplification of (the traditionally) imaginary numbers in geometry. Here he came close to providing an early geometric representation of the complex numbers.[79]

8

Wallis's drive for mathematical generality helps explain his endorsement of the negative and imaginary numbers as well as other major aspects of his mathematical work. The push for general results was evident not only in the symbolical style he adopted and the generalized concepts he employed but also in his choice of methods. The very last chapter of the *Treatise* contained a defense of induction, written in reply to Fermat's attack on the inductive proofs of *Arithmetica infinitorum*. In this final chapter Wallis argued that induction, which was for him essentially mathematical generalization based on the study of a few specific cases (and not what would eventually come to be defined as "mathematical induction"), was a valid mathematical method. Beginning to answer Fermat's main charge that induction was a method of investigation but not demonstration, he observed that, after using induction to reach certain conclusions, he could in turn demonstrate these conclusions geometrically. On the other hand, Wallis claimed that induction alone sometimes constituted a sufficient demonstration of a general result. "I look upon *Induction*," he wrote,

> as a very good Method of *Investigation*; as that which doth very often lead us to the easy discovery of a General Rule; or is at least a good preparative to such an one. And where the Result of such Inquiry affords to the view, an obvious discovery; it needs not (though it may be capable of it,) any further Demonstration. And so it is, when we find the Result of such Inquiry, to put us into a regularly orderly Progression (of what nature soever,) which is observable to proceed according to one and the same general Process; and where there is no ground of suspicion why it should fail, or of any case which might happen to alter the course of such Process.[80]

Citing the example of an inductive derivation of the binomial theorem through consideration of the second, third, and higher powers of the expression $a + e$, he suggested that students could either accept the

---

79.   Ibid., chap. 67. For evaluation of Wallis's efforts at the geometric representation of complex numbers, see, e.g., Scott, *Mathematical Work of John Wallis*, 162.

80.   Wallis, *Treatise of Algebra*, 305–306.

derivation as demonstrative after "some few Steps" or "continue the Process (by continual Multiplication into $a + e$,) as far as they please; and then content themselves (instead of the general) with a particular conclusion (for they prove no more,) that *it holds true as to so many steps*; and rest there." The former was the more mathematical course of action, he implied, as he noted that "most Mathematicians that I have seen . . . are satisfied (from such evidence,) to conclude universally."[81]

More tellingly, Wallis explained his belief that mathematicians seemed to have little choice but to accept induction as a valid method since, "without this, we must be content to rest at particulars (in all such kind of Process,) without proceeding to the Generalls."[82] He, for one, was unwilling to settle for particulars. Therefore, consistent with his penchant for generalization and "venturous" mathematical style, he gladly pushed beyond the traditional canons of mathematical argument into the shaky area of inductive proofs.

9

In summary, in his influential *Treatise* Wallis wholeheartedly endorsed the symbolical style, probed the arithmetic foundations of algebra, carefully distinguished "pure algebra" from its applications, intimated that algebra was superior to geometry, argued for the negative and complex numbers, and even defended a premature form of mathematical induction. As he elaborated a philosophy of algebra, Wallis took great care to synthesize the most important results of early modern equation theory. Thus, assigning major credit to Harriot as he himself filled the gaps in the *Praxis*, he covered the canonical form of equations, the reduction of equations to binomial factors, the fundamental theorem of algebra, the solutions of quadratic and cubic equations, and so on. Specifically, he corrected Descartes's rule of signs, showing by example that it did not apply to equations with imaginary roots.[83]

Although he made of the new equation theory a respectable part of English algebra (almost in spite of its significant debt to Descartes), he perhaps exacerbated the problem of the negative numbers even as he went far beyond any of his English predecessors in affirming the imaginary numbers. He stressed that the negatives, which had been well on

81. Ibid., 308. 82. Ibid.
83. For a list of the results Wallis attributed to Harriot, see ibid., 198–200. On Descartes's rule of signs, see ibid., 158–159. On Wallis's contributions to equation theory, see Scott, *Mathematical Work of John Wallis*, 156–160.

their way to acceptance in the major English algebras published prior to his *Treatise,* were as imaginary as numbers involving $\sqrt{-1}$. Indeed, his "coattails argument" settled neither the problem of negative numbers nor that of (the traditionally) imaginary numbers.

Finally, in his quest to resolve the problems of the negative and imaginary numbers, Wallis recognized new criteria for algebraic objects: he emphasized precedent, utility, physical applicability, and geometric representation over "strict notions" or traditional definitions. At least implicitly, he also enunciated an early version of mathematical freedom: certainly not explicitly claiming that mathematicians were free to create the objects of mathematics, he nevertheless implied that they were free (1) to choose to work with mathematical objects that arose from standard arithmetic and algebraic operations but still defied traditional justification and (2) to extend the rules of arithmetic to the manipulation of these objects. He recognized that mathematicians as a group were free to set mathematical standards. According to his view of the mathematical process, in the absence of traditional definitions, a few venturous algebraists had chosen to use the negative and imaginary numbers; some of them then tried to persuade their colleagues to follow suit; and eventually the questionable numbers were on their way to integration into the algebraic universe. Wallis himself took some credit for pioneering the use of imaginary numbers as well as the reduction of surds. Referring to his early work with surds, he wrote: "since that time it is grown more common, and I perhaps have somewhat contributed thereunto." On his early reluctance to use an imaginary number, he added: "Since that time I have been more venturous, and I find now that others do not scruple to use it as well as I."[84] Indeed, Wallis was the most "venturous" of the early modern English algebraists!

84. Wallis to Collins, 6 May 1673, 2:578.

6

English Mathematical Thinkers Take Sides on Early Modern Algebra

Thomas Hobbes and Isaac Barrow against John Wallis

In late-seventeenth-century England there developed a geometric and synthetic backlash to the algebraic tradition established by Oughtred, Harriot, Pell, Kersey, and Wallis. Wallis, in particular, became entangled in an extended debate over the relative ranks of geometry and arithmetic, the appropriateness of using algebra to solve geometric problems, and the legitimacy of reasoning on symbols. The restive, pensive mood of the Scientific Revolution, combined with the lack of clear disciplinary lines, made such a debate possible and permitted its participants to roam freely over the mathematical, philosophical, and educational aspects of the new algebra.

The debate peaked prior to the publication of Wallis's *Treatise*, with critiques of his early writings by Thomas Hobbes (1588–1679) and Isaac Barrow (1630–1677). It was perhaps significant that Hobbes and Barrow were royalists, opponents of the Puritan movement with which Wallis had been so intimately connected. Moreover, both men craved certainty, in real life as well as in mathematics, and both men found that certainty in traditional geometry. Both saw themselves simultaneously as defenders of an enduring mathematical tradition with roots back to ancient Greece and as enlightened thinkers of the new age of science. Favoring geometry over arithmetic and algebra, and synthesis over analysis, they helped to lay the foundations for the preoccupation with geometry that was to mark British mathematical education through the early nineteenth century.

Still, neither thinker was a naive defender of the old mathematics. Hobbes's mathematical views were shaped by seventeenth-century respect for experience as a source of knowledge and his nominalist inclinations. Ironically, despite his animadversion toward early modern algebra, his nominalist approach potentially nudged algebra in the symbolical direction. Barrow's views were also shaped partially by an empiricism that lauded geometry as a sense-based science. Subordinating arithmetic to geometry, he explained numbers as mere signs of magnitude, a view on which George Berkeley would later build a philosophy of arithmetic and algebra as sciences of signs.[1]

I

Thomas Hobbes – born in 1588 in Malmesbury, England, and destined to become one of the major philosophers of his country – was the son of a minor clergyman and a yeoman mother, who, as Hobbes oft repeated, gave premature birth to him as she feared an imminent invasion by the Spanish Armada. As a teenager Hobbes acquired a good grounding in Greek and Latin from a young clergyman, Richard Latimer. When Hobbes was sixteen, his father struck a fellow parson and fled Malmesbury to avoid prosecution. Hobbes then came under the protection of a prosperous uncle, who paid his way through Magdalen Hall (later Hertford College), Oxford. As an undergraduate Hobbes read widely outside the scholastic curriculum. Receiving his bachelor's degree in 1608, he took a position as tutor to the son of William Cavendish, Baron Hardwick, later the first earl of Devonshire,[2] and service to and preferments from England's aristocracy became the warp and woof of his life. This was a logical path for a man who, as his family's second son, inherited little from his uncle and who never pursued an ecclesiastical or academic career.

With his appointment to the Cavendish household, Hobbes was introduced to the cultured world of the English aristocracy.[3] In 1610 he and his pupil embarked on a continental tour. Through his employer he came to know Francis Bacon, for whom he briefly served as secretary, as

1. This chapter expands on my earlier article: Helena M. Pycior, "Mathematics and Philosophy: Wallis, Hobbes, Barrow, and Berkeley," *Journal of the History of Ideas* 48 (1987): 265–286, esp. 265–277.
2. There is a biography of Hobbes in John Aubrey, *"Brief Lives," Chiefly of Contemporaries, Set Down by John Aubrey, between the Years 1669 & 1696*, ed. Andrew Clark, 2 vols. (Oxford: Clarendon Press, 1898), 1:321–403. Later biographies include George Croom Robertson, *Hobbes* (Edinburgh: William Blackwood, 1886); Samuel I. Mintz, "Thomas Hobbes," *Dictionary of Scientific Biography*, ed. Charles C. Gillispie, 18 vols. (New York: Scribner's Sons, 1970–1990), 6:444–451; Miriam M. Reik, *The Golden Lands of Thomas Hobbes* (Detroit: Wayne State University Press, 1977); Arnold A. Rogow, *Thomas Hobbes: Radical in the Service of Reaction* (New York: Norton, 1986); and Richard Tuck, *Hobbes* (Oxford: Oxford University Press, 1989).
3. Mintz, "Thomas Hobbes," 444.

well as the more scientific branch of the Cavendish family, William Cavendish, later duke of Newcastle, and his younger brother, Sir Charles Cavendish. According to Aubrey, the latter "was a little, weake, crooked man, and nature having not adapted him for the court nor campe, he betooke himselfe to the study of the mathematiques, wherein he became a great master."[4] This Cavendish and his associates, Walter Warner and John Pell, awakened in Hobbes a strong scientific interest in optics.[5]

Curiously, according to the standard accounts, neither his studies at Oxford nor the influence of Charles Cavendish and his associates turned Hobbes toward pure mathematics, which later became for him an avocation and, one might say, passion. Only around the age of forty did he encounter Euclid's *Elements* – a copy of which lay open before him in a gentleman's library – and thereupon, like so many great minds before and after him, he fell "in love with geometry." As Aubrey told the story, the open pages contained proposition 47 of book I (the Pythagorean theorem), to which he initially responded: "By G – . . . this is impossible!" By following the proof of the proposition, however, Hobbes "was demonstratively convinced of that truth." This chance encounter with Euclid thus made a mathematical devotee of Hobbes, who from that point on was "wont to draw lines on his thigh and on the sheetes, abed." A stretching of the truth (since Hobbes probably knew some mathematics before the age of forty), Aubrey's dramatic retelling of his friend's mathematical awakening seems to have been crafted at least partially to help explain Hobbes's mathematical deficiencies. Had Hobbes studied mathematics earlier in his life, Aubrey speculated, he "would have made great advancement in" the subject and "would not have layn so open to his learned mathematicall antagonists."[6]

Aubrey's account also reinforced the image of geometry as certain knowledge. As Wallis's reading of Oughtred's *Key* had seemed to seal his fate as an analytic mathematician, Hobbes's encounter with Euclid turned him into a geometric and synthetic partisan. The *Elements* brought him face to face with the synthetic method by means of which

---

4. Aubrey, *Brief Lives*, 1:153. Aubrey's account of Cavendish comes from John Collins.

5. Mintz, "Thomas Hobbes," 445–446; Rogow, *Thomas Hobbes*, 108–109. See also Jean Jacquot, "Sir Charles Cavendish and His Learned Friends," *Annals of Science* 8 (1952): 13–27.

6. Aubrey, *Brief Lives*, 1:332–333. Some of Hobbes's biographers have questioned his total ignorance of geometry prior to this experience. Also, later biographers have placed the event in 1629–1630. See, e.g., Rogow, *Thomas Hobbes*, 100–103.

clear definitions and self-evident premises led to necessary conclusions. According to modern scholars, this experience was of personal as well as philosophical importance to Hobbes, who – at least partially because of the circumstances of his birth, which seemed to haunt him into adult life, and his early abandonment by his father – had hitherto known a great deal of fear and uncertainty. Reportedly, even his health began to improve after the encounter.[7]

He came to value geometry especially as a model for political theory. Indeed: "We may hypothesize that one root of Hobbes's interest in mathematics was a need, traceable to the insecurities and uncertainties of his childhood, to establish a scientific politics of authority and stability."[8] Thus a note to the reader preceding *Human Nature, on the Fundamental Elements of Policy*, which Hobbes published in 1640, stated "that *Mr. Hobbes* hath written a body of philosophy, upon such principles and in such order as are used by men conversant in demonstration: this he hath distinguished into three parts . . . each of the consequents beginning at the end of the antecedent, and insisting thereupon, as the later Books of *Euclid* upon the former." In the dedication to the same work, Hobbes explained that there were two kinds of learning: mathematical, which is "free from controversy and dispute," and dogmatical, in which "there is nothing undisputable." It was his purpose here, he stressed, to develop a doctrine of justice and policy, which would build in degrees from first principles (which "passion, not mistrusting, may not seek to displace") to "inexpungable" conclusions.[9]

In adapting the synthetic method to political theory, Hobbes helped lay the foundations for a strong geometric influence on European rational thought of the late seventeenth and early eighteenth centuries. Generally, the entry of "the mathematical, and particularly the geometrical, model . . . into the wider intellectual consciousness" was "a major event in early modern intellectual history."[10] Hobbes himself and his disciples

7. For analyses along these lines, see, e.g, Richard Peters, *Hobbes* (Baltimore: Penguin, 1967), 20; Richard M. Pearlstein, "Of Fear, Uncertainty, and Boldness: The Life and Thought of Thomas Hobbes," *Journal of Psychohistory* 13 (1986): 309–324.

8. Rogow, *Thomas Hobbes*, 102.

9. F. B., "To the Reader," and Thomas Hobbes, "Dedication" of *Human Nature* to the Earl of Newcastle, May 9, 1640, in *The English Works of Thomas Hobbes of Malmesbury*, ed. Sir William Molesworth, 11 vols. (1839–1845; reprint, Aalen: Scientia, 1962), 4:n.p.

10. G. A. J. Rogers, "Hobbes's Hidden Influence," in *Perspectives on Thomas Hobbes*, ed. G. A. J. Rogers and Alan Ryan (Oxford: Clarendon Press, 1988), 189–205, on 195.

François Du Verdus and William Petty judged his geometric approach to politics to be his most fruitful insight.[11] Du Verdus, a student of Roberval, generalized that, in order to be a philosopher, one had to be a geometrician.[12] As the Englishman Petty, who respected algebra as well as geometry and wrote his own *Political Arithmetick,* put it: "I think that the best Geometricians were the most sagacious men, or that the most Sagacious men did ever make the Best Geometricians."[13] Few other English scholars of the period publicly claimed an intellectual affiliation with Hobbes, and yet some quietly adapted his geometric manner to their nonmathematical studies. For instance, although Henry More, the Cambridge Platonist of Christ's College, had attacked Hobbes's atheism, he wrote his *Account of Virtue* in a loosely geometric style. In the *Account* of 1668, More defined ethics and offered twenty-three "noemata" or axioms pertaining to virtue, which were supposed to be clear and self-evident truths forming the basis of his system.[14]

2

Hobbes also produced an innovative philosophy of mathematics, which, like his geometry of politics, perhaps exerted a "hidden influence" on the philosophy of the next century, especially on Berkeley's mathematical philosophy. Since Hobbes was a materialist[15] who believed that all human knowledge was derived through sense impressions, his mathematical philosophy had a strong materialist component, with which geometry fit well and arithmetic, less well. As George Croom Robertson

11. Quentin Skinner, "Thomas Hobbes and His Disciples in France and England," *Comparative Studies in Society and History* 8 (1966): 153–167, on 161. Hobbes met Du Verdus (François Bonneau, sieur de Verdus) and Petty while he was in exile in Paris.

12. Ibid., 161. Skinner quotes from a letter from Du Verdus to Hobbes of December 1655, which elaborates this point.

13. Petty to Robert Southwell, 21 Sept. 1685, *The Petty-Southwell Correspondence, 1676–1687,* ed. The Marquis of Landsdowne (London, 1928; reprint, New York: Augustus M. Kelley Publishers, 1967), 158.

14. Rogers, "Hobbes's Hidden Influence," 199, 200–202.

15. Hobbes held that: "All that *really* exists . . . is body. Those things which are usually considered to be immaterial, such as space and time, or thought, or logical relations, are attributes of the mind; they are 'phantasms' of the mind, and mind . . . is a material phenomenon" (Samuel I. Mintz, *The Hunting of Leviathan* [Cambridge: Cambridge University Press, 1969], 63).

140 Symbols, Numbers, and Geometric Entanglements

summarized a century ago: "The prominence he [Hobbes] assigns to Geometry, as if it were equivalent to Mathematics, follows from his original position that nothing but extended body exists as the subject of science."[16] According to Hobbes, extended body was the subject of geometry, and, as he emphasized in his *Six Lessons to the Professors of the Mathematics* of 1656, geometry was demonstrable – that is, based on knowledge of causes rather than effects – precisely because "the lines and figures from which we reason are drawn and described by ourselves."[17]

The preceding statement can be taken literally, since Hobbes's philosophy of geometry drew as heavily on his rejection of general abstract ideas as on his emphasis on sense experience. He maintained that the geometer reasoned on specific lines and figures, or on their universal names, and not on general abstract ideas of such. This approach fit in with the nominalist leanings expressed in his *Elements of Philosophy: The First Section, concerning Body* of 1656, but it also put him at odds with the traditional Platonic interpretation of Euclid's *Elements*. In his *Elements* Hobbes turned to universal names rather than general abstract ideas to explain the human ability to generalize. General abstract ideas appeared as his sixth example of "erring, falsity, and captions." "In the sixth manner they err," he declared, "that say the *idea of anything is universal*; as if there could be in the mind an image of a man, which were not the image of some one man, but a man simply, which is impossible; for every idea is one, and of one thing; but they are deceived in this, that they put the *name* of the *thing* for the idea thereof." Later, in his discussion of triangles, he reinforced his point that generality in reasoning comes by means of universal names:

> [I]f any man, by considering a triangle set before him, should find that all its angles together taken are equal to two right angles, and that by thinking of the same tacitly, without any use of words either understood or expressed; and it should happen afterwards that another triangle, unlike the former, or the same in different situation, should be offered to his consideration, he would not know readily whether the same property were in this last or no, but would be forced, as often as a different triangle were brought before him (and the difference of triangles is infinite) to begin his contemplation anew; which he would have no need to do if he had the use of

16. Robertson, *Hobbes*, 104–105.
17. Thomas Hobbes, *Six Lessons to the Professors of the Mathematics, One of Geometry, the Other of Astronomy . . . in the University of Oxford,* in *English Works,* 7:184.

names, for every universal name denotes the conceptions we have of infinite singular things.[18]

According to Hobbes, then, the name "triangle" stands for all those singular plane figures with such distinctive properties as having an interior angle-sum equal exactly to 180 degrees. The name permits mathematicians to reach general conclusions about all triangles.[19]

Rejection of general abstract ideas made a geometric maverick of Hobbes. In the heated quarrel in which the two thinkers debated various mathematical issues, Wallis endorsed Euclid's definitions of a point and a line, whereas Hobbes argued for new definitions, which better fit his materialism and nominalism. Specifically, Wallis defended the definitions of a point as "that whereof there is no part" and of a line as "length which hath no breadth" – definitions that, according to Hobbes, referred to "neither substance nor quality" and thus stood for "nothing." Hobbes, in turn, insisted that a point was a physical "body whose quantity is not considered" and a line was "a body whose length is considered without its breadth."[20] In brief, geometry dealt not with the abstract ideas of point and line, but with sensible points and lines whose parts or breadth were ignored by mathematicians. Generalizations about such points and lines depended on the use of the universal names "point" and "line" to denote specific points and lines.

3

Hobbes's rejection of general abstract ideas and his emphasis on sense observations also shaped his philosophy of arithmetic. But this philosophy was not as well developed and clear-cut as that of geometry. He seemed torn between viewing arithmetic as dependent on geometry, on the one hand, and developing it as a somewhat independent theory of names, on the other. This tension was evident in his *Elements of Philosophy*. In the work's major section on quantity, he avoided the question of

18. Thomas Hobbes, *Elements of Philosophy. The First Section, Concerning Body,* in *English Works,* 1:60, 79–80. The *Elements of Philosophy* is the English translation of Hobbes's *De corpore*.

19. For a deeper discussion of this theory of names as well as its deficiencies, see Frithiof Brandt, *Thomas Hobbes' Mechanical Conception of Nature* (London: Librairie Hachette, 1928), 231–239.

20. For Hobbes's summary of the controversy over these Euclidean definitions, see *Six Lessons,* in *English Works,* 7:200–202.

the nature of number and, working from his materialist conception of science, asked rather how number could be "exposed" or "set before" the senses. "Quantity exposed," he explained, "must be some standing or permanent thing, such as is marked out in consistent or durable matter; or at least something which is revocable to sense." Number could be exposed in two ways, through geometric representation or recitation of the names of numbers:

> either by the exposition of points, or of the names of number, *one, two, three,* &c.; and those points must not be contiguous, so as that they cannot be distinguished by notes, but they must be so placed that they may be *discerned* one from another; for, from this it is, that number is called *discrete quantity*. . . . But that number may be exposed by the names of number, it is necessary that they be recited by heart and in order, as one, two, three, &c.[21]

The first method of exposition implied a dependence of arithmetic on geometry. Although Hobbes did not explicitly identify numbers as geometric points ("bod[ies] . . . whose quantity is not considered"), he at least suggested that such points could serve the senses as representatives of numbers. But still, he proposed a second, nongeometric way of exposing number. This method depended on a view of numbers as names and required their recitation in the right order, with the saying and hearing of numbers apparently providing their sensible realization.

There is evidence in another section of the *Elements* that the problem of zero and negative numbers pushed him toward the view of numbers as names. Whereas whole numbers were susceptible of representation as points, he could find no physical referents for the number zero or negative numbers. A philosophy of numbers that provided for names that stood for no specific extended bodies, then, seemed essential for explanation of zero and the negatives. "Nor, indeed," he opined,

> is it at all necessary that every name should be the name of something. . . . this word *nothing* is a name, which yet cannot be the name of any thing: for when, for example, we substract 2 and 3 from 5, and so nothing remaining, we would call that substraction to mind, this speech *nothing remains,* and in it the word *nothing* is not unuseful. And for the same reason we say truly, *less than nothing* remains, when we substract more from less; for the mind feigns such remains as these for doctrine's sake, and desires, as often as is necessary, to call the same to memory.[22]

Denying abstract general ideas, unable to find physical representations for all numbers, and yet mindful of the usefulness of zero and the

---

21. Hobbes, *Elements of Philosophy,* in *English Works,* 1:140–141.
22. Ibid., 17–18.

negative numbers, Hobbes thus toyed with a new approach to arithmetic as a theory of names. Here, interestingly, he agreed with his mathematical nemesis Wallis and Descartes that the negative numbers were "not unuseful" (the term used by both Hobbes and Wallis) and that mathematicians "imagined," "supposed," or "feigned" negative numbers.[23] Hobbes, however, seems not to have elaborated a philosophy of numbers as names. Moreover, in his *Examinatio et emendatio mathematicae hodiernae* of 1660 – where he argued against Wallis's thesis of the superiority of arithmetic over geometry – he claimed that arithmetic was a part of geometry.[24]

4

The *Examinatio* was part of a nasty quarrel in which, from the 1650s through 1670s, Wallis attacked Hobbes's eccentric geometry and Hobbes lambasted Wallis's algebra. Paralleling the two men's opposite mathematical preferences were fundamental political and religious differences. As Wallis had supported and benefited from the advance of the parliamentarians in the 1640s, Hobbes – whose political inclinations were royalist – had fled his homeland in 1640 to begin what proved to be an eleven-year exile in France. Too, Hobbes's religious views, which have been characterized as deistic through agnostic to atheistic, put him at odds with Wallis as well as with the majority of his influential contemporaries. As others have noted, Wallis and Robert Boyle worried especially about the popular effects of the religious opinions and anti-clericalism that Hobbes expressed in his *Leviathan* of 1651. Since Hobbes seemed to link an understanding of mathematics to an understanding of religion, Wallis and Boyle concluded that "the most effective way of nullifying *Leviathan* and undermining Hobbes's influence . . . was not a refutation of *Leviathan* chapter by chapter but a successful effort to discredit Hobbes as a mathematician and scientist."[25]

23. Hobbes and Descartes viewed the imagination differently. See Dennis L. Sepper, "Imagination, Phantasms, and the Making of Hobbesian and Cartesian Science," *Monist* 71 (1988): 526–542. Sepper does not consider negative or imaginary numbers.
24. On this aspect of the *Examinatio,* see Douglas M. Jesseph, "Of Analytics and Indivisibles: Hobbes on the Methods of Modern Mathematics," *Revue d'Histoire des Sciences et de leurs Applications* 46, 2–3 (1993), 168–169.
25. Rogow, *Thomas Hobbes,* 202–203. Hobbes originally quarreled with both Wallis and Seth Ward. In discussing the controversy over condensation

Entangled in politics and religion and made all the more bitter thereby, the public quarrel between Hobbes and Wallis touched on serious issues within the philosophy of mathematics, including the nature and legitimacy of early modern algebra as well as algebra's relationship with geometry. Hobbes almost universally condemned the new algebra. Geometry and its subordinate, arithmetic, were, in his opinion, sciences; algebra, which he saw essentially as symbolical reasoning, was an art of invention. As we have seen, in responding to the Renaissance claim that mathematics did not adhere to the highest type of syllogistic reasoning, Wallis had argued essentially that all mathematics was certain or demonstrative. Hobbes's response was quite different: since geometry alone considered causes, it alone conformed to the high standards of demonstration. Thus mathematics was demonstrative, but – Hobbes said in so many words – mathematics was geometry and synthetic geometry at that.[26] Algebra was not demonstrative. In a striking way, then, Hobbes and Wallis stood as prototypes of two kinds of British thinkers who would debate the direction of mathematics for more than a century: Hobbes speaking for the geometric traditionalist and the general scholar; Wallis representing the algebraic or analytic mathematician and, secondarily, the emerging professional mathematician.

Hobbes criticized algebra in a section of his *Elements* dealing with geometric analysis. Here he expressed a preference for synthesis as the sole demonstrative method over analysis; he distinguished algebra from analysis, making it something even less than geometric analysis; and he attacked the algebraist's dependence on symbols. First of all, he defined synthesis as "the art itself of demonstration" and analysis as the art leading from suppositions to "known propositions," which could, in

and rarefaction in which the three men engaged, Schaffer has explored interrelated scientific, mathematical, religious, and political issues (Simon Schaffer, "Wallification: Thomas Hobbes on School Divinity and Experimental Pneumatics," *Studies in the History and Philosophy of Science* 19 [1988]: 275–298). On the origins of Hobbes's quarrel with the Savilian professors, see also Siegmund Probst, "Infinity and Creation: The Origin of the Controversy between Thomas Hobbes and the Savilian Professors Seth Ward and John Wallis," *British Journal for the History of Science* 26 (1993): 271–279.

26. On Hobbes's response to the Renaissance debate, see Paolo Mancosu, "Aristotelian Logic and Euclidean Mathematics: Seventeenth-Century Developments of the *Quaestio de Certitudine Mathematicarum*," *Studies in History and Philosophy of Science* 23 (1992): 241–265, esp. 255–258; on Hobbes's association of geometry with causes and demonstration, see, e.g., Jesseph, "Of Analytics and Indivisibles," 174–177.

turn, be used for demonstration.[27] As such, synthesis was the superior mathematical method, even though, as he admitted, there was precedent for the use of analysis in the geometry of Pappus and Archimedes. Ignoring the inventive aspects of geometry as he focused on the demonstrative and pedagogical, he declared that "no man can ever be a good analyst without being first a good geometrician; nor do the rules of analysis make a geometrician, as synthesis doth. . . . For the true teaching of geometry is by synthesis, according to Euclid's method." A student could thus learn geometry from Euclid alone, as he presumably had, but not from any of the geometric analysts.[28] In his own twist on the relationship between analysis and algebra, he then dismissed algebra as an almost useless shorthand for analysis. "[A]lgebra, or the analytics specious, symbolical, or cossick . . . are, as I may say, the *brachygraphy* of the analytics, and an art neither of teaching nor learning geometry, but of registering with brevity and celerity the inventions of geometricians." He ended by claiming to wonder ("I know not") whether such "discourse by symbols . . . deserve to be thought very profitable, when it is made without any ideas of the things themselves."[29]

Similarly, in his *Six Lessons* directed against Wallis and Ward, he promoted synthetic geometry as the paradigm of demonstrative reasoning and harshly attacked the symbolical style. Appealing to authority, he noted that the ancients used analysis but not symbols, and asked:

> Had Pappus no analytics? or wanted he the wit to shorten his reckoning by signs? Or has he not proceeded analytically in a hundred problems (especially in his seventh book), and never used symbols? Symbols are poor unhandsome, though necessary, scaffolds of demonstration; and ought no more to appear in public, than the most deformed necessary business which you do in your chambers.[30]

As Hobbes saw it, symbols were useful, perhaps "necessary," for mathematical invention, but useless for the teaching of mathematics. In invention, symbols provided a shorthand to record the mathematician's progress. Symbolical formulation, however, prolonged rather than shortened the communication of mathematical results. Communication required translation of symbols into the "things" they represented. Thus, concerning Wallis's algebraic reformulations of some of the geometric results of Hobbes, the latter retorted: "you show me how you could demon-

27. Hobbes, *Elements of Philosophy,* in *English Works,* 1:310.
28. Ibid., 1:314.
29. Ibid., 1:316–317. "Cossick" refers to the Italian use of "co" ("cosa") for the unknown.
30. Hobbes, *Six Lessons,* in *English Works,* 7:248.

strate the . . . articles a shorter way. But though there be your symbols, yet no man is obliged to take them for demonstration. And though they be granted to be dumb demonstrations, yet when they are taught to speak as they ought to do, they will be longer demonstrations than these of mine."[31]

In a later passage, he elaborated a two-stage process of translation of algebraic symbols. Referring again to Wallis's application of algebra to geometry, he remarked that "the conception of the lines and figures (without which a man learneth nothing) must proceed from words either spoken or thought upon. So that there is a double labour of the mind, one to reduce your symbols to words, which are also symbols, another to attend to the ideas which they signify."[32] Thus a symbolical demonstration was a waste of time: to be understood, its symbols had to be translated, first into words or names and then into the particular ideas for which the words stood. For example, a person intent on understanding a demonstration involving *a,* where *a* represented a line, would move through two stages: first, association of *a* with the universal name "line" and, second, recall of the particular bodies covered by "line."

Hobbes thus trivialized the symbolical style, which Oughtred, Harriot, Ward, and Wallis had seen as the heart of early modern algebra. Whereas the latter had stressed the economy, clarity, fertility, and even concreteness of algebraic symbolism – which permitted the eye to see complex mathematical relationships at a glance – Hobbes could view algebraic symbols as no more than a shorthand, which mathematicians used in the heat of invention to record their ideas. His very term "brachygraphy" betrayed a myopic view that associated algebraic symbols more with the shorthand ("brachygraphy") movement of sixteenth- and seventeenth-century England than with the new algebra.[33] Whereas in his *Treatise of Algebra* Wallis had carefully traced the cumulative steps by which Diophantus, Viète, Oughtred, Harriot, and others had formulated the symbolical style, Hobbes asked if any reasonable man could assume that Pappus "wanted the wit to shorten his reckoning by signs?"

Failing to see the fertility of the symbolical style, he turned the full

31. Ibid., 7:281–282.   32. Ibid., 7:329.
33. According to Salmon, "shorthand . . . remained an exclusively English phenomenon from its invention by Timothy Bright in 1588 until the middle of the 17th century." And: "*Brachigraphy* was one of the most popular names for shorthand." On shorthand and its relationship to the universal language movement, see Vivian Salmon, *The Study of Language in 17th-Century England* (Amsterdam: John Benjamins, 1979), 157–175 (quotations from p. 157).

power of his prose against the new algebra. He used vivid figures of speech to characterize symbolical algebra as an arbitrary, superfluous art. One simile emphasized algebraic tenuousness while implying that algebra resembled running about in a circle. "The symbols," he declared, "serve only to make men go faster about, as greater wind to a windmill." In another passage, he pretended to misspeak: he accused Wallis of "writing . . . not in language, but in *gambols*; I mean . . . symbols." In another, he likened symbolical mathematics to an animal's random activity. He claimed to find in such mathematics "no knowledge, neither of quantity, nor of measure, nor of proportion, nor of time, nor of motion, nor of any thing, but only of certain characters, as if a hen had been scraping there."[34]

In the *Six Lessons* he attacked Wallis on other mathematical grounds. He lambasted the methods used in *Arithmetica infinitorum* and *Conic Sections*. Foreshadowing Berkeley's famous attack on the calculus, he claimed to accept Wallis's results while rejecting both his proofs by induction and his handling of indivisibles. He quoted Wallis on induction, and then countered: "Egregious logicians and geometricians, that think an induction, without a numeration of all the particulars sufficient, to infer a conclusion universal, and fit to be received for a geometrical demonstration!"[35] He stressed apparent inconsistencies in Wallis's theory of indivisibles – a theory with roots to Nicole Oresme, Kepler, and Galileo, which had been developed geometrically by Bonaventura Cavalieri and then algebraically by Wallis. According to this theory, plane figures were viewed as consisting of an infinite number of parallel straight lines or infinitesimal parallelograms ("indivisibles"), which were used in the calculation of their areas.[36] Hobbes claimed that Wallis regarded an indivisible alternately as "nothing" and as "something."[37] Still, although the theory of indivisibles was riddled with such inconsistencies, it led to true conclusions. Regarding some specific propositions from the *Arithmetica infinitorum* and *Conic Sections*, Hobbes stressed, for example, "though your lemma be true, and by me . . . demonstrated; yet you did not know why it is true."[38]

In the *Seven Philosophical Problems* of 1662, Hobbes argued specifi-

34. Hobbes, *Six Lessons*, in *English Works*, 7:188, 247, 330.
35. Ibid., 7:307–308 (quotation from p. 308).
36. On the background to Wallis's theory of indivisibles, see J. F. Scott, *The Mathematical Work of John Wallis, D.D., F.R.S. (1616–1703)* (London: Taylor and Francis, 1938), 18–19.
37. Hobbes, *Six Lessons,* in *English Works*, 7:300–301, 308–309.
38. Ibid., 7:308 (quotation), 310.

cally against analytic geometry, maintaining that "impossibilities do necessarily follow the confounding of arithmetic and geometry." Earlier he had published an anonymous work claiming to duplicate the cube (that is, to construct, with compass and straightedge only, a cube whose volume is twice that of a given cube), and Wallis, in turn, had used algebraic calculations to expose the inadequacies of Hobbes's construction. In the *Philosophical Problems*, Hobbes dismissed Wallis's calculations as arithmetically valid but lacking geometric sense. At the heart of the problem, he maintained, was Wallis's mixing of various numbers without any concern for their geometric interpretations, as in the equation $2\sqrt{2} = \sqrt{8}$. "Here the root of 8 is put for the cube of the root of 2," he observed. "Can a line be equal to a cube?" Responding negatively, he concluded that "the calculation in numbers is right, though false in lines."[39]

He then developed a general argument against meaningful application of arithmetic or algebra to geometry. He reiterated his claims that numbers were exposed by points, and that points, as bodies, had length. "In geometry there are . . . three dimensions, lengths, superficies, and body. In arithmetic there is but one, and that is number or length which you will." Analytic geometry could produce results inconsistent with those of synthetic geometry because unidimensional arithmetic was unable to encompass all three dimensions of geometry. Claiming a complete victory over Wallis, Hobbes concluded that he had

> wrested out of the hands of our antagonists this weapon of algebra, so as they can never make use of it again. Which I consider as a thing of much more consequence to the science of geometry, than either of the duplication of the cube, or the finding of mean proportionals, or the quadrature of a circle, or all these problems put together.[40]

He could not have expressed his geometric bias more powerfully. Standing in awe of and deriving personal and intellectual comfort from the certainty of synthetic geometry, Hobbes saw algebra as no more than a "weapon" in the modern war on geometry.

5

As Hobbes defended ancient mathematics while yet exploring a nominalist approach to the subject, so Isaac Barrow intermingled old and

39. Hobbes, *Seven Philosophical Problems and Two Propositions of Geometry*, in *English Works*, 7:65–66.
40. Ibid., 7:67–68.

new mathematical ideas. The son of a well-to-do linendraper, Barrow (like Wallis) received a firm grounding in Greek and Latin under Martin Holbeach at Felsted, Essex.[41] A royalist and an Arminian,[42] Barrow had the misfortune of coming of age at the outbreak of the Civil War and spending his early adult years in an England dominated by Oliver Cromwell, who in fact occupied Cambridge University in 1642.

Family tradition and scholarly aspirations nevertheless pushed Barrow toward Cambridge, and by December 1643 his academic career seemed to be on track as he was appointed a foundation scholar at Peterhouse, Cambridge. This appointment, which had apparently been arranged by Barrow's namesake uncle, then a fellow of Peterhouse, was withdrawn the next month when the parliamentary visitors removed the senior Isaac from Cambridge. With the help of friends the younger Isaac weathered the mid-1640s away from Cambridge, and then entered Trinity College, Cambridge, in 1646. Here he found an unlikely patron in Thomas Hill, the master of the college and a staunch Presbyterian who "by the late 1640s . . . had sufficiently mellowed to tolerate even Arminians if exceptionally learned." After a successful undergraduate career and receipt of his M.A. in 1652, Barrow was, in fact, in line for the Regius professorship of Greek at the university. Cromwell, however, personally intervened to see that the professorship went to Ralph Widdrington, a popular tutor from Christ's College and brother of Sir Thomas Widdrington, speaker of the House of Commons and one of Cromwell's favorites.[43]

Barrow then obtained a traveling fellowship from Trinity and supposedly also sold his books to help pay for a foreign tour, which took him throughout continental Europe and kept him out of England until late

41. Biographical studies of Barrow include Aubrey, "Isaac Barrow," *Brief Lives,* 1:87–94; Abraham Hill, "Some Account of the Life of Dr. Isaac Barrow to the Reverend Dr. Tillotson, Dean of Canterbury," *The Works of the Learned Isaac Barrow, D.D.,* ed. John Tillotson (London, 1683), I; preface to *The Mathematical Works of Isaac Barrow, D.D.,* ed. William Whewell (Cambridge: Cambridge University Press, 1860); and D. T. Whiteside, "Isaac Barrow," *Dictionary of Scientific Biography,* 1:473–476. The most detailed studies are Percy H. Osmond, *Isaac Barrow: His Life and Times* (London: Society for Promoting Christian Knowledge, 1944), and Mordechai Feingold, "Isaac Barrow: Divine, Scholar, Mathematician," in *Before Newton: The Life and Times of Isaac Barrow,* ed. Mordechai Feingold (Cambridge: Cambridge University Press, 1990), 1–104.

42. Opponents of Calvinism, the Arminians rejected predestination. On Barrow's Arminianism, see John Gascoigne, "Isaac Barrow's Academic Milieu: Interregnum and Restoration Cambridge," in *Before Newton,* ed. Feingold, 250–290, on 258–262.

43. Feingold, "Isaac Barrow," 4–6, 37–38 (quotation from p. 6).

1659, a year after Cromwell's death and shortly before the Restoration of Charles II. Taking holy orders on his return, he received the Greek professorship that had been denied him earlier. In 1662 Barrow – now supported by John Wilkins – succeeded Laurence Rooke as the professor of geometry at Gresham College, London.[44] In 1663 he was appointed to the Lucasian professorship at Cambridge. During the next six years he gave three series of Lucasian lectures, and possibly also encouraged Newton's mathematical studies.

As the first Lucasian professor, he stood as the symbol of the university's formal acceptance of mathematics as an independent academic subject. No longer confined to the colleges, mathematics now had its own university professor, its own mandated university lectures, and assumedly its own student following. The symbolism of the professorship did not escape Barrow. Exaggerating the mathematical poverty of Cambridge up until Henry Lucas's endowment of the mathematical chair, he declared in his inaugural Lucasian lecture that he did not know "whether it is more to be accounted a Matter of Wonder, or of Grief, that it [mathematics] had hitherto obtained no Place, no assigned Reward, no allowed privilege in the University."[45]

Still, his own career seemed somewhat to belie the rewards and significance of mathematics in late-seventeenth-century England. During a midlife crisis he worried about having betrayed his clerical calling in the pursuit of mathematics. Around this time he praised John Tillotson, a practicing clergyman, for "deliver[ing] the mysteries of sacred truth to the people," and added: "I . . . am so unhappy as to be fixed to the books which you see here [his geometric lectures], and so waste my time and my power."[46] True to what he saw as his higher religious calling, he resigned the Lucasian professorship in 1669 and became royal chaplain in London the next year. He excelled as a royal chaplain and then as college preacher at Trinity, with many of his sermons being published

44.   Ibid., esp. 47, 55, 61; John Ward, *The Lives of the Professors of Gresham College* (London, 1740; reprint, The Sources of Science, no. 71, New York: Johnson Reprint, 1967), 158–160. Gresham College's role in the development of English mathematics needs to be studied in more detail.

45.   Isaac Barrow, *The Usefulness of Mathematical Learning Explained and Demonstrated: Being Mathematical Lectures Read in the Publick Schools at the University of Cambridge,* trans. John Kirkby (London, 1734), xiii. This English translation of Barrow's *Lectiones mathematicae* is hereafter cited as the *Mathematical Lectures.*

46.   Quoted in William Whewell, "Barrow and His Academical Times," in *The Theological Works of Isaac Barrow, D.D.,* ed. Alexander Napier, 9 vols. (Cambridge, 1859), 9:xxxviii–xxxix.

posthumously.[47] In 1673 he returned to Cambridge as master of Trinity, and two years later he became vice-chancellor of the university. He died in 1677 of a "malignant fever," possibly exacerbated by the opium with which he treated himself.[48]

6

Although a productive mathematician, Barrow was, first of all, a humanist. His interests encompassed the Greek language and its literature, theology, optics, medicine, and mathematics. As a mathematician, he was biased toward geometry and against algebra and even an independent arithmetic. In a very real sense it was Barrow who laid the formal foundations for the much touted geometric tradition that dominated the undergraduate curriculum at modern Cambridge into the nineteenth century.

For Barrow, as for Hobbes, a careful reading of Euclid's *Elements* seems to have been the rite of passage into mathematics. According to an account written shortly after his death, while reading Joseph Scaliger he had "perceived the dependence of Chronology on Astronomy, which put him on the study of *Ptolomy's Almagest,* and finding that Book and all Astronomy to depend on Geometry, he applied himself to *Euclide's Elements,* not satisfied till he had laid firm foundations; and so he made his first entry into the Mathematicks."[49] Once he had decided to study mathematics seriously – a step that, according to Feingold, he may have taken as early as 1648 or 1649 – he found at Cambridge other scholars who encouraged his pursuit of the subject, including John Ray, who became the Trinity College lecturer in mathematics in 1653.[50] Barrow seems to have ranged widely in his mathematical readings; by the time of his death his library included ancient classics as well as such mainstays of early modern mathematics as the collected works of Viète (1646),

47. Barrow's religious convictions and charity shone through his sermons and private life, and his sermons "rank as masterpieces of pulpit artistry in an age of great preaching" (Richard S. Westfall, *Science and Religion in Seventeenth-Century England* [New Haven: Yale University Press, 1958], 157). See also Irène Simon, "The Preacher," in *Before Newton,* ed. Feingold, 303–332.
48. According to Aubrey, Barrow died from an overdose of opiates, which he began taking as a young man in Turkey while waiting out the English Civil War (Aubrey, *Brief Lives,* 1:91).
49. Hill, "Some Account of the Life of Dr. Isaac Barrow," n.p.
50. Feingold, "Isaac Barrow," 19, 40.

Oughtred's *Key* (the Latin edition of 1652), Wallis's *Arithmetica infinitorum* (1656), some of Wallis's writings against Hobbes, Schooten's second Latin edition of Descartes's *Géométrie* (1659–1661), and the English edition of Rahn's *Algebra*.[51] Although it is impossible to determine if and when Barrow read each of the mathematical books he owned, it is clear that by the early 1650s he was familiar with Euclid's *Elements,* Oughtred's *Key,* the major Euclidean commentaries, and the more advanced works of Archimedes and possibly Apollonius.[52]

In 1655 he published *Euclidis Elementorum libri XV breviter demonstrati,* a work containing glimmers of his future role as a staunch defender of geometry's superiority among the mathematical subdisciplines. Combining old and new elements, the book was an epitome of Euclid's *Elements* and was "designed as a quadrivium undergraduate text, with emphasis on its deductive structure rather than its geometrical content."[53] Barrow promised, as his first objective, to "conjoin the greatest Compendiousness of Demonstration with as much Perspicuity as the quality of the subject would admit." He boasted that his edition of the *Elements* was "conveniently portable" even though, unlike other editions (he mentioned specifically the one André Tacquet had published just a year earlier), his included all Euclid's original books. Furthermore, the edition met his second objective of presentation in the symbolical style "in favour of Their desires who more affect Symbolical than Verbal Demonstrations." Observing that "most of our own Nation are accustomed to the Notes of Mr. *Oughtred,*" he followed his preface with a chart, "The Explication of the Signes or Characters," which included Oughtred's symbols for equality, greater than, less than, square *(Q* and *q),* cube *(C* and *c),* and so on.[54]

Barrow's *Elements,* however, betrayed but a superficial commitment to the symbolical style. In referring to "*Their* desires who more affect Symbolical ... Demonstrations," Barrow seemed to distance himself from the new style. Instead of using the occasion of his preface to explore the style's advantages, he criticized the excesses to which it had been carried by Pierre Hérigone, a Parisian mathematician. In the six volumes of *Cursus mathematicus,* which was published between 1634

51. Mordechai Feingold, "Isaac Barrow's Library," in *Before Newton,* ed. Feingold, 333–372.
52. Whiteside, "Isaac Barrow," 474.
53. Ibid.
54. Isaac Barrow, *Euclide's Elements; The Whole Fifteen Books Compendiously Demonstrated* (London, 1660), preface. Barrow's symbol for less than did not exactly match Oughtred's.

and 1642 and included the first six books of Euclid's *Elements*, Hérigone had aimed at a compendium of elementary and intermediate mathematics presented in a completely symbolical fashion, with symbols for logical as well as mathematical terms.[55] For the period, this was symbolism carried to the extreme, which, according to Barrow, led to "difficulty and occasion of doubting." Thus worrying about the excesses of symbolism rather than reveling in its lucidity and inventiveness, Barrow now reserved for himself the right to intermingle "Words and Signes at discretion" and seemed in his *Elements* proper to use symbols as mere abbreviations. Tellingly, at the bottom of his introductory chart on "Signes or Characters," he alluded to "Other Abbreviations of words, where ever they occurr, [that] the Reader will without trouble understand of himself."[56]

In his preface, he also argued for appreciation of Euclid's geometric approach to arithmetic. Noting that Tacquet had limited his edition of the *Elements* to eight books (I–VI, XI, and XII), he accused him of "slighting or undervaluing the other Seven as lesse relating to the Elements of Geometry." Then he entered into a spirited defense of Euclid's books VII through X, which cover the theory of numbers and incommensurables. "No man that ha's arriv'd to any measure of skill in Geometry," he declared, "is ignorant how exceedingly usefull they [books VII–X] are in Geometricall matters, aswell in regard of the very neer alliance between Arithmetick and Geometrie, as for the knowledge of Commensurable and Incommensurable Magnitudes."[57] That is, as his more analytic countrymen were starting to pull algebra apart from geometry and found it on arithmetic, he argued for an intimate relationship between arithmetic and geometry. This relationship would in his later writings be refined into the subordination of arithmetic to geometry and would rule out his accepting an algebra that was independent of geometry and founded on arithmetic.

Unlike his more advanced geometric writings, which covered Archimedes' work, Apollonius's *Conics,* and Theodosius's *Spherics,* Barrow's pocket-sized *Elements* enjoyed a wide readership. An English edition of the work appeared in 1660 and was reissued throughout the early

55. Per Stromholm, "Pierre Hérigone," *Dictionary of Scientific Biography,* 13:299. On the various editions of Euclid, including Hérigone's (1639), see Thomas L. Heath, "Principal Translations and Editions of the Elements," in *The Thirteen Books of Euclid's Elements,* trans. and ed. Thomas L. Heath, 3 vols. (New York: Dover, 1956), 1:91–113, 108 (Hérigone).
56. Barrow, *Euclide's Elements,* preface.
57. Ibid. See also, Feingold, "Isaac Barrow," 40–42.

eighteenth century. Shortly after succeeding Barrow as the Lucasian professor, Newton corrected the Latin edition, which was eventually combined with Barrow's epitome of Euclid's *Data* and *Lectio . . . in qua theoremata Archimedis De sphaera & cylindro per methodum indivisibilium investigata exhibentur* in a posthumous work of 1678.[58]

7

As the first Lucasian professor, Barrow continued to lay the underpinnings of a strong geometric tradition at Cambridge. According to the terms of the professorship, its recipient was to deliver regular lectures to senior students on "Geometry, Arithmetic, Astronomy, Geography, Optics, Statics or some other Mathematical Discipline."[59] For his first set of lectures, the *Mathematical Lectures* of 1664–1666, Barrow discoursed on the foundations and usefulness of mathematics; his second set of lectures concerned geometry; and his third, optics.

The decision to begin his Lucasian lectures with a probing analysis of the foundations and usefulness of mathematics evidenced as much about Barrow the humanist as about Barrow the teacher. As he wrote Collins in an excessively modest passage: "that little study I have employed upon mathematical businesses, . . . [was] never designed to any other use than the bare knowledge of the general reasons of things, as a scholar, and no further."[60] In his earliest Lucasian lectures Barrow assumed, then, what was for him the comfortable role of the humanist bent on penetrating the "general reasons" of mathematics rather than advancing the discipline in a more technical way. Practical considerations also motivated his choice of subject. He believed that many Cambridge men had little regard for mathematics or lacked the background necessary to understand the subject's more advanced aspects. "That I may humour, or rather oblige those who come less prepared to such Studies," he stated, "[I] shall first of all touch upon certain *general Things* belonging thereto."[61]

58. Whiteside, "Isaac Barrow," 474–475.
59. Quoted in "General Introduction," *The Mathematical Papers of Isaac Newton*, ed. D. T. Whiteside, 8 vols. (Cambridge: Cambridge University Press, 1967–1980), 3:xviii.
60. Isaac Barrow to [John Collins], n.d., *Correspondence of Scientific Men of the Seventeenth Century*, ed. Stephen Jordan Rigaud, 2 vols. (Oxford, 1841; reprint, Hildesheim: Georg Olms, 1965), 2:33.
61. Barrow, *Mathematical Lectures*, xxiii, 1.

In his *Mathematical Lectures* he made an elaborate case for geometry as the premier mathematical science. Correspondingly, he rejected the subordination of geometry to arithmetic, the thesis that Wallis had begun to develop in his *Mathesis universalis* of 1657. Like Hobbes, he stressed geometry's connection with the sensible. Going beyond Hobbes, however, he consistently denied arithmetic an existence independent of geometry and, when discussing the new algebra, confused it with geometric analysis and dismissed it as a part of logic. Furthermore, as the problem of sensible referents for whole numbers had moved Hobbes to suggest a theory of numbers as names, Barrow came to argue that numbers were nothing but words with arbitrary signification.[62]

Influenced by Aristotle, the Cambridge Platonists, and the early English empiricists, Barrow has been described as a "transitional figure" in epistemology.[63] The mathematical philosophy developed in his *Lectures* did not assume Platonic ideas but rather involved a largely sensationalist account of the origin of mathematical universals.[64] As he explained, "there is no need, at least in Speculative Sciences, for supposing any *physical Anticipations, common Notions, or congenite Ideas.*" In place of these philosophical assumptions, he offered a theory of universal ideas that assumed the accuracy of sense observations and saw ideas as products of human reason applied to those observations, while minimizing the distance between the sensible and the universal. Evidencing his faith in the senses, he asked: "Perhaps I am ignorant of the Manner of perceiving by my Senses, but do I not therefore see what is before mine Eyes?" He soon thereafter declared that "the *Sense* then, when right and perfect, as it is naturally in most Men of a sound Constitution, discerns many Objects *certainly.*" The human reason, he continued, produced universal ideas from accurate sense observations: "The Mind, from the Observation of the Things objected, takes Occasion of framing like Ideas, which, as soon as it clearly perceives to agree with the Things that may exist, it affirms and supposes; then appropriating Words to them forms Definitions."[65]

Barrow seemed wary of reasoning exclusively on universal ideas and warned against setting up too great a distance between the sensible

62. On these points, see Pycior, "Mathematics and Philosophy," 274–277, and Michael S. Mahoney, "Barrow's Mathematics: Between Ancients and Moderns," in *Before Newton*, ed. Feingold, 185–191, 200–202.
63. Gascoigne, "Isaac Barrow's Academic Milieu," 266–273 (quotation from p. 272).
64. Pycior, "Mathematics and Philosophy," 274–275.
65. Barrow, *Mathematical Lectures*, 115, 69, 70, 115.

and the intelligible. Taking geometry as his example, he attributed its excellence at least partially to the accuracy of the sense perceptions on which it was based, and urged mathematicians to keep particular geometric objects before their eyes even as they reasoned on universal geometric ideas. The mathematical sciences were ones of "excellent demonstration," he wrote, precisely "because we do clearly conceive, and readily obtain distinct ideas of the Things which these Sciences contemplate; they being Things the most simple and common, such as lie exposed to Senses." Asking in illustration what right lines, triangles, squares, circles, and the like were, he responded: "Things which we perceive clearly and distinctly."[66] He suggested, moreover, that mathematicians should permit sensible objects to inform their reasonings on geometric ideas. Somewhat blurring the distinction between the particular and the universal, he explained:

> [I]t is a very weak and slippery Foundation to depend upon, that the *Mathematics* are conversant about Things *intelligible* and Things *sensible,* because in reality every one of its Objects are at the same time both intelligible and sensible in a different respect; intelligible as the Mind apprehends and contemplates their universal . . . Ideas, and sensible as they agree with several particular Subjects occurring to the Sense: For who does not view with the Eye and feel with the Hand all the particular dimensions of Bodies? But there is no reason why the Doctrine of Generals should be separated from the Consideration of Particulars, since the former entirely includes and primarily respects the latter. . . . Why ex. gr. should one Science treat of an intelligible Sphere, and another of a sensible one? when these, as to the Verity of the Thing, are altogether the same, and as to the Action of the Mind subordinate; nor can any thing be attributed to the intelligible Sphere (i.e. one understood universally) which does not perfectly agree with the sensible (i.e. with every particular one).[67]

In short, Barrow held, geometry was a science of universal ideas which were derived by the mind from sensible objects and which, according to the preceding quotation, were possibly little or no more than particular objects "understood universally."

Geometry was the superior mathematical science precisely because, unlike arithmetic, it dealt with sensible objects or, more specifically, with

66. Ibid., 53–54.
67. Ibid., 19. (Barrow refers here to Aristotle's *Metaphysics* [II, 3] and *Analytics Posteriori* [I, 24].) Barrow's remarks also challenge the traditional distinction between "mixed mathematics" (e.g., astronomy and optics) and the "pure mathematics" of geometry and arithmetic. On Barrow's rejection of the latter distinction, see Mahoney, "Barrow's Mathematics," 185–186.

magnitude or quantity realized in nature. "There is really no Quantity in Nature different from what is called *Magnitude* or *continued Quantity*," Barrow declared, "and consequently . . . this alone ought to be accounted the Object of Mathematics."[68] Numbers were not quantities; rather, he argued, they were mere "notes" or "signs" of magnitude and hence the study of numbers, or arithmetic, was subordinate to the study of magnitude, or geometry. Barrow's treatment of numbers was a curious one: it was conservative in its emphasis on geometry over arithmetic, to the extent of recognizing geometry as perhaps the only, certainly the superior, mathematical science; it was progressive in its foreshadowing of Berkeley's view of numbers as arbitrary characters.

Barrow specifically attacked Wallis's assertion that "because a line of two feet added to a line of two feet makes a line of four feet, it does not follow that two and two make four; on the contrary, the former follows from the latter."[69] He challenged Wallis to consider the addition of a line of two feet to a line of two palms (the palm being a unit of measurement based on the length of the human palm). As Barrow noted, the sum here was not "a Line of four Feet, four Palms, or four of any Denomination." Now, every student of arithmetic ought to know that the units of measurement must be the same for such additions to work. Barrow admitted as much, and used the admission to make his general point that numbers were not independently existing entities. "No Number of itself," he wrote, "signifies any thing distinctly, or agrees to any determinate Subject, or certainly denominates any thing. For every Number may with equal Right denominate any Quantity: and in like manner any Number may be attributed to every Quantity." For example, "Any Line A may be indifferently called *One, Two, Three, Four,* or any other Number . . . as it remains undivided, may be cut into, or compounded of, two, three, four or any other Number of Parts." Specifically, the line of two feet used in Wallis's example may be called one, if viewed as undivided; two, if divided into feet; twenty-four, according to inches; and so on. In short, the number assigned to a given body depends simply on an arbitrary unit of measurement. Denied Platonic ideal backing and absolute sensible referents, numbers became, then, in Barrow's philosophy of mathematics mere "words" with "arbitrary signification." "I say that a *Mathematical Number*," he summarized,

> has no Existence proper to itself, and really distinct from the Magnitude it denominates, but is only a kind of *Note* or *Sign* of Magnitude

68. Barrow, *Mathematical Lectures,* 20.
69. John Wallis, *Mathesis universalis* (Oxford, 1657), 69.

considered after a certain Manner. . . . For in order to expound and
declare our Conception of Magnitude, we design it by the *Name* or
Character of a certain Number, which consequently is nothing else
but the *Note* or *Symbol* of such Magnitude as taken.[70]

Seeing geometry as the science of magnitude and numbers as mere
signs of magnitude, Barrow referred in his *Lectures* to "the Identity of
Arithmetic and Geometry." As he elaborated, "Arithmetic and Geometry
are not conversant about different Matters, but do both equally demonstrate Properties common to one and the same Subject." This "identity"
meant, above all, that there was no independent science of arithmetic. "I
note," he wrote, "that Numbers of themselves can neither be added to
nor subtracted from one another." The attributes of number follow "not
from the abstract Reason of the Numbers, but from the Condition of the
Matter to which they are applied." Taking Wallis's example of addition,
Barrow interpreted "two" and "four" as mere "words" which through
"arbitrary signification" bore the relationship "2 + 2 = 4." Furthermore, the formula "2 + 2 = 4" did not hold in all cases – the point
Barrow originally made – for such formulas were fundamentally dependent on the "Condition of the Matter to which they . . . [were] applied."
The formula covered only cases involving homogeneous magnitudes or
those with a "like Nature and Denomination." In short: "whatsoever
Attributes among Arithmeticians are proved to agree with Numbers,
they agree not with Numbers taken abstractly and of themselves, but
concretely according to the Condition of the Things they are attributed
to."[71] And thus formed the wide gap between Wallis's arithmetico-
algebraic mathematics and Barrow's geometric: for Wallis, the arithmetic
formula "2 + 2 = 4" was independent of and prior to any geometric
application; for Barrow, the formula was quite arbitrary and limited in
applicability by the conditions of magnitude. For Wallis, arithmetic ruled
geometry; for Barrow, geometry included arithmetic.

70. Barrow, *Mathematical Lectures,* 37, 34–35, 41. The sources of Barrow's
philosophy of arithmetic need to be probed. A possibility is Cardano's *De
uno,* in which Cardano, following Nicholas of Cusa's *Trialogus de possest,*
wrote: "Number is a fiction of the human mind. Its principle is the One.
'Numerical unity' is therefore a fiction of the soul." On Cardano's statement (but not on any connection to Barrow), see Markus Fierz, *Girolamo
Cardano, 1501–1576: Physician, Natural Philosopher, Mathematician, Astrologer, and Interpreter of Dreams,* trans. Helga Niman (Boston: Birkhäuser, 1983), 61 (quotation), 82–83.
71. Barrow, *Mathematical Lectures,* 30, 35–38.

8

If positive rational numbers were mere signs of magnitude, what of the irrational, negative, and imaginary numbers – or of numbers applied to objects without any consideration of magnitude? Committed to early modern as well as traditional mathematics, Barrow elaborated his philosophy of arithmetic to include the irrationals and, just barely, the negatives; (as far as I have been able to determine) his lectures did not mention the imaginaries. He wrote of the irrationals in two different contexts: he used them to buttress the case for his philosophy of arithmetic and also to attack Wallis's algebraic approach to proportion. First, he stressed the ease with which his philosophy explained the irrational or surd numbers – such as $\sqrt{2}$, which seemed a problem since "there is no Number, Integer or Fracted, which being drawn into itself produces 2." He noted that this lack of whole number or fractional values for the irrationals had led other mathematicians to cast these numbers out of arithmetic and to consign them to algebra. But he objected to the classification and to algebra in general. Instead he saw the irrationals as the "noblest and most profitable Number[s]" of arithmetic and put them forward as prime evidence for his philosophy. As he argued, the irrational $\sqrt{2}$ was not an independently existing entity that required reduction to whole numbers or their fractions; like *all* numbers, it was simply and fundamentally the sign of a magnitude, in this case, the diagonal of a square with a side of one unit. "These [surd] Numbers," he concluded, "since they cannot even in thought itself be abstracted from all Magnitude, do make it sufficiently evident that Numbers differ nothing in reality from Magnitude."[72] In short, the irrationals were both explained by and lent support to his philosophy of arithmetic.

Barrow focused again on the irrationals as he argued against Wallis's algebraization of proportion. "From hence a most learned . . . Man [Wallis]," he began, "takes occasion to say that all *Geometrical Reasons* [*ratios*] of whatsoever Quantities are homogeneous to one another, *because all are in the Genus of Number.*"[73] Additionally: "what he [Wallis] infers seems foreign to the Truth, viz. that the whole doctrine of Reasons suits more with the Speculation of Arithmetic than Geometry." For his part, Barrow argued that a ratio or "Reason is, and is acknowl-

72. Ibid., 44–45.
73. Ibid., 328. Here Barrow referred specifically to Wallis, and used the term "Reason" for ratio.

edged to be a pure perfect Relation." "Neither," he added, "in my
Opinion are *Reasons* Quantities, nor capable of Quantity." Still, he
conceded, some ratios could be "exhibited" in numbers because numbers "are Symbols most accommodate to our Capacity for determining
the Measures of Things." Thus, the ratio of a line of twelve feet to a line
of four feet could be exhibited as 12/4 or 3/1. But not every ratio was
"subject to *Numeration* . . . [or] agreed with Numbers, or might be
expressed by Numbers." Even Wallis, Barrow scoffed, did not "himself
deny but there are certain *Surd Reasons* and such as are inexplicable by
Numbers." Implicitly rejecting Wallis's extension of the concept of number to surds (by means of algebra) – and insisting on his own view of
surds as signs of magnitude – he added: "but there is no *Reason* at all
which cannot be exhibited in Magnitudes of every Kind. Therefore
Number is unsuitably assumed for the adequate Subject of *Reason*; and
Magnitude is much more suitable."[74]

Barrow defended negative numbers, which he called "negative terms,"
in a less direct way. He referred to them in his explanation of "*Arithmetical Reason* or . . . the Relation of two Quantities as to Difference, i.e. as
one exceeds or is short of the other." Arithmetical reason involved
subtraction, he began, but subtraction led to "*two Differences,* the one
Positive shewing an *Excess,* the other *Negative* signifying a *Defect.*" "*To
lack* and *to exceed,*" he emphasized, "are not the same, and therefore
ought not to be denoted by the same Sign." In illustration, he compared
the "Difference between 1 and 2 . . . [or] − 1, as signifying 1 to lack one
of 2" and the "Difference between 2 and 1 . . . [or] + 1, as signifying 2
to exceed 1 by one."[75]

But what was a negative difference, such as − 1, when the number 1
was no independent entity but merely a sign of magnitude? Barrow, as
so many of the English mathematicians before and after him, admitted
that it "perhaps seems difficult to conceive, *Something* less than *Nothing*
itself." Still, he maintained, "nor does any Thing hinder but there may
be such an improper *Subduction of a greater from a lesser.*" And thus,
bereft of a suitable traditional definition of negative numbers, he began
his own cumulative defense of negative terms – a defense that appealed
to analogy, usefulness, and geometric exhibition and a defense that was

74. Ibid., 374, 368, 328, 336. Although Barrow referred to book V of Euclid's
Elements (ibid., 374), he did not simply accept the Euclidean theory of
proportion. On Barrow's final theory of proportion, see Chikara Sasaki,
"The Acceptance of the Theory of Proportion in the Sixteenth and Seventeenth Centuries," *Historia Scientiarum* 29 (1985): 83–116, esp. 95–98.
75. Barrow, *Mathematical Lectures,* 321, 325.

thus similar to that used by Wallis for negative and imaginary numbers. After all, Barrow observed, mathematicians accepted "the *Division of a lesser by a greater*," although it was no easier "to conceive how often a *greater is contained in a lesser*" than it was to conceive "Something less than Nothing itself." Thus: "It is enough that this Kind of *Subduction* is subject to Arithmetical Laws as well as that Kind of Division, and is not devised or applied without Cause." Furthermore: "It may be added that this Method of *Subduction* is very useful." For example, work on progressions that decreased from unity was facilitated by the designation of the increasing terms by "Positive Exponents" and the decreasing terms, by "Negative Exponents." Finally, he "notice[d], that these *Negative Differences,* or *Terms less than Nothing* are easily exhibited in Geometry." Intent on preserving as much of the modern mathematics, including negative numbers and exponents, as he could in an empiricist framework, he then launched into what was becoming the standard example of lines taken in a forward direction versus those taken in a backward direction. His positive and negative terms were signs of magnitude taken forward and backward.[76]

Although thus successful with the irrational and negative numbers, Barrow's philosophy of arithmetic failed to explain those applications of the whole numbers involving no unit of measurement. As he specifically acknowledged, his philosophy did not provide for the assigning of whole numbers to collections of objects such as men, angels, and mountains. Yet he saw this failure as no serious drawback. Exhibiting a readiness to separate scientific from metaphysical concerns and seeming to fall back on a Platonic view of ideas for the metaphysical, he argued that some numbers were mathematical whereas others were "metaphysical" or "transcendental." According to his scheme, metaphysical numbers – which applied to "Things agreeing together no otherwise than by a certain generical Ratio" – lay outside mathematics and hence outside his philosophy of arithmetic. Although not strictly mathematical, however, such numbers were prior and essential to mathematics, for "Geometry as well as Arithmetic borrow those transcendental Numbers it uses in framing its Definitions and delivering its Argumentations."[77] For example, in defining a triangle, geometers use the metaphysical number "three," since their definition refers to the general term "straight line" rather than any unit of measurement.

76. Ibid., 326.   77. Ibid., 38–39.

9

If arithmetic emerged subservient to geometry in Barrow's scheme of mathematics, algebra fared worse. The Lucasian professor generally ignored algebra, and, in his few references to the subject, he presented it as an instrument of logic rather than a science. He characterized algebra as "yet . . . no Science"[78] in his explanation of the irrationals; in another lecture he proved unable to distinguish algebra from the method of analysis. "I am wholly silent about that which is called *Algebra* or the *Analytic Art*," he commented,

> . . . this was not done unadvisedly. Because indeed *Analysis*, understood as intimating something distinct from the Rules and Propositions of *Geometry* and *Arithmetic*, seems to belong no more to *Mathematics* than to *Physics*, *Ethics*, or any other Science. For this is only a Part or Species of *Logic*, or a certain Manner of using Reason in the Solution of Questions, and the Invention or Probation of Conclusions, which is often made use of in all other Sciences. Wherefore it is not a Part or Species of, but rather an Instrument subservient to Mathematics: No more is *Synthesis*, which is the manner of demonstrating Theorems in Contradistinction to *Analysis*.[79]

On the one hand, Barrow's muddling of algebra with the method of analysis was a convenient trivialization of early modern algebra, a subject that he rejected perhaps because of his lack of real appreciation of the symbolical style and his strong attachment to geometry. On the other hand, he exposed the foundational confusion that underlay the claims of Viète and his early English disciples that algebra was the art of analysis. After all, Oughtred's *Key*, which Barrow had studied in his formative years of the early 1650s, spoke of the "Analyticall art" rather than algebra and defined that art as that "in which by taking the thing sought as knowne, we finde out that we seeke." Barrow, in turn, maintained that analysis was no more than a method, which was equally applicable to all sciences and certainly no separate science in and of itself. In his later *Treatise of Algebra* Wallis would concede as much, but without mentioning Barrow: as we have seen, Wallis wrote of "algebra" and treated the subject as no simple "analytic art" but rather as the science of symbols or notes, based on arithmetic, which "manageth Proportion" in an abstract fashion. More generally, from the late seventeenth century on, English mathematicians would usually distinguish the science of algebra from the method of analysis.

78. Ibid., 44.   79. Ibid., 28.

Barrow however permitted himself a double standard with respect to analysis and synthesis. Even as he abjured discussion of analysis on the grounds that it was "only a Part or Species of Logic," he extolled the synthetic method of geometry. Going beyond Oughtred, who in the early 1630s had promoted mathematics as an academic discipline, Barrow as Lucasian professor argued for geometry as a part of the undergraduate curriculum on the basis of its pedagogical usefulness in training the human mind. He not only listed the various ways in which geometry improved the mind and character, but also explicitly presented geometry as a paradigm of logic and therefore an essential component of a liberal education. He thus formally sketched the argument that would be used repeatedly into the early nineteenth century to solidify the role of mathematics at Cambridge University.

His remarks on the pedagogical usefulness of mathematics came primarily in his inaugural oration and in later sections of the *Mathematical Lectures* that were specifically entitled "Excellence of Mathematical Demonstration." In the oration he reminded students that "Mathematics . . . depends upon Principles clear to the Mind, and agreeable to Experience; . . . [and] draws certain Conclusions." The study of such deductive mathematics, he continued, helped students to learn "to turn aside the Strokes of true Arguments, and warily decline the Blows of false ones; to dispute strenuously as well as judge solidly." Mathematics also taught diligence, stabilized the fancy, sharpened the wit, restrained the headstrong, and roused the dull.[80] Perhaps revealing some influence of Cambridge Platonism on his thought or perhaps primarily trying in this inaugural oration to reach as many of his Cambridge colleagues as possible, he also briefly and eloquently lapsed into a seemingly Platonic vision of mathematics as the study of "pure Forms" associated with the "Beauty of Ideas" and "a more divine Contemplation,"[81] a vision that (as we have seen) would not consistently inform his *Lectures*.

In the later lectures on the "Excellence of Mathematical Demonstration," he inched ever so carefully beyond these rather general arguments for university mathematics to the specific argument that mathematics actually taught logic, and was perhaps the best instrument for teaching logic. When elaborated by his successors at the university, this argument would prove of the utmost importance in assuring mathematics a firm place within the undergraduate curriculum. Logic was after all the liberal

---

80. Ibid., xxviii.
81. Ibid., xxx–xxxi. Gascoigne cites this in support of a Platonic "imprint" in Barrow's mathematical writings (Gascoigne, "Isaac Barrow's Academic Milieu," 272).

art that, according to the Elizabethan statutes, was supposed to be the main study of Cambridge men of the second and third years.[82] Arguing that mathematics was indeed logic and (frequently) that geometry was the best logic, Barrow's successors would by the end of the eighteenth century focus the undergraduate curriculum on mathematics, to the point where mathematics emerged as the sole path to honors at the university.[83]

Meanwhile, a full century earlier, in a role befitting the first Lucasian professor, Barrow tentatively and diplomatically sketched the argument for geometry as logic. Speaking of "Demonstration" or deductive reasoning, he began: "But it is neither my Office nor Intention to handle this Matter at large, since it properly belongs to *Logic.*" He then referred students to Aristotle, the "famous Inventor and Author" of demonstration. "But it may suffice for my Purpose," he continued, "only to take Notice of his [Aristotle's] Sayings and Prescripts as they accord with the Mathematical Way of Reasoning." Then, appealing to Aristotle, he argued that: "Demonstration agrees properly and peculiarly with these [mathematical] Sciences only, and cannot with equal Justice be supposed to agree with other Disciplines." Even Aristotle, he reminded his audience, admitted that scholars could not strictly demonstrate moral truth "because the Rigour of Demonstration is only to be found in the Mathematical Sciences, and not in that Kind of Learning"; that they should no more ask for demonstrations from an orator than persuasive arguments from a mathematician; and that they could not expect mathematical exactness from a natural philosopher. Offering additional support for the acceptance of mathematics as logic, Barrow cited "that excellent Logician *Jac. Zabarella* . . . [who] diligently pursued *Euclid's* Elements over and over, that he might the better understand and explain the Nature of Demonstration."[84]

So the major themes of Barrow's experience as a university mathematician came together. As he had confided to Collins, in his mathematical work he was interested primarily in the "general reasons of things, as a scholar." But no facet of mathematics could so attract a general scholar as its certainty, as Hobbes's library experience had shown. Perhaps at

82. These statutes are quoted in Mordechai Feingold, *The Mathematicians' Apprenticeship: Science, Universities and Society in England, 1560–1640* (Cambridge: Cambridge University Press, 1984), 24–25.

83. On the focus on mathematics at eighteenth-century Cambridge, see John Gascoigne, *Cambridge in the Age of the Enlightenment: Science, Religion and Politics from the Restoration to the French Revolution* (Cambridge: Cambridge University Press, 1989), 270–299.

84. Barrow, *Mathematical Lectures*, 52–53.

the beginning of his mathematical career, Barrow, as a royalist and Arminian, had needed some intellectual reassurance in the confused times of the Interregnum. Perhaps not coincidentally he had highlighted this theme of mathematical certainty in a "prayer" that he had written on the front page of his manuscript on the first four books of Apollonius's *Conics,* completed in 1653 but not published until 1675. "In other Arts and Sciences," he had confided to God (the "great Geometrician"), "our Understanding is able to do almost nothing; and, like the Imagination of Brutes, seems only to dream of some uncertain Propositions: Whence it is that in so many Men are almost so many Minds." But mathematics was different, a bedrock of certainty amid intellectual dissension for "in these Geometrical Theorems all Men are agreed: In these the Human Faculties appear to have some real Abilities, and those Great, Wonderful and Amazing. . . . Thee therefore do I take hence occasion to Love, Rejoice in, and Admire."[85]

Barrow thus brought mathematics to Cambridge University with more emphasis on its deductive method and foundations than on its content. He laid the groundwork for the defense of mathematics as a logic and helped shape Cambridge's obsession with geometry. After all, he had argued that arithmetic was only a part of geometry; he had dismissed algebra as analysis; and, above all, he had extolled geometry as the perfect logic. In short, the mathematical legacy of the first Lucasian professor, which was to prove an enduring one, spoke of a mathematics whose principal (and, to a certain extent, only) branch was the logically paradigmatic geometry.

10

Major seventeenth-century English thinkers reached no consensus on algebra. Early in the century Oughtred and Harriot strongly advocated the symbolical style. Later Wallis argued for the symbolical style, an expanded algebraic universe, and arithmetic foundations for the new algebra. But Wallis's theory of algebra was stymied on two fronts. Internally, the negative and imaginary numbers refused strictly arithmetic legitimation, and seemed instead to require elaborate cumulative justification. Externally, Hobbes and Barrow, two of the period's most reflective thinkers, proved unreceptive and even hostile to the new algebra. The employment of algebraic symbolism, in particular, was a major

85. Quoted in Feingold, "Isaac Barrow," 54; *The Theological Works of Isaac Barrow, D.D.,* ed. Alexander Napier, 9 vols. (Cambridge, 1859), 1:xlvii.

stumbling block for Hobbes; even Barrow seemed to see the symbolical style as a system of abbreviations rather than an instrument of discovery. Too, the empiricist inclinations of the English Scientific Revolution affected mathematics, with Hobbes and Barrow arguing for the primacy of the sense-based geometry over arithmetic and algebra. As these inclinations fostered disciplines like medicine, they threatened to stifle the more abstract science of algebra. Nevertheless, in a curious twist of the Scientific Revolution, rejection of Platonism pushed Barrow and, on at least one occasion, Hobbes to begin to reconstruct the foundations of arithmetic in what would prove a modern direction.[86]

Although Hobbes and Barrow were on the losing side of one of the most important mathematical battles of their century – that pitting the new algebraists against mathematicians linked more closely to the geometric tradition of Western mathematics – their ideas exerted an enduring influence on British mathematics and, in particular, mathematical education. The new algebra survived and by the end of the eighteenth century the very best of European mathematics was analytic. But through the early nineteenth century geometry remained the core of the mathematical curriculum at Cambridge University; undergraduates were supposed to master the first two books of Euclid's *Elements*. Bolstering Cambridge's mathematical and, in particular, geometric focus were arguments for the pedagogical merits of geometry, similar to those that Barrow offered, as well as arguments for the applicability of the geometric method to nonmathematical disciplines, at least partially a legacy of Hobbes.

Ultimately, however, neither Wallis, on the one hand, nor Hobbes and Barrow, on the other, determined England's mathematical destiny. As in the history of nearly all the exact sciences of early modern England, the towering figure of Isaac Newton intervened at this crucial mathematical turning point.

---

86. English empiricism also helps to explain the geometric optics developed by Barrow, Newton, and James Gregory. See Antoni Malet, "Studies on James Gregorie (1638–1675)" (Ph.D. diss., Princeton University, 1989), 100.

7

The Mixed Mathematical Legacy of Newton's Universal Arithmetick

Isaac Newton was born in 1642, more than a decade after the publication of the algebraic works of Oughtred and Harriot. Thus he was not of the mathematical generation that brought early modern algebra to England, but rather an immediate beneficiary of that generation's algebraic efforts. As an undergraduate, he read and built on many of the mainstays of early modern algebra (principally the algebraic writings of Viète, Oughtred, and Descartes, as well as the early analytic writings of Wallis). As the second Lucasian professor, he wrote his own lectures on algebra, which were first published in Latin in 1707 and later in English under the title of *Universal Arithmetick*.

There was irony in the standing of *Universal Arithmetick* as the most popular of Newton's mathematical works in the eighteenth century. He agreed to publication of the first edition of the book only under pressure from his Cambridge colleagues. The lectures on which the book was based were delivered relatively early in his tenure at Cambridge, and they were neither predominantly original nor as "venturous" or philosophical as Wallis's *Treatise on Algebra*. In *Universal Arithmetick*, Newton, like Pell and Kersey, treated roots involving $\sqrt{-1}$ essentially as signs of impossibility, and, on that basis, he argued against the applicability of Cardano's solution of the cubic equation to the irreducible case. Still, Newton's work offered the best of the early modern notation, some technical originality (including an extension of Descartes's rule of signs), and, above all, economy.

If early-eighteenth-century English thinkers had any lingering doubts about the validity of algebra as an academic subject, Newton's publication of his algebra textbook helped to quell them – his contemporaries said as much. But *Universal Arithmetick* was no unqualified endorsement of an algebra that was "pure" and superior to geometry, the vision of algebra that had evolved in England from Oughtred and Harriot through Wallis. The very contemporaries who argued that Newton's textbook enhanced the "esteem" of algebra also wrote that he had "condescended to handle" the subject.[1] In *Universal Arithmetick* Newton mixed Cartesian geometric algebra with the arithmetic algebra that Wallis supported, and he hinted at a preference for classical geometry over analytic geometry. Even as *Universal Arithmetick* solidified the standing of algebra as an academic subject, then, it left open questions about the nature of algebra and, particularly, its relationship with geometry. The premier English mathematician of his period, an expert on the modern mathematics, Newton refused to let himself or his countrymen forget the enduring merits of classical geometry.

I

Born on Christmas Day, Newton (1642–1727) never knew his father, who had died a few months before his birth.[2] At the age of three, when his mother married the Reverend Barnabas Smith, he was sent to live with his maternal grandmother. Reunited with his mother upon Smith's death in 1653, he entered the Free Grammar School at Grantham two years later. He was called home in 1659 to learn to manage his estate; a failure at rural pursuits, he was returned to Grantham the next year; and he entered Trinity College, Cambridge, in 1661.

Like so many of his mathematical contemporaries, he was largely a mathematical autodidact. According to Abraham De Moivre's account, Newton happened upon the serious study of mathematics in a way similar to Barrow's journey from Scaliger to Euclid. He was supposed to have bought a book on astrology at the Stourbridge fair of 1663; the book, in turn, led him toward trigonometry, then Euclid's geometry, and, finally, early modern mathematics.[3] The appeal of the analogy with

1. Anon., "To the Reader," in Isaac Newton, *Universal Arithmetick: or, A Treatise of Arithmetical Composition and Resolution*, trans. Mr. Ralphson and Rev. Mr. Cunn, 2d ed. (London, 1728), reprinted in *The Mathematical Works of Isaac Newton*, ed. Derek T. Whiteside, 2 vols. (New York: Johnson Reprint, 1964–1967), 2:i (4). Subsequent references to *Universal Arithmetick* are to this edition, which followed Newton's Latin edition of 1722. The reference to i (4) is to p. i of the original work and p. 4 of the reprint.
2. For a masterful portrait of Newton, see Richard S. Westfall, *Never at Rest: A Biography of Isaac Newton* (Cambridge: Cambridge University Press, 1980). On the Newton biographical tradition and the chronology of the "Life, Career and Works of Newton," see Derek Gjertsen, *The Newton Handbook* (London: Routledge & Kegan Paul, 1986), 74–83, 314–319.
3. For De Moivre's account, see D. T. Whiteside, "Introduction" to part 1, *The Mathematical Papers of Isaac Newton*, ed. D. T. Whiteside, 8 vols.

Barrow notwithstanding, both Newton's preparatory and undergraduate educations seem to have provided occasions for his learning some mathematics. As a student at Grantham, he may have benefited from the reform movement of the mid-seventeenth century that brought arithmetic, English, and modern history to the English grammar schools. Thus Whiteside has "tentatively suppose[d] that . . . Newton's schoolmaster John Stokes . . . drummed into his pupils a basic familiarity with (if not understanding of) standard methods of addition, subtraction, multiplication and division, reduction of fractions, and the rule of proportion and their application to elementary weight and money problems, perhaps even simple techniques of casting accounts."[4] Moreover, as an undergraduate at Cambridge from 1661 to 1665, Newton found a somewhat more supportive mathematical environment than Wallis had at the university thirty years earlier and Barrow, less than twenty years earlier. The very year that Newton supposedly began his serious study of mathematics, 1663, was in fact the year that Barrow was appointed the Lucasian professor. Newton attended some of Barrow's Lucasian lectures, or at least he later implied as much. Even if Barrow was neither his tutor at Cambridge nor his close mentor, it is then possible that the early Lucasian lectures were the spark igniting Newton's mathematical interest.[5] Furthermore, Barrow seems to have been the first to recognize Newton's mathematical genius.[6]

In addition to enjoying the real and symbolical benefits of Barrow's tenure as Lucasian professor, Newton was able to take advantage of Cambridge's newly increased tolerance of independent reading by its seniors. In 1664 he began to study contemporary scientific and philosophical works, including those of Robert Boyle, Descartes, Pierre Gassendi, and Hobbes, as well as traditional and early modern math-

(Cambridge: Cambridge University Press, 1967–1981), 1:5–6. As all modern scholars of Newton's mathematics, I am heavily indebted to Whiteside. The research for this chapter was made possible by his publication of Newton's mathematical papers and his valuable annotations. The synthetic tracing of Newton's evolving views on algebra is of course my own, and any errors should be attributed to me and not Whiteside.

4. D. T. Whiteside, "Isaac Newton: Birth of a Mathematician," *Notes and Records of the Royal Society of London* 19 (1964): 53–62, on 54.

5. The latter argument comes from Westfall, *Never at Rest*, 99.

6. In 1669, e.g., Barrow described Newton as having "a very excellent genius to those [mathematical] things" (Barrow to Collins, 20 July 1669, *The Correspondence of Isaac Newton,* ed. H. W. Turnbull, J. F. Scott, A. R. Hall, and Laura Tilling, 7 vols. [Cambridge: Cambridge University Press, 1959–1977], 1:13).

ematics.[7] As in his astronomical and physical studies, Newton the mathematician would stand on the shoulders of giants. It was the writings of these giants (Viète and Descartes) and some first-rung mathematicians (Oughtred, Wallis, and Barrow) as well – some of whose writings were only recently composed, others only recently widely available in England – that he devoured in 1664 and 1665.

He purchased a copy of Barrow's 1655 edition of Euclid's *Elements* and by early 1664 he had annotated books V, VII, and X (on proportion and number theory),[8] and thus two of the books that Barrow had made a point of including in order to impress upon students the "very neer alliance between Arithmetick and Geometrie." In the same year Newton studied both the third Latin edition of Oughtred's *Key* (1652) and Schooten's second Latin edition of Descartes's *Géométrie*. In December he bought a copy of the latter work (in its first or second edition) as well as Schooten's *Exercitationum mathematicarum libri quinque*, and borrowed Wallis's *Opera mathematica* (1656–1657), a collection of Wallis's early mathematical lectures, including his *Arithmetica infinitorum*. He also studied Viète's researches in *Francisci Vietae opera mathematica*, which Schooten had edited in 1646.[9]

Newton absorbed the corpus of early modern mathematics at an astonishing pace. He compared the notations, techniques, methods, and mathematical styles it offered as he struggled to find his own mathematical way. For example, he studied the Pythagorean theorem from both Barrow's edition of Euclid's *Elements* and Oughtred's *Key*. As Whiteside has noted, in entering the theorem into his notebook (tentatively dated late 1664) he first gave a citation to *The Key* but then replaced it with one to the *Elements*.[10] Significantly, he gave the theorem without a diagram and in algebraic form. Among the earliest of his surviving mathematical notes, this inscription thus captures in a simplified way the mathematical ambidexterity – at times, schizophrenia between the new

7. Whiteside, "Isaac Newton: Birth of a Mathematician," 57–58.
8. On Newton's early acquaintance with Euclid, see D. T. Whiteside, "Sources and Strengths of Newton's Early Mathematical Thought," in *The "Annus Mirabilis" of Sir Isaac Newton, 1666–1966*, ed. Robert Palter (Cambridge, Mass.: MIT Press, 1970), 69–85, esp. 71–72.
9. For an analysis of Newton's mathematical readings of 1664 and 1665, as well as the bibliographical details of the works he read, see Whiteside, "Newton's Early Mathematical Thought," 72–74, and Whiteside, "Bibliographical Note on the Works Annotated by Newton," in *Mathematical Papers of Newton*, 1:19–24.
10. For the notebook entry, see *Mathematical Papers of Newton*, 1:25. See also Whiteside, "Newton's Early Mathematical Thought," 74.

and old, the algebraic and geometric – that would distinguish Newton the mathematician from the beginning through the very end of his career and would explain both his strengths and limits.

Moreover, he seems to have evaluated rapidly the various mathematical notations used by his immediate predecessors. At the beginning of his notes on some of Viète's writings, probably written also in late 1664, he used Oughtred's simplified notation, translating, for example, *A quadratum* to *Aq*. By the end of these notes, however, he was already experimenting with the Cartesian notation according to which, for example, *Lc* suddenly became *l*³.[11] Newton's predecessors thus spared him the notational trials they had endured: his basic notation came from Descartes's *Géométrie,* and his notation for negative exponents built on Wallis's *Arithmetica infinitorum.*

He drew more than a commodious notation from the existing corpus of early modern mathematics. Building on Wallis's treatment of series in *Arithmetica infinitorum,* he arrived by interpolation at the general binomial theorem. Whereas the purely algebraic binomial theorem for expanding $(a + b)^n$ for an integer n had long been appreciated, and proved by Blaise Pascal in 1654, Newton generalized the theorem to real exponents in 1665 and made it a "keystone" of his calculus.[12] But even more than Wallis's works, Schooten's second Latin edition of Descartes's *Géométrie* "direct[ed] the future course of Newton's mathematical development."[13] Indeed, the study of Descartes's *Géométrie* (and the commentaries of Schooten, Hudde, and Heuraet) fixed his major mathematical research on analytic geometry and calculus.[14]

2

Descartes's *Géométrie,* moreover, played a leading role in shaping Newton's algebraic style and interests. In May 1665 Newton began a manu-

11. For the original and Whiteside's commentary on the notations Newton employed, see *Mathematical Papers of Newton,* 1:63–71, esp. 63, 71.
12. D. T. Whiteside, "Newton's Discovery of the General Binomial Theorem," *Mathematical Gazette* 45 (1961): 175–180, at 175; see M. Pensivy, "The Binomial Theorem," in *Companion Encyclopedia of the History and Philosophy of the Mathematical Sciences,* ed. I. Grattan-Guinness, 2 vols. (London: Routledge, 1994), 1:492–498.
13. Whiteside, "Newton's Early Mathematical Thought," 74.
14. For an introduction to Newton's early work in both areas, see Whiteside, "Newton's Early Mathematical Thought," 74–82.

script that referred specifically to Descartes and followed his lead in finding geometric constructions for the roots of algebraic equations.[15] Still, sections of book III of *La géométrie* could take a seventeenth-century mathematician beyond geometric algebra, or algebra mixed with geometry, and to the outskirts and even the heart of what Wallis called "pure algebra." The book encouraged the young Newton to explore key algebraic relationships between roots and coefficients of equations. In research notes probably dating from 1665–1666, he gave the coefficients of an equation in terms of the symmetric functions of its roots. (A symmetric function – a modern term that Newton did not use – is a function of two or more variables that is unaltered when any two of its variables are interchanged.) Writing here "of the Nature of Equations," he took as his example:

$$x^8 + px^7 + qx^6 + rx^5 + sx^4 + tx^3 + vxx + yx + z = 0.$$

He noted in succession that the sum of the eight roots of the equation (a symmetric function) is equal to $-p$, or the negative of the coefficient of the equation's second term; the sum of the products of the roots taken two at a time (another symmetric function) is $+q$; the sum of the products of the roots taken three at a time (also symmetric), $-r$; and so on. Too, the sum of the squares of the roots (another symmetric function) is $+pp - 2q$; the sum of the cubes is $-p^3 + 3pq - 3r$; and so on.[16]

In this same set of notes, he extended Descartes's rule of signs, according to which the algebraist finds the number of positive and negative roots of an equation, or at least upper limits to those numbers, by counting the number of times that the signs of the coefficients change. He extended the rule to incomplete equations, or those lacking some term(s), as

$$x^9 * + ax^7 * * - bx^4 * * * -c = 0,$$

---

15. Newton, "The Geometrical Construction of Equations," 30 May 1665, *Mathematical Papers of Newton*, 1:492–502.

16. Newton, "Researches in the Theory of Equations" [1665–1666], *Mathematical Papers of Newton*, 1:517–539, at 519. Newton was probably unfamiliar with the *Invention nouvelle en l'algebre* of 1629, in which Albert Girard had already expanded an equation's coefficients through symmetric functions of its roots (ibid., 518 n. 12). On symmetric functions and "Newton's Formulae for Sums of Powers of Roots," see Florian Cajori, *An Introduction to the Modern Theory of Equations* (New York: Macmillan, 1904), 13–14, 84–86.

which he found to have one positive and eight negative roots.[17] He also clarified an ambiguity in the rule. "[I]f any roots bee imaginary," he opened, "this rule soe far admitts of exception." He proceeded with an example, taking the equation

$$x^3 - pxx + 3ppx - q^3 = 0,$$

which the rule "show[s] . . . to have three true roots." Multiplying the given equation by $x + 2p = 0$, he obtained:

$$x^4 + px^3 + ppxx + (6p^3 - q^3)x - 2pq^3 = 0.$$

If the original equation has three positive roots, then the new equation should have three positive roots and one negative root, $x = -2p$. But, assuming once again that Descartes's rule gives the exact number of positive and negative roots and applying it directly to the new equation, Newton was led to the (different) conclusion that the new equation has one positive root and three negative roots (since indeed there is but one change in signs between coefficients). With penetrating mathematical insight, he "conclude[d] therefore that the two roots wch in ye one case appeare true, & in ye other false are neither, but imaginary; & that of ye other two roots one is true ye other false."[18]

Other seventeenth-century mathematicians might have expostulated here on the mysteries surrounding imaginary roots and hence the inapplicability of Descartes's rule to such equations, but Newton left no evidence here of indulging in what might be called the philosophy of the imaginary numbers. Rather he set about extending Descartes's rule to cover equations with imaginary roots. He wrote matter-of-factly that "it appeares y^t to know ye particular constitution of any Equation it is cheifely necessary to understand w^t imaginary roots it hath." From realization of the need to enumerate imaginary roots, he moved quickly to rules for doing so. He developed, but did not prove, a quite complicated "incomplete" rule for enumerating imaginary roots[19] – the likes of which rule appears to have eluded even Wallis, who in his *Treatise of*

17. Newton, "Researches in the Theory of Equations," 1:517. According to Descartes's rule, an equation *can* have as many positive roots as the number of times that the signs of its coefficients change from + to − or − to +.
18. Ibid., 1:520–521. To make sense of the multiplication of the original equation by $x + 2p = 0$, I entered a few corrections in Newton's manuscript (as reproduced here).
19. Ibid., 1:521–531 (quotation from p. 521).

Algebra of roughly twenty years later would also give an example to show the "uncertainty" of applying Descartes's rule to equations with imaginary roots but would state only that the subject was "capable of farther Improvement."[20] Published later in *Universal Arithmetick*, Newton's rule would be elaborated by George Campbell, Colin MacLaurin, and Edward Waring in the eighteenth century and proved in its complete form by J. J. Sylvester in the nineteenth.[21]

The influence of Descartes seems to have been crucial in Newton's accepting imaginary roots and embedding them firmly in equation theory. Newton's French mentor had after all written that "Neither the true nor the false roots are always real; sometimes they are [only] imaginary."[22] In his notes of 1665–1666, Newton recorded: "Now of these rootes some are true some false & some imaginary." Soon after, adopting Descartes's strong version of the fundamental theorem, he stated: "Every Equation hath soe many roots as dimensions of wch some may be true some false & some imaginary or impossible." Moreover, an addendum to this section of the manuscript ("because imaginary roots are properly neither true nor false"), which Newton wrote but then canceled,[23] showed that he was here not merely following Descartes's views but refining them as well.

In addition, a belief in limits on human reasoning may have helped the young Newton avoid struggling with the philosophy of the imaginaries. He indirectly suggested as much in the late 1660s, when he attributed mathematicians' inability to understand infinite series and the irrational numbers to such innate limits. His lean statement on these limits appeared in *De analysi per aequationes numero terminorum infinitas* (Of Analysis by Equations of an Infinite Number of Terms), which he showed Barrow in 1669 but which appeared in print only in

20. John Wallis, *A Treatise of Algebra, Both Historical and Practical, Showing the Original, Progress, and Advancement Thereof, From Time to Time, and by What Steps It Hath Attained to the Height at Which Now It Is* (London, 1685), 158–159.

21. For Whiteside's sketch of the history of Newton's rule through Sylvester's proof, see *Mathematical Papers of Newton*, 1:524–525 n. 40.

22. *The Geometry of René Descartes with a Facsimile of the First Edition*, trans. David Eugene Smith and Marcia L. Latham (New York: Dover, 1954), 174–175. Descartes here wrote: "mais quelquefois seulement imaginaires" (174).

23. For Newton's original manuscript and Whiteside's comments (including a reference to the canceled remark), see Newton, "Researches in the Theory of Equations," 1:519–521.

$1711.$[24] In the way of justification for the techniques used in *De analysi*, including expanding functions into infinite series that he then integrated term by term, he stated simply:

> And whatever the common Analysis performs by Means of Equations of a finite Number of Terms (provided that can be done) this can always perform the same by Means of infinite Equations: so that I have not made any Question of giving this the Name of *Analysis* likewise. For the Reasonings in this are no less certain than in the other; nor the Equations less exact; albeit we Mortals whose reasoning Powers are confined within narrow Limits, can neither express, nor so conceive all the Terms of these Equations, as to know exactly from thence the Quantities we want: Even as the surd Roots of finite Equations can neither be so exprest by Numbers, nor any analytical Contrivance, that the Quantity of any one of them can be so distinguished from all the rest, as to be understood exactly.[25]

This is a striking passage. Newton here implied an ignorance or rejection of Barrow's explanation of irrational numbers and appealed to limits on human reason as an argument for his doing cutting-edge mathematics without worrying overmuch about some of its philosophically fine points. Also, by using the term "analysis" here and elsewhere to describe a calculus in which he seemed to generalize algebra to cover infinite series and then applied those series to the calculation of areas, he helped to confuse anew the terms "algebra" and "analysis."[26]

It is possible that Newton adapted the argument about the limits of human reason from Barrow. In his discussion of the divisibility of magnitude in the *Mathematical Lectures* of 1664–1666, Barrow had referred to "the Imperfection of the Mind of Man and the Smallness of our Capacities." Specifically, he had argued for the "perpetual Divisibility . . . of Quantity" but admitted that "we do . . . perceive it to be of that kind of Things, which cannot be comprehended by our Minds, as being finite."[27] An expanded version of the argument, moreover, appeared in

24. On the circulation of the paper beyond Collins and its eventual printing, see introduction to *Mathematical Works of Newton*, 1:xi–xiii, xvi–xvii.
25. Newton, "Of Analysis by Equations of an Infinite Number of Terms," *Mathematical Works of Newton*, 1:340 (22).
26. On Newton's application of the term "analysis" to infinite series and calculus, see Carl Boyer, "Analysis: Notes on the Evolution of a Subject and a Name," *Mathematics Teacher* 47 (1954): 450–462, esp. 456.
27. Isaac Barrow, *The Usefulness of Mathematical Learning Explained and Demonstrated: Being Mathematical Lectures Read in the Publick Schools*

Robert Boyle's "Discourse of Things above Reason" of 1681. Here Boyle wrote that "we have no such clear and symmetrical conceptions [of surds] as we have of many other things that are of a nearer and more intelligible order." He attributed the difficulty with the surds to the limits imposed on human reasoning by God:

> I shall not scruple to acknowledge that, partly by my own experience, and partly by the confessions of others and by their unsuccessful attempts, I am induced to think that God, who is a most free agent, having been pleased to make intelligent beings, may perhaps have made them of differing ranks or orders whereof men may not be of the principal; and that, whether there be such orders or no, he hath at least made us men of a limited nature (in general), and of a bounded capacity.

According to Boyle, humans could aspire only to "competent knowledge of as much truth as God thought fit to allow . . . [their] minds in their present (and perchance lapsed) condition, or state of union with their mortal bodies."[28] At least for some of the key English mathematical thinkers of the pre-Enlightenment period, then, a modest assessment of the human mind tempered the criterion of preeminent clarity for mathematical objects.

3

However large the gap in enthusiasm for early modern algebra that separated Newton and Barrow, especially after Newton's substantial analytic readings and reflections of 1664–1666, and however minimal the early personal connection between the two men, in 1669 Barrow seems to have arranged for Newton to succeed him as Lucasian professor.[29] In the same year Barrow directed Newton's attention to Nicholas Mercator's Latin translation of Gerard Kinckhuysen's *Algebra*, an introductory textbook that built on the work of Descartes and Schooten and originally appeared in 1661. Collins had probably first told Barrow, one

at the *University of Cambridge*, trans. John Kirkby (London, 1734), 152, 162. I found the reference to Barrow's use of the argument in Antoni Malet, "Studies on James Gregorie (1638–1675)" (Ph.D. diss., Princeton University, 1989), 241.

28. Robert Boyle, "A Discourse of Things above Reason, Enquiring Whether a Philosopher Should Admit There Are Any Such," *Selected Philosophical Papers of Robert Boyle,* ed. M. A. Stewart (Manchester: Manchester University Press, 1979), 209–242, on 238–239.

29. Westfall, *Never at Rest,* 206–208.

of his regular correspondents, of his intention to publish the translation with some accompaniments. On the brink of stepping down from the Lucasian professorship and, in any case, not favorably inclined toward algebra, Barrow seems to have directed Collins to see Newton for help with the proposed book.[30] By doing so, Barrow set in motion a curious series of circumstances that led ultimately to publication of Newton's *Universal Arithmetick.*

Meeting for the first time around the beginning of December 1669, Collins and Newton seem to have agreed that the new Lucasian professor would annotate Kinckhuysen's *Algebra.* By February, however, Newton was less than enthusiastic about the project; he evaluated Kinckhuysen's lean textbook as "a good introduction ... [but] not worth the paines of a formall comment" and told Collins that he intended to confine his notes to essential corrections to the original text.[31] Somehow Collins then persuaded Newton to expand his notes. On 11 July Newton sent him the new set of "Observations" with an obliging letter in which he wrote: "I know not whither I have hit your meaning or noe but I have added & altered those things wch I thought convenient to bee added or altered, & I guesse that was your desire I should doe. All & every part of what I have written I leave wholly to your choyse whither it shall bee printed together with your translation or not."[32]

Initially, Collins had singled out for improvement Kinckhuysen's brief treatment of radicals (roots of numbers and algebraic quantities) and, in particular, his extraction of the cube roots of binomials, essential for the algebraic solution of some cubic equations. He had recommended Ferguson's *Labyrinthus algebrae* to Newton for its exemplary treatment of cube roots, and even suggested that Newton incorporate Ferguson's ideas into his annotations. Newton, however, found Ferguson's treatment of the cube roots of binomials defective, and he tried to explain the deficiencies to Collins (as Wallis would similarly do in 1673). Ferguson, Newton told Collins, had indeed given the real roots of some irreducible

30. On Barrow's bringing Newton and Collins together for work on Kinckhuysen's *Algebra,* see ibid., 222–223, and Whiteside, "Introduction" to part 3, *Mathematical Papers of Newton,* 2:280.

31. Newton to Collins, 6 Feb. 1669/1670, *Correspondence of Newton,* 1:24. For a discussion of the correspondence between Collins and Newton on Kinckhuysen's *Algebra,* see Whiteside, "Introduction" to part 3, 2:281–291. A sketch of this correspondence, Kinckhuysen's book, and Newton's comments appears in Christoph J. Scriba, "Mercator's Kinckhuysen-Translation in the Bodleian Library at Oxford," *British Journal for the History of Science* 2 (1964): 45–58.

32. Newton to Collins, 11 July 1670, *Correspondence of Newton,* 1:30.

cubic equations but without explaining how to extract the cube roots of complex binomials. That there was a method for extracting such cube roots and that he knew it, Newton assured Collins. But, he wrote on 11 July: "I think it not worth ye inserting into Kinckhuyson, yet if you think it convenient . . . I will send you it done in my next letter."[33]

Although Collins's reply to Newton of two days later was warmly enthusiastic ("you have much obliged the young Students of Algebra," he began), he pressed Newton to expand Kinckhuysen's section on radicals, still seemed confused about the merits of Ferguson's textbook, and asked Newton to send his method for extracting cube roots of complex binomials.[34] By 16 July Newton was agreeing to the additions Collins suggested, but he seemed a little defensive. "I sometimes thought," he opened his letter, "to have altered & enlarged Kinkhuysen his discourse upon surds [radicals] but judging those examples I added would in some measure supply his defects I contented my selfe wth doing that onely."[35] Newton and Collins were perhaps here engaged in a pedagogical dispute. In his section on radicals Kinckhuysen had discussed the extraction of square roots of algebraic quantities at some length but his coverage of cube roots was scanty. Newton, who here and elsewhere stressed the value of examples in mathematical education, had supplemented the original discussion with examples involving higher roots. Newton had noted, for example, that $\sqrt{C:2a^3b + a^4} = a\sqrt{C:a + 2b}$.[36] Collins wanted a more general treatment.

As Newton offered hints of his pedagogical philosophy in this revealing letter, he also assumed the role of patient instructor when trying to convince Collins of Ferguson's inadequacies in extracting the cube roots of complex binomials. He analyzed Ferguson's solution of the equation

$$x^3 = 6x + 4,$$

which had first of all brought Ferguson to the point where he needed to extract the cube roots of $2 + \sqrt{-4}$ and $2 - \sqrt{-4}$. Newton quoted Ferguson's next directive: "Multiply ye binomium [$2 + \sqrt{-4}$] by 1000, put it in pure numbers & c." As Newton explained: "$2 + \sqrt{-4}$ in 1000 makes $2000 + \sqrt{-4000000}$, but to put this in pure numbers is impossible for $\sqrt{-4000000}$ is an impossible quantity & hath noe pure

33. Ibid., 30–31.
34. Collins to Newton, 13 July 1670, ibid., 1:32–33.
35. Newton to Collins, 16 July 1670, ibid., 1:34.
36. Newton, "Observations on the Algebra of Gerard Kinckhuysen," [1669–1670], *Mathematical Papers of Newton*, 2:364–447 (example from p. 370).

number answering to it. His rule therefore failes."[37] By 19 July Collins seemed to concede Ferguson's limitations, explaining that he had viewed his examples "cursorily" and "scrupled his rootes of negative quadratick quantities, and imagined that they expunged one another being affected with contrary Signes."[38]

As he continued to work on Kinckhuysen's *Algebra* from July to September 1670, Newton kept rethinking the exact form that the published version of his work ought to take. In mid-July he voiced concerns about the appropriateness of attaching his already lengthy commentary to Kinckhuysen's book and suggested that at the least the title page should note the work had been "enriched by another Author."[39] In September he confessed to having thought about writing his own "compleate introduction to Algebra."[40] His confidence in himself as a potential textbook author had clearly grown during the summer as he composed a brief and lucid piece on "reducing problems to an aequation" – a topic which, he had stressed in his pedagogically rich letter of 16 July, was "the most requisite & desirable doctrine to a Tyro & scarce touched upon by any writer unles in generall circumstances."[41] This piece was meant as a prelude to part 3 of Kinckhuysen's *Algebra,* where algebra was used to solve specific problems. Whereas in his introduction Newton gave general guidelines for writing equations for problems, he complained that Kinckhuysen solved his problems "not by any generall Analyticall method but by particular & contingent inventions, wch though many times more concise then a generall method would allow, . . . are lesse propper to instruct a learner." Kinckhuysen's examples, he claimed, were as valuable to a new student of algebra as a study of "Acrostick's & such kind of artificiall Poetry" would be to one intent on mastering Ovidian poetry.[42]

In the 1670s Newton neither wrote his own algebra textbook nor saw his "Observations" appear in print. Considering that several algebra

37. Newton to Collins, 16 July 1670, 1:35.
38. Collins to Newton, 19 July 1670, *Correspondence of Newton,* 1:37. Collins also recommended Ferguson's extraction process to Wallis, who in 1673 tried to convince him of Ferguson's deficiencies (see Chapter 5). According to Whiteside, Collins's claims for Ferguson had some validity, since the latter found some real roots of irreducible cubics by working with conjugate complex quantities (*Mathematical Papers of Newton,* 2:418–420 n. 101).
39. Newton to Collins, 11 July 1670, 1:31.
40. Newton to Collins, 27 Sept. 1670, *Correspondence of Newton,* 1:43.
41. Newton to Collins, 16 July 1670, 1:35.
42. Newton to Collins, 27 Sept. 1670, 1:43–44.

textbooks had already been published and that he had other academic commitments, he decided against an independent algebra and dutifully completed his "Observations."[43] Then, Kinckhuysen's *Algebra,* supplemented by Newton's "Observations," fell a victim of the vicissitudes of the English publishing industry. In July 1671 Collins reported that Moses Pitts, the publisher who had the rights to the Latin translation of Kinckhuysen's *Algebra,* was "not desirous as yet to put the Introduction to Algebra to the Presse." Tentatively, Collins now suggested that Newton be given as the author of what he had started calling "the Introduction." Under Newton's name, the textbook would "find the better entertainement, and more Speedy Sale."[44] Newton took the suggestion well. He was working on *Tractatus de methodis serierum et fluxionum* (A Treatise of the Methods of Series and Fluxions), an "ambitious exposition of his fluxional calculus" that drew on *De analysi*.[45] He now proposed that *De methodis* be attached to the "Introduction to Algebra," reasoning: "[I]f I must helpe to fill up its title page, I had rather annex somthing wch I may call my owne, & wch may bee acceptable to Artists as well as ye other to Tyros."[46]

In general, in 1671 and 1672 Newton seemed to stand on the brink of sharing his hitherto unpublished researches with the larger mathematical and scientific communities. Having completed his "Observations," he had been moved by Barrow to begin *De methodis* in the winter of 1670–1671, and, the next winter, again at Barrow's suggestion, he started revising his Lucasian lectures on optics for publication.[47] But most of the proposed publications of this period remained unprinted until the next century.[48] The failure of *De methodis,* the optical lectures, and even, to a certain extent, the "Introduction to Algebra" to reach early print can be attributed somewhat to Newton's bad experience with his first published work. It was only in February 1672 that Newton, nearly thirty years old, became a published author when his paper on the theory of colors appeared in the *Philosophical Transactions.* Less than

43. Ibid. According to Whiteside, the "Observations," as reproduced in the *Mathematical Papers of Newton,* was completed by Christmas 1670 ("Introduction" to part 3, *Mathematical Papers of Newton,* 2:287).
44. Collins to Newton, 5 July 1671, *Correspondence of Newton,* 1:66.
45. Westfall, *Never at Rest,* 226.
46. Newton to Collins, 20 July 1671, *Correspondence of Newton,* 1:68.
47. Westfall, *Never at Rest,* 226, 231, 238.
48. The *Opticks* was published only in 1704; Newton's algebra was first published in Latin as *Arithmetica universalis* in 1707; and *De methodis,* in 1736. For publication details on Newton's major works, see Gjertsen, *The Newton Handbook,* 614–623.

two weeks after the paper was read to the Royal Society, he received a harsh and condescending critique from Robert Hooke. Hooke thereby began what was to become for Newton an extremely painful debate on the nature of light. As Westfall has explained, the inability of the paper's critics "to recognize the force of his demonstrations quickly drove Newton to distraction. He was unprepared for anything except immediate acceptance of his theory. The continuing need to defend and explain what he took to be settled plunged him into a personal crisis."[49] The pressure wore on him so heavily that he tried to isolate himself from science and other scientists; he abandoned some of his major publication projects. In May 1672 he thanked Collins for his offer to assist with the publishing of the optical lectures, but stated: "I have now determined otherwise of them [that is, not to publish them]; finding already by that little use I have made of the Presse, that I shall not enjoy my former serene liberty till I have done with it; wch I hope will be so soon as I have made good what is already extant on my account."[50]

The latter was perhaps a reference to his "Observations," which he now reminded Collins was completed and "at . . . [his] command." Indicating that at this point he was still quite interested in publishing his algebraic work, he stated that, should nothing come of the attempt to publish the expanded version of Kinckhuysen's *Algebra,* he himself might "possibly hereafter" publish the "Observations" with *De methodis.* That he found some temporary comfort in algebra as he retreated from optics seems indisputable. In this letter he reminded Collins that his "Observations" included the "discourse concerning invention or the way of bringing Problems to an Æquation," in which he obviously continued to take special pride.[51] Furthermore, in the fall of 1673 he changed the topic of his Lucasian lectures from optics to algebra.

But Newton's continued interest in Kinckhuysen's *Algebra* was not enough to assure its publication. Having received copies of the synopsis of Kersey's *Algebra,* which Pitts was publishing, Newton seemed to question Pitts's commitment to Kinckhuysen's work and offered to see if a Cambridge publisher would purchase Pitts's rights to the latter. Perhaps beginning in July 1672 to back away from publishing his "Observations," he wondered if Pitts thought the *Algebra* ought to be published with or without his notes.[52] In August Collins informed him that Pitts

49. For an elaborate account of this episode, see Westfall, *Never at Rest,* chap. 7, 238–280 (quotation from p. 239).
50. Newton to Collins, 25 May 1672, *Correspondence of Newton,* 1:161.
51. Ibid.
52. Newton to Collins, 13 July 1672, ibid., 1:215.

was "willing to take 3£ for his Interest in K[inckhuysen's] Introd[uction]."[53] Newton himself bought the rights to the work, but failed to find a publisher for it, with or without his "Observations."[54] In any case, in June 1673, deeply hurt by the controversy over light, Newton declared that he "intend[ed] to be no further sollicitous about matters of Philosophy."[55] By then he had turned his research from optics and mathematics to chemistry, alchemy, and theology.[56] He left the "Observations" unprinted, and *De methodis,* unfinished. Furthermore, by the next year he ceased writing to Collins, who in 1675 reported to James Gregory that Newton had embarked on chemical studies and Newton, with Barrow, had begun "to think mathematical speculations to grow at least dry, if not somewhat barren."[57]

4

Despite Newton's inability to find a publisher, his comments on Kinckhuysen's *Algebra* displayed technical prowess and, at times, a distinct pedagogical style. The "Observations" testified to the steady progress that textbook algebra – at the hands of Newton and intermediate authors like Kinckhuysen and Ferguson – had made after the 1640s, when Wallis felt "venturous" in reducing $\sqrt{12}$ to $2\sqrt{3}$. In part 1 of his *Algebra* Kinckhuysen had reduced, added, subtracted, multiplied, and divided (largely) square roots, involving numbers (such as $\sqrt{75}$) and algebraic quantities (such as $\sqrt{a^3 + aab}$).[58] Newton added examples of the reduction of cube roots and gave examples of and some algorithms for the addition, subtraction, multiplication, and division of cube and higher

53. Collins to Newton, 1 Aug. 1672, ibid., 1:226. Pitts also asked for ten copies of the work, if printed.
54. On Newton's purchase and his final attempts to publish the algebra, see Whiteside, "Introduction" to part 3, *Mathematical Papers of Newton,* 2:290–291.
55. Newton to Henry Oldenburg, 23 June 1673, *Correspondence of Newton,* 1:294–295.
56. Westfall, *Never at Rest,* 281–334.
57. Collins to Gregory, 19 Oct. 1675, *Correspondence of Scientific Men of the Seventeenth Century,* ed. Stephen Jordan Rigaud, 2 vols. (Oxford, 1841; reprint, Hildesheim: Georg Olms, 1965), 2:280.
58. For the Latin translation of Kinckhuysen's original, see "Mercator's Latin Version of Kinckhuysen's 'Algebra,'" in *Mathematical Papers of Newton,* 2:295–364; 307–312 (on radicals).

roots, as well as examples of the basic operations on "fractional radicals."[59]

Moreover, he replaced the section on extracting the cube roots of binomials at the end of part 1. In one of the more creative sections of his "Observations" and one that, as Whiteside has concluded, "was the fruit of much hard thought,"[60] Newton developed satisfactory processes for extracting the cube roots of surd binomials (binomials with an irrational term, such as $26 + 15\sqrt{3}$ – one of his examples) and complex binomials (such as $-2 + \sqrt{-121}$ – another of his examples). Whereas Wallis claimed that in his early days as an algebraist he had struggled alone to invent rules for such extractions, Newton here faced a different task, the evaluation of two rules for the extraction of the cube roots of surd binomials: Ferguson's and Descartes–Waessenaer's (a rule due to Descartes but first published by Jacob van Waessenaer, a version of which Schooten had included in his Latin edition of *La géométrie*).[61] Newton realized, however, that neither of these rules for surd binomials had been formulated in a fashion to include complex binomials. After some wavering, he opened his section on the "Rule for Extracting the Cube Root of Binomials" with Ferguson's rule. Then, a few pages later he began to develop his own algorithm, which he described as "neater and more universal." His algorithm was "universal" since, as he soon showed, it could be extended to complex binomials. "Indeed, the roots of complex binomials may be extracted by this method," he explained, "and, since some may perhaps judge it pertinent to the store of analysis inasmuch as by its means all cubic equations come out reducible by Cardan's rules, I will not be reluctant to elucidate it."[62]

Thus the new rule for extracting cube roots of complex binomials put Newton in a position to reject Kinckhuysen's implication that Cardano's formula for the cubic equation did not apply to the irreducible case and that solution of the irreducible case required Descartes's geometric, "trisection" method.[63] In a later section, Newton reiterated this point:

59. Newton, "Observations," 2:370–373. A fractional radical has a fractional radicand.
60. Whiteside, *Mathematical Papers of Newton*, 2:376 n. 17.
61. On Descartes–Waessenaer's rule, see ibid., 2:314–315 n. 39.
62. For Newton's "Rule for Extracting the Cube Root of Binomials," as well as Whiteside's indispensable explanations, see Newton, "Observations," 2:376–397 (quotations from pp. 383, 393). Note Newton's use of Oughtred's metaphor of "the analytical store."
63. For Kinckhuysen's statement and Whiteside's comments, see "Mercator's Latin Version of Kinckhuysen's 'Algebra,'" 2:350–352.

"if $1/4q^2$ be not greater than $1/27p^3$, the root specified by these rules is in fact impossible; and yet by the rule for extracting the [cube] roots out of impossible binomials delivered above real roots may be obtained from them." Using the latter rule, he then showed specifically that the equation

$$x^3 = 15x - 4$$

has three real roots. Cardano's formula gave:

$$x = \sqrt[3]{-2 + \sqrt{-121}} + \sqrt[3]{-2 - \sqrt{-121}}.$$

By Newton's extraction rule,

$$x = -2 + \sqrt{-1} - 2 - \sqrt{-1} = -4.$$

Reducing the given equation by the factor $x + 4$, he obtained a quadratic equation that generated two additional real roots.[64]

Besides displaying such technical prowess (he also qualified Kinckhuysen's statement of Descartes's rule of signs along the lines of his notebook entries of 1665),[65] Newton evidenced genuine concern and talent for the pedagogy of algebra. His concern shone in some revisions to part 2 of Kinckhuysen's *Algebra,* where for example he defined an equation and carefully described how to put equations into Descartes's standard form,[66] and especially in the new introduction he wrote for part 3. This was Newton's discourse on reducing problems to equations, to which he had referred at least three times in his correspondence with Collins. This discourse in 1670 seemed to tempt him to undertake a "compleate introduction to Algebra," and in the spring of 1684 he would highlight it as one of his original pedagogical achievements.[67]

Here Newton was concerned particularly with inspiring students to deeper algebraic studies. "After the novice has exercised himself some little while in algebraic computation . . . ," he explained, "I judge it not unfitting that he test his intellectual powers in reducing easier problems

---

64. Newton, "Observations," 2:420–421. Whiteside notes that Newton made an arithmetic mistake, found the wrong quadratic equation, and so gave two erroneous real roots (ibid., n. 104).

65. Ibid., 2:412–415. Newton, however, did not give his full rule for enumerating complex roots.

66. Ibid., 2:396–399.

67. Newton to Collins, 16 July 1670, 1:35; Newton to Collins, 27 Sept. 1670, 1:43–44 (quotation on p. 43); Newton to Collins, 25 May 1672, 1:161; and Newton, "The 'First Book of Universal Arithmetick,' " [Spring 1684?], *Mathematical Papers of Newton,* 5:538–621, on 565.

to an equation, even though perhaps he may not yet have attained their resolution." The objective of setting students at an early stage on the writing of equations was to give them an incentive to study the more difficult aspects of equation theory. Having learned to translate problems into equations, they would "with greater profit and enjoyment contemplate the nature and properties of equations and learn their algebraic, geometrical and arithmetical resolutions."[68]

The merits of this approach thus sketched, Newton followed with an introduction to the writing of equations that explained in an elementary and still masterly way the key role of the symbolical style in algebra. He presented the process of reducing prose problems to symbolical equations as one of "translation" – "the translation from latin or any other language in which the problem is proposed into that of algebra . . . , that is, into characters suitable to denote our concepts of quantitative relationships." Generally, he (here following Kinckhuysen) advised students to try to designate all the conditions of a given problem by an equal number of equations. Then, evidencing his special pedagogical commitment to examples, he set two exemplary problems and patiently worked toward the equations corresponding to each. To ease the student's transition from prose to algebraic symbols, he divided each "question expressed verbally" into its constituent parts and next to each part gave "the algebraic expression of the same."[69]

The first problem, which Newton took from Kinckhuysen but worked in his more pedagogically transparent fashion, "required three continuously proportional numbers whose sum is 20 and the sum of their squares 140." He proceeded as follows:

Left column
line 1: The question expressed verbally
line 2: There are required three numbers subject to these conditions:
line 3: they must be continuously proportional,
line 4: their sum total must be 20,
line 5: and the sum of their squares 140.

Right column
line 1: The algebraic expression of the same
line 2: x, y, z.
line 3: $x{:}y = y{:}z$ or $xz = y^2$.
line 4: $x + y + z = 20$.
line 5: $x^2 + y^2 + z^2 = 140$.

68. Newton, "Observations," 2:422–425.
69. Ibid., 2:424–427.

Since he aimed here solely at illustrating the translation of a problem
into algebraic equations, he did not solve the given problem. Rather he
laid down an additional guideline for the translation process: "But it
should be noted that the solutions to questions come out the more
speedily and skilfully the fewer the unknown quantities supposed at the
beginning." Using the preceding problem to illustrate this point, he
noted that $z = y^2/x$, and thus the problem and its equations could be
restated:

Left column
line 1: The question expressed verbally
line 2: There are sought three numbers in continued proportion,
line 3: whose sum is 20
line 4: and the sum of their squares 140.

Right column
line 1: The algebraic expression of the same
line 2: $x, y, y^2/x$.
line 3: $x + y + y^2/x = 20$.
line 4: $x^2 + y^2 + y^4/x^2 = 140$.[70]

After his second example, which centered around a merchant's capital,
Newton offered some final advice on translating from prose into algebraic symbols. Sometimes, he observed, for prose expressions there
seemed to be no equivalent algebraic terms. In those cases, students had
to be aware of the idioms peculiar to different prose languages, and to
translate from the "sense" of the question and not from its literal terms.
His general advice thus completed, he turned students loose on the
problems, equations, and solutions, with which Kinckhuysen had ended
the original *Algebra*. He added: "[S]o that I may impress an intimate
understanding of this method of reducing problems of this sort to equations and illustrate it, and because craft skills are more easily learnt by
example than by precept, it seems appropriate to adjoin the solutions of
the following problems."[71]

Besides highlighting Newton's technical and pedagogical abilities, the
"Observations" testified to his interest in applying algebra to geometry
and his willingness to address such applications in an algebra textbook,
but in a limited way. Although in his *Algebra* Kinckhuysen seems to
have purposely set only (what Newton termed) "Algebraick problems,"

70. Ibid.   71. Ibid., 2:426–429.

Newton added eight "geometrick ones."[72] He added his geometric problems right after the arithmetic ones and with no explanation. That is, he did not choose to cordon off geometric applications from arithmetic and algebra per se. Still, unlike Descartes, Newton posed only a few geometric problems, translated them into equations, and solved those equations algebraically. Thus the geometric content of the "Observations" was minimal.

5

Although, according to Collins, Newton's "Observations" would "have much obliged the young Students of Algebra"[73] of the 1670s, it is likely that only a few close colleagues and those students who attended the Lucasian lectures of 1673 to 1683 learned anything of his early algebraic work. How many students Newton taught during this decade, and exactly what he lectured on, remain open questions. By 1683, despite the Lucasian statutes, he had given only his optical lectures of 1670–1672 to the Cambridge University Library. When his algebra lectures were drawing to a close, he finally composed a manuscript on the subject. His "Lectures on Algebra" – which was an elaboration and, in some respects, significant revision of his "Observations" and which would form the core of *Universal Arithmetick* – was probably written during winter of the academic year, 1683–1684,[74] and was deposited in the library that spring. Perhaps he was moved to compose the "Lectures" by the circulation in 1683 of Wallis's "Proposal about Printing a Treatise of Algebra." More definitely, as Westfall has pointed out, Newton found himself in 1683 with an amanuensis, Humphrey Newton, who eased the burden of formalizing his second series of Lucasian lectures.[75] Soon after depositing the "Lectures," Newton began to revise the manuscript, seemingly with publication in mind.[76]

---

72. Ibid., 2:428–445 (quotation from p. 429). Whiteside shows that some of these geometric problems were inspired by Descartes and Schooten; Scriba notes that Newton added geometric problems to Kinckhuysen's algebraic ones (Scriba, "Mercator's Kinckhuysen-Translation," 49).
73. Collins to Newton, 13 July 1670, 1:32.
74. Whiteside, "Introduction" to part 1, *Mathematical Papers of Newton*, 5:5.
75. Westfall, *Never at Rest*, 398.
76. Here Newton addressed the "reader" (Whiteside, "Introduction" to part 2, *Mathematical Papers of Newton*, 5:535).

During 1683–1684 he twice reworked his much valued discourse on reducing problems to equations, once for the "Lectures" and, again, for the revised manuscript of 1684.[77] In both cases he continued to reflect on language, algebra, and the parallels between the two. By the revised manuscript he was writing that in the "language [of algebra] quantities fill the rôle of words and equations that of sentences." Repeating his claim to pedagogical novelty, he was boasting that the "finding of equations . . . [is] the part of this [algebraic] art which is by far the most difficult and yet is explained by no one."[78]

If in his earlier "Observations" he had attended primarily to the finding of equations for arithmetic problems, his "Lectures" made it clear that by the late 1670s and early 1680s geometric problems were as much, or more, on his mind than arithmetic ones and that he was struggling with the question of the relative merits of algebra and geometry in solving problems that were geometrically motivated. For the "Lectures" Newton wrote two significantly new and lengthy sections that betrayed these concerns: "How Geometrical Questions Are to Be Reduced to an Equation" (designed to follow the general discourse on reducing problems to equations) and "The Linear Construction of Equations."[79] The former section began with what he described as an "easy" example, a geometric question that could be reduced to an equation "by the same procedures as those propounded in regard to abstract quantities." The question was proposition 11 of book II of Euclid's *Elements*, the very problem with which Oughtred had introduced chapter 19 of *The Key* ("Examples of Analytical Æquations, for inventing of Theoremes, and resolving of Problemes").[80] Following this example, Newton explained that the setting of equations for geometric questions was frequently more difficult: "in geometrical situations which more often occur they tend so to be dependent on a variety of positions of lines and their complex inter-relationships as to need further skilful manipulation to be reducible to algebraic terms." Explicitly trying to "smooth the way for the beginning student," he noted that the route to

77. Newton, "Newton's Lectures on Algebra during 1673–1683," [Winter 1683–1684?], ibid., 5:54–517, on 128–135, and Newton, "First Book of Universal Arithmetic," 5:564–567.
78. Newton, "First Book of Universal Arithmetic," 5:565.
79. Newton, "Lectures on Algebra," 5:158ff. and 5:420ff. Some of the latter section was adapted from Newton, "Problems for Construing Æquations," [1670?], *Mathematical Papers of Newton*, 2:450–517.
80. Ibid., 5:159; William Oughtred, *The Key of the Mathematicks, New Forged and Filed* . . . (London: Thomas Harper, 1647), 81.

equations for some geometric problems was through the relationship between the sides of similar triangles (the ratio between the sides giving the equation); in others, the Pythagorean theorem; and so on.[81] Having offered his advice for writing equations for "problems in the geometry of straight lines," he turned to "geometrical curves," even providing an inadequate sketch of Descartes's analytic geometry.[82] Whereas the geometric content of the early "Observations" had been minimal, that of the "Lectures" was then extensive, prominent, and, as will be discussed, confusing.

Whatever his plan for the revised version of the "Lectures" was in the spring of 1684, the manuscript, like so many of Newton's other projects, was pushed aside as he found a new focus for his genius. According to the frequently told story, in August of the year Edmond Halley visited Newton and asked him what the curve of a planet would be if the force holding the planet in its orbit was inversely proportional to the square of the distance from the sun. After replying "an ellipse," Newton threw himself into the research and writing that would lead in 1687 to the publication of the *Principia*.[83]

In the long run, some of Newton's colleagues at Cambridge wanted more than a library copy of his algebraic lectures. During the English elections of 1705, when he was master of the Mint and in the midst of a gloomy campaign for a seat in Parliament, these colleagues pledged their support for his candidacy in return for a contribution of sixty pounds from Newton to Trinity College and his agreement that William Whiston be permitted to publish his "Lectures on Algebra."[84] Although he ran last in a field of four, he was still bound to let his lectures appear in print. According to his own accounts, he neither sanctioned nor participated in preparation of the lectures for publication. Rather Whiston, his handpicked successor as Lucasian professor, reworked the deposited "Lectures on Algebra" – taking some notice of the revised version of 1684 but largely following the "Lectures" – into the anonymous *Arithmetica universalis* of 1707, and from a distance Newton criticized Whiston's efforts. When the book was in press, Newton told David Gregory that he had "not seen a sheet of it," could not "well remember the contents of it" (presumably his "Lectures on Algebra"), and intended, when it was printed, to go to Cambridge to buy up the whole edition if "it did

81. Newton, "Lectures on Algebra," 5:158–167 (quotations from p. 159).
82. Ibid., 5:180–185. Whiteside notes that Newton here failed to establish basic coordinate systems (ibid., 5:180–181 n. 206).
83. Westfall, *Never at Rest*, 402–403.
84. Ibid., 624–626.

not please him."[85] Although Newton professed dissatisfaction with the published work, he did not fulfill his threat. Rather, in 1722, following publication of an English translation of the text two years earlier, he prepared his own Latin edition of the work, ostensibly to correct the many errors that he accused Whiston of introducing into the original edition. But, as Whiteside has noted, the revisions that Newton incorporated into this second Latin edition were relatively minor: he corrected obvious misprints, inserted new running heads, and reordered the geometric problems of the book's fourth section.[86]

6

Newton's *Universal Arithmetick,* which thus nearly escaped publication but achieved great popularity and influence, evidenced a concern for the practice of algebra rather than for lengthy exploration of its foundations. In comparison with the algebras of Kersey and especially Wallis, Newton's was ahistorical and aphilosophical. It offered sparse explanations of key concepts, succinct rules, and multiple examples – all solved in careful detail. Thus the book's prefatory note "To the Reader," which appeared in later editions, stressed that

> it was but small in Bulk, and yet ample in Matter, not too much crowded with Rules and Precepts, and yet well furnished with choice Examples, (serving not only as Praxes on the Rules, but as Instances of the great Usefulness of the Art it self; and, in short, every Way qualified to conduct the young Student from his first setting out on this Study).

Newton himself defended its style on pedagogical grounds: "For in learning the Sciences Examples are of more Use than Precepts. Wherefore I have been the larger on this Head."[87]

Given his pedagogical philosophy and pragmatic mathematical orientation, Newton generally confined treatment of the larger questions

85. Whiteside, "Introduction" to part 1, *Mathematical Papers of Newton,* 5:9–14 (quotation of Gregory's account from pp. 9–10). Whiteside suggests that Newton may have had a hand in the *Ad Lectorem* to the work, which stressed that it was based on lectures delivered almost thirty years before and not intended for publication (ibid., 5:12–13). See Anon., "To the Reader," in Newton, *Universal Arithmetick,* ii (5).
86. Whiteside, "Introduction" to part 1, *Mathematical Papers of Newton,* 5:13–14.
87. Anon., "To the Reader," in Newton, *Universal Arithmetick,* iii (5); Newton, *Universal Arithmetick,* 189 (100).

about the nature of early modern algebra to two introductory paragraphs and an introductory section, "Of the Signification of Some Words and Notes." Here his foundational reflections were synthetic and somewhat pedestrian rather than original and profound, and few and succinct rather than many and expansive. Following the arithmetic tradition that had come through Oughtred and Harriot to Wallis and himself, he subsumed arithmetic and algebra under the general heading of "the Science of Computing," and declared that they were "both built on the same Foundations, and aim at the same End." The chief differences concerned degrees of generality and methods. Echoing Viète and his early followers, he noted that arithmetic computation used numbers whereas algebraic computation used species, and that the ends of arithmetic were definite and particular whereas those of algebra were indefinite and universal (hence the name "universal arithmetick"). He explained furthermore that the method of analysis – to which he did not refer by name – played a special role in algebra:

> Algebra is particularly excellent in this, that whereas in Arithmetick Questions are only resolv'd by proceeding from given Quantities to the Quantities sought, Algebra proceeds in a retrograde Order, from the Quantities sought, as if they were given, to the Quantities given, as if they were sought, to the end that we may some way or other come to a Conclusion or Equation, from which one may bring out the Quantity sought.

He, like Wallis, explained that through analysis "the most difficult Problems are resolv'd, the Resolutions whereof would be sought in vain from only common Arithmetick."[88]

Thus went Newton's tight and rather prosaic introduction, which was followed by an eight-page essay in which he defined a few key terms and described the basic symbols of algebra. It was here that he established not only whole numbers but fractions, surds, and (what he called) "affirmative and negative quantities" as algebraic objects. He opened with a paragraph on the definition of number, which in content was somewhat reminiscent of Barrow's view of the subject but in length (two sentences) spoke tellingly of Newton's avoidance of the mathematical philosophizing in which his Lucasian predecessor had seemed to revel.[89] Although he mentioned nowhere the view that numbers were mere signs (or that arithmetic was subservient to geometry), he wrote that "By

88. Newton, *Universal Arithmetick*, 1–2 (6–7).
89. Ibid., 2 (7). Behind Newton's brevity there was possibly the understanding that those interested in the philosophy of number could read Barrow's lectures.

Number we understand not so much a Multitude of Unities [Euclid's definition], as the abstracted Ratio of any Quantity, to another Quantity of the same Kind, which we take for Unity. And this is threefold; integer, fracted, and surd." According to Newton, if a quantity could be measured by unity, then its ratio was an integer; if it could be measured only by a "submultiple" part of unity, its ratio was a fraction; and, if it could not be measured, its ratio was a surd.[90]

There followed paragraphs on the symbolical style and on "affirmative and negative quantities." In three sentences he told students to denote quantities by species or letters if they were "unknown, or look'd upon as indeterminate"; unknown quantities were to be denoted by the last letters of the alphabet and known quantities, by the initial letters. Then, his mathematical pragmaticism shining through, he introduced the negatives, with no apologies and no philosophical reflections. Unlike Kersey, for example, he did not describe negative numbers as "fictitious"; unlike Wallis, he did not describe them as "imaginary." He simply defined these terms, and offered practical examples of each. He wrote: "*Quantities are either* Affirmative, *or greater than nothing*; *or* Negative, *or less than nothing*." Three examples followed:

> Thus in humane Affairs, Possessions or Stock may be call'd affirmative Goods, and Debts negative ones. And so in local Motion, Progression may be call'd affirmative Motion, and Regression negative Motion; because the first augments, and the other diminishes the Length of the Way made. And after the same Manner in Geometry, if a Line drawn any certain Way be reckon'd for Affirmative, then a Line drawn the contrary Way may be taken for Negative.[91]

There was, on the surface, little new in Newton's treatment of the negatives except his ignoring of the difficulties that these numbers had posed for earlier algebraists. The terms "affirmative" and "negative" had already appeared in Pell's English edition of Rahn's *Algebra* and in Kersey's *Algebra*. Furthermore, his definitions of affirmative and negative quantities were similar to those that Kersey had imported into English algebra from Descartes's *Géométrie*;[92] even Barrow had alluded to these definitions when he distinguished between positive and negative terms. Neither were Newton's examples original. Barrow had used both

90. Ibid.   91. Ibid., 3 (7).
92. See John Kersey, *The Elements of That Mathematical Art Commonly Called Algebra, Expounded in Four Books,* 2 vols. (London: William Godbid, 1673–1674), 1:3, 269.

geometric lines and motion to exemplify positive and negative terms; Cardano and Kersey had used debts to exemplify the negatives, with Kersey's explaining explicitly that " + 5 *l*. may represent five pounds in money, or the Estate of some person who is clearly worth five pounds; so − 5 *l*. may represent a Debt of five pounds."[93]

Still, the author of *Universal Arithmetick* was "the great Sir Isaac" and so his definitions and examples enjoyed an inflated significance. Interestingly, he perhaps complicated the case for the negatives by a slight modification in Kersey's definition: whereas Kersey had written of − 5 as a "*number* less than nothing by 5,"[94] Newton wrote of "*quantities* less than nothing." Although he so defined the negatives, Newton perhaps meant otherwise. As his examples, like Kersey's, imply, he thought not so much of inherently affirmative or negative quantities (or quantities actually greater than or less than nothing), but rather of quantities that may "be call'd" or may "be reckon'd" affirmative or negative. But Newton's definition captured nothing of this subtlety, and so, for the rest of the century, some of his literalist English students were forced to try to make some sort of philosophical peace with the idea of *quantities* that really were "less than nothing."

Newton unobtrusively completed his endorsement of the negative numbers by discussing in turn the signs + and −, subtraction leading to a negative result, and the law of signs. First he carefully explained that there were two uses of the signs + and −. "A *negative* Quantity is denoted by the Sign −; the Sign + is prefix'd to an *affirmative* one." But, if the same signs appear in an "Aggregate of Quantities," he continued, then "*the Note + signifies, that the Quantity it is prefix'd to, is* to be added, *and the Note* −, *that it is* to be subtracted." In providing illustrations of the second use of these signs, he introduced subtraction involving a subtrahend greater than the minuend. "And 5 − 3, or 5 less 3, denotes the Difference which arises by subducting 3 from 5, that is 2: And − 5 + 3 signifies the Difference which arises from subducting 5 from 3, that is − 2." Thus, what earlier algebraists, including Wallis, had highlighted as an "impossible subtraction" was for Newton just another example of subtraction. Later in the book, he gave the rule of signs, again without any apology or justification but with multiple examples.[95]

93. Ibid., 1:3.
94. Ibid. The emphasis is mine.
95. Newton, *Universal Arithmetick*, 3 (7), 16 (14).

7

Once the basic definitions and symbols were established, the body of *Universal Arithmetick* broke down into five main sections. The five sections and their ordering came from the "Lectures," the sixth section of which had however been transformed by Whiston into an appendix to *Universal Arithmetick*. The first section offered the rules for the basic algebraic operations as well as for what Newton termed "the invention of divisors," the reduction of fractions to a common denominator, and the reduction of radicals.[96] The book's second main section dealt with the standardizing and reducing of equations. Newton opened it with a definition of equations as "Ranks of Quantities either equal to one another, or, taken together, equal to nothing." He then explained that the root(s) of an equation "may be the more easily discovered, . . . [if] the Equation . . . be transformed most commonly various Ways, until it becomes the most Simple that it can." He laid down the standard guideline for ordering the terms of an equation according to the "Dimensions of the unknown Quantity," and then seven, rather standard, rules for reducing an equation.[97] Finally, Newton gave rules for the "Transformation of Two or More Equations into One," which constituted an early systematic attempt to solve simultaneous linear equations through elimination.[98]

The third section featured the revised discourse on the setting of equations that Newton had written for his "Lectures on Algebra," replete with the detailed verbal and algebraic analyses of the two exemplary problems that he had first used in his "Observations."[99] The fourth section concerned the translating of geometric problems into equations. It took its title and, with some rewriting, its content from the

96. Ibid., 9–54 (10–33).
97. Ibid., 55–60 (33–36) (quotations from p. 55 [33]). In annotating Newton's "Lectures on Algebra," Whiteside notes that these rules derive, some more loosely than others, from Kinckhuysen's *Algebra* (*Mathematical Papers of Newton*, 5:112–113 n. 92).
98. Newton, *Universal Arithmetick*, 60–67 (36–39). See also Carl Boyer, "Cartesian and Newtonian Algebra in the Mid-Eighteenth Century," *Actes du XIᵉ congrès internationales d'histoire des sciences* (Warsaw, 1968), 3:195–202, on 201.
99. Newton, *Universal Arithmetick*, 67–85 (39–48), esp. 67–70 (39–41). Newton's 1684 revision of this section, offering verbal and algebraic analyses of additional problems ("First Book of Universal Arithmetick," 5:564ff.), was ignored.

section on "How Geometrical Questions Are to Be Reduced to an Equation" that Newton had composed for the "Lectures." As had the "Lectures," *Universal Arithmetick* offered many more geometric problems than arithmetic ones: sixteen "arithmetic questions" stretched across sixteen pages of its third section and sixty-one "geometrical questions," across nearly ninety pages of the fourth section.[100]

In the last major section of *Universal Arithmetick* ("How Equations Are to Be Solved," also taken from the "Lectures on Algebra"), Newton again displayed the technical skill that had shone through the examples of the third and fourth sections. He showed here, too, that in the seventeenth and early eighteenth centuries even towering genius was insufficient to assure that a mathematician would make peace with the imaginaries of the newly expanded algebraic universe. For preliminaries, he defined the term "Root" as "a Number which being substituted in the Equation for the Letter or Species signifying the Root, will make all the Terms vanish." He gave examples showing that specific (both affirmative and negative) roots satisfied specific equations, and, using geometric examples, he explained why "the same Equation may have several Roots." Then, quickly moving to the more advanced points of equation theory, he stated (without proof) the weak version of the fundamental theorem: *"an Equation may have as many Roots as it has Dimensions, and not more." "But the Roots are of two Sorts,"* he continued, "Affirmative ... *and* Negative.... *And of these some are often* impossible."[101] Here, curiously, the mature Newton appeared more conservative on the issue of imaginary numbers than the young Newton, who in 1665–1666 had, following Descartes, stated the strong version of the fundamental theorem: "Every Equation hath soe many roots as dimensions of wch some may be true some false & some imaginary or impossible."[102] The change in the verb form used in the theo-

100. For the fourth section, including the "geometrical questions," see Newton, *Universal Arithmetick*, 86–189 (49–100); for the "arithmetic questions," see ibid., 70–85 (41–48). For the many printed sources from which Newton took his geometric examples and technical explanations of the examples, see Whiteside's notes on Newton's "Lectures on Algebra" (*Mathematical Papers of Newton*, 5:184–336). Newton seems to have added examples to the "Lectures" "piecemeal at whim, without system or regard to any grading by relative difficulty." They were so published by Whiston but reordered by Newton for the second Latin edition of *Universal Arithmetick* (ibid., 5:184–185 n. 216).

101. Newton, *Universal Arithmetick*, 190–192 (101–102).

102. Newton, "Researches in the Theory of Equations," 1:520–521.

rem – "may have" in place of "hath" – and the exclusive use of the term "impossible" in place of "imaginary" to describe roots involving $\sqrt{-1}$ spoke of a growing tentativeness toward such roots.

Although Newton seems not to have explained why by the time of the Lucasian lectures he came to see these roots as basically impossible rather than imaginary, his manuscripts – (1) the algebra notes of 1665–1666, (2) the "Observations" on Kinckhuysen's *Algebra,* and (3) the "Lectures on Algebra," from which *Universal Arithmetick* was taken – permit a partial reconstruction of the evolution of his position on imaginary numbers through his long mathematical career. His previously quoted statement of 1665–1666 on the fundamental theorem was the strongest of his career. In the "Observations," moreover, he let stand Kinckhuysen's close translation of Descartes's strong version of the fundamental theorem. In *La géométrie* Descartes had written: "Au reste tant les vrayes racines que les fausses ne sont pas tousiours reeles; mais quelquefois seulement imaginaires; c'est a dire qu'on peut bien tousiours en imaginer autant que iay dit en chasque Equation; mais qu'il n'y a quelquefois aucune quantité, qui corresponde a celles qu'on imagine."[103] In the "Observations" Newton echoed: "Both true and false roots are not always real but on occasion merely imaginary, that is, there may in any equation be conceived to be in imagination as many roots as we mentioned just now, but it may happen on occasion that there is no quantity which accords with those concepts."[104] In short, in his algebra notes of 1665–1666 and "Observations," Newton seemed largely content to deal with roots involving $\sqrt{-1}$ in Cartesian terms: he let stand the term "imaginary roots"; he associated such roots with the exercise of the human imagination; and he gave strong versions of the fundamental theorem.

Yet, even in the late 1660s and early 1670s he was no slave to Descartes's view of the imaginaries, as proved by a few suggestive additions he made to Descartes's and Kinckhuysen's expositions of the subject. In the notes of 1665–1666 and the "Observations," he referred to roots involving $\sqrt{-1}$ as "impossible" as well as "imaginary."[105] And in the "Observations" he explicitly qualified Descartes's reference to imaginary roots as positive and negative: "those [imaginary] roots, properly

103. Descartes, *The Geometry,* 174.
104. Newton, "Observations," 2:412–413.
105. Kinckhuysen had referred to these roots as "imaginary" ("Mercator's Latin Version of Kinckhuysen's 'Algebra,' " 2:323); Newton referred to them as "imaginary or impossible" (Newton, "Observations," 2:412–413, e.g.).

speaking, are neither true nor false but may be assumed indifferently to be one or the other." Neither true nor false, such roots appeared as true or false when Descartes's unqualified rule of signs was applied to certain equations – the latter, a point made by Newton in the "Observations" as well as his earlier algebra notes.[106]

In composing the manuscript of his Lucasian lectures on algebra, Newton recast his views on roots involving $\sqrt{-1}$. He did not completely shake off Descartes's and Kinckhuysen's influence, but the revisions to his earlier "Observations" brought him more in line with the conservative view of such roots that had gained momentum in England during the late 1660s and early 1670s. In a nutshell, he now gave a weaker version of the fundamental theorem and, like Pell and Kersey, presented roots involving $\sqrt{-1}$ as fundamentally impossible. "An equation can, in truth," he began, "have as many roots as it possesses dimensions, and no more." He then wrote that roots "are of two groups" – affirmative and negative (he no longer used Descartes's terms "true" and "false") – and that "[o]f these [roots], indeed, several not infrequently prove to be impossible." This employment of the term "impossible" to the exclusion of the term "imaginary" was a key change from his "Observations." Testifying further to his solidifying view of such roots as signs of impossibility, he here (in a paragraph that had no equivalent in the "Observations") explained that the two roots, $a + \sqrt{a^2 - b^2}$ and $a - \sqrt{a^2 - b^2}$, of the equation $x^2 - 2ax + b^2 = 0$

> are assuredly real when a^2 is greater than b^2, but when a^2 is less than b^2 they prove to be impossible ones, because $a^2 - b^2$ will then prove to be a negative quantity and the square root of a negative quantity is impossible. For every possible root, be it positive or negative, if it be multiplied by itself will produce a positive square; and consequently one which has to produce a negative square will be impossible.

Although $\sqrt{a^2 - b^2}$, where a was less than b, was impossible, he noted, the entities $a + \sqrt{a^2 - b^2}$ and $a - \sqrt{a^2 - b^2}$ were still roots of the given equation, since, when each was substituted in the original equation, it made "all its terms mutually destroy one another."[107]

Taking account of his new emphasis on the impossibility of such roots,

106. Newton, "Observations," 2:412–413.
107. Newton, "Lectures on Algebra," 5:340–343. Note that in the "Lectures" Newton wrote of quantities that were "affirmativæ" and "negativæ"; Whiteside translates the former as "positive." Also, although Newton gave both terms "imaginary" and "impossible" in a section of his incomplete, revised version of the "Lectures," he still presented such roots as basically impossible. Thus, roots "may on occasion lose their possibility,

Newton changed the language but not the content of his discussion of the rule of signs as he moved from the "Observations" to the "Lectures." In the latter he reiterated that the original rule did not apply to equations with impossible roots; that such roots were "neither negative nor positive [affirmative]"; but that "impossible roots lie hidden among the positive [affirmative] roots or among the negative ones"; and that his rule for enumerating impossible roots could ascertain if they were among the affirmative or the negative roots.[108]

To a certain extent, in his "Lectures" Newton seemed to want his cake and to eat it too, or to want to accept impossible roots and yet to exclude them from the universe of algebraic objects. Equations did and did not have as many roots as their dimensions, he seemed to say. The entity $a + \sqrt{a^2 - b^2}$, where a was less than b, was a root of the equation $x^2 - 2ax + b^2 = 0$, since, as mandated by the definition of root, it satisfied the given equation. But, as a root, it was impossible, since "every possible root ..., if it be multiplied by itself will produce a positive square." Even his earlier concession to the limits of human reasoning seemed not to cover the case of such roots, since (as he, like Kersey, indirectly emphasized) such roots were seen as not merely inexplicable but contradictory as well – they did not fit with the law of signs. Why, then, Newton asked himself, did equations generate impossible roots? What was their purpose? "It is fair, however," he concluded, "that the roots of equations are often impossible; otherwise they would display as possible cases of problems which are often impossible."[109]

Following Newton's "Lectures," *Universal Arithmetick* offered (as noted earlier) a weak version of the fundamental theorem. In addition, the textbook merely repeated (with a few insignificant changes) the explanation of impossible roots that had been included in the "Lectures." Thus in *Universal Arithmetick* Newton began with the example of the equation $xx - 2ax + bb = 0$, worked his way to the statement that the appearance of roots involving $\sqrt{-1}$ was "just" since some problems were impossible, and gave his expanded rule of signs.[110]

At the end of the textbook's fifth section, after dealing with the transmutation of equations (including transforming negative roots into affirmative roots and vice versa) and the limits of equations, Newton had a final confrontation with impossible roots. Having referred to the demands of brevity, he presented a short, somewhat cryptic explanation

and then they are *usually said to be* imaginary or impossible" (Newton, "First Book of Universal Arithmetick," 5:562–563; my emphasis).

108. Newton, "Lectures on Algebra," 5:346–353.
109. Ibid., 342–343.
110. Newton, *Universal Arithmetick*, 191–198 (101–105).

of the algebraic solution of the cubic equation. "[T]here are some Reductions of *Cubick Equations* commonly known, which, if I should wholly pass over," he wrote, "the Reader might perhaps think us deficient." He carefully explained that any cubic could be reduced to the form:

$$x^3 + qx + r = 0.$$

He then supposed that the root of this equation was $x = a + b$, and, pursuing a variant of Cardano's formula, found

$$a^3 = -\frac{1}{2}r \pm \sqrt{\frac{1}{4}rr + \frac{q^3}{27}} \text{ and } b = -\frac{q}{3a}.$$

"And after this Way," he explained, "the Roots of all Cubical Equations may be extracted wherein q is Affirmative; or also wherein q is Negative, and $\frac{q^3}{27}$ not greater than $\frac{1}{4}rr$, that is, *when two of the Roots of the Equation are impossible.*" There remained of course the irreducible case, which Newton now declared irresolvable by Cardano's formula:

But where q is Negative, and $\frac{q^3}{27}$ at the same time greater than $\frac{1}{4}rr$, $\sqrt{\frac{1}{4}rr - [sic] \frac{q^3}{27}}$ becomes an impossible Quantity; and so the Root of the Equation x or y will, in this Case, be impossible, *viz.* in this Case *there are three possible Roots,* which all of them are alike with respect to the Terms of the Equations q and r, and are indifferently denoted by the Letters x and y, and consequently all of them may be extracted by the same Method, and expressed the same Way as any one is extracted or expressed; *but it is impossible to express all three by the Law aforesaid.*[111]

With respect to the imaginary numbers, then, Newton's public legacy was a mixed bag: in some respects, it was progressive; in others, conservative; and, undoubtedly, it was confusing for the many eighteenth-century students who labored through *Universal Arithmetick.* He recognized the utility of roots involving $\sqrt{-1}$: he stressed that they indicated when problems were impossible and that their enumeration entered into the determination of the number of affirmative and negative roots of equations. He gave, after all, his rule for determining the number of complex roots of a given equation, arguably "the most remarkable passage" in *Universal Arithmetick.*[112] Yet, like Kersey, Newton pre-

111. Ibid., 220–222 (116–117); for the parallel argument, see Newton, "Lectures on Algebra," 5:410–411.
112. H. W. Turnbull, *The Mathematical Discoveries of Newton* (London: Blackie and Son, 1945), 49.

sented the imaginaries basically as impossible roots. He appeared to see these roots as neither numbers nor quantities. He ignored the imaginaries in his introductory section on the "Signification of Some Words and Notes." In the body of his textbook, he referred repeatedly to "impossible roots," but never to an impossible number and only rarely to "an impossible Quantity."[113]

On top of this, at the end of the main text of *Universal Arithmetick* he left algebra students with but a garbled account of the algebraic solution of the irreducible case of the cubic equation. Instead of correcting the implications of some earlier algebraists, including Kinckhuysen, that Cardano's solution did not cover the irreducible case – an implication that (as we have seen) Newton had earlier challenged in his "Observations" – he now argued this very point. Quite naturally, Newton's changing view of roots involving $\sqrt{-1}$ had affected his appreciation of Cardano's solution. As he prepared his "Observations," he, like Descartes, had seen these roots as entities to be imagined, and, armed with his algorithm for extracting the cube roots of complex binomials, he had carefully illustrated how Cardano's formula led to real roots of irreducible cubic equations. But as he came to see roots involving $\sqrt{-1}$ as signs of impossibility, which was the view of the matter he took in his "Lectures" and hence in *Universal Arithmetick,* he could no longer apply Cardano's formula to the irreducible case. He could, in fact, no longer use his algorithm for extracting the cube roots of complex binomials, and he did not include the algorithm in his textbook. As the previously quoted passage explained, he could no longer pretend to extract the real or possible from the impossible. He said not only that $\sqrt{\frac{1}{4}rr - \frac{q^3}{27}}$ "becomes an impossible Quantity" but also that "so the Root of the Equation x or y will . . . be impossible."

8

Newton's remarks on the relationship between algebra and geometry probably also bewildered and tantalized students rather than clarifying and settling the issue for them. He revealed mixed sympathies – with

---

113. Newton, *Universal Arithmetick,* 221 (116). Here, in the only example of the latter that I have found, the reader is told that a radical "becomes an impossible Quantity." The parallel section of his "Lectures" states that "the quantity . . . becomes impossible" (Newton, "Lectures on Algebra," 5:410–411).

arithmetic algebra, on the one hand, and geometric algebra,[114] on the other. Further complicating the issue, he hinted in *Universal Arithmetick* and other later writings at a growing attraction toward ancient geometry and possibly away from early modern mathematics, toward traditional geometric analysis in place of analytic geometry and even geometric synthesis in place of all other mathematics.

Much of *Universal Arithmetick* was an exercise in the arithmetic algebra that Wallis had endorsed in his *Treatise of Algebra*. In his introduction Newton explained that arithmetic and algebra formed "one perfect *Science of Computing*," and that he would "therefore . . . explain them both together."[115] Still, unlike his earlier "Observations," Newton's "Lectures" and hence *Universal Arithmetick* contained a significant amount of geometric algebra. Of course, even Wallis had not totally separated algebra and geometry: in his *Treatise* he had given geometric constructions of the roots of quadratic, cubic, and quartic equations. As Boyer has emphasized, however, Wallis "looked upon these as really only applications of algebra to geometry, and not as a part of algebra proper."[116]

Newton "intangled" algebra with geometry. The fourth main section of *Universal Arithmetick*, "How Geometrical Questions May Be Reduced to Equations," was the textbook's longest section and, in content and language, it bore Descartes's imprint.[117] But numerous examples of algebra applied to geometry, and even the paraphrasing of some of Descartes's language, did not prove Newton a geometric algebraist. More revealing of his mixed algebraic sympathies was the appendix to *Universal Arithmetick*, entitled "On the Linear Construction of Equations" and containing some examples from his manuscript on the topic dating from the early 1670s. With the confused explanation of Cardano's solution of the cubic equation at the end of the textbook's fifth main section, the first paragraph of the appendix explained:

> It remains now only to shew, how, after Equations are reduced to their most commodious Form, their Roots may be extracted in Numbers. And here the chief Difficulty lies in obtaining the two or three first Figures; which may be most commodiously done by either

114. As in Chapter 3, the term "geometric algebra" is adapted from Boyer, "Cartesian and Newtonian Algebra," 195–202. It is used in Chapter 3 and here in the sense of algebra mixed with geometry.
115. Newton, *Universal Arithmetick*, 2 (7).
116. Wallis, *Treatise of Algebra*, 268ff., 273ff.; Boyer, "Cartesian and Newtonian Algebra," 198.
117. On similarity of ideas and language between this section and *La géométrie*, see Boyer, "Cartesian and Newtonian Algebra," 198.

the Geometrical or Mechanical Construction of an Equation. Wherefore I shall subjoin some of these Constructions.[118]

Like Descartes, Newton now used geometric constructions to solve algebraic equations. In particular, in the appendix he sketched geometric constructions of the different types of cubic equations.[119] Through the body and appendix of *Universal Arithmetick*, then, Newton taught not only that algebra was to be applied to geometry, a statement with which Wallis could have agreed, but also that geometry was to be applied to algebra – and at a crucial algebraic moment, that is, at the point of the solution of an equation.

Still, the preceding historical analysis does not fully capture what Newton said in *Universal Arithmetick* about the relationship between algebra and geometry. In mathematics as in so many other subjects, his complicated mind seemed to pull him constantly in multiple directions. In the cracks of the body of his textbook and especially in its appendix, he aired doubts about the very enterprise of using algebra in geometry and seemed to begin to argue for a return to classical geometry. Thus, he solved some of the geometric problems of the fourth section in a classical rather than modern way. "And some [questions] which occurred as I was putting down the rest," he explained at the section's end, "I have given their Solutions without using Algebra, that I might shew that in some Problems that at first Sight appear difficult, there is not always Occasion for Algebra."[120]

The second long, dense paragraph of the appendix contained his oft-quoted statement of concern that analytic geometry was compromising classical geometry. Writing as an advocate of the traditional separation between arithmetic and geometry, which Viète had muddied and Descartes had discarded, he cautioned his readers that

> Equations are Expressions of Arithmetical Computation, and properly have no Place in Geometry, except as far as Quantities truly Geometrical (that is, Lines, Surfaces, Solids, and Proportions) may be said to be some equal to others. Multiplications, Divisions, and such sort of Computations, are newly received into Geometry, and that unwarily, and contrary to the first Design of this Science. For whoever considers the Construction of Problems by a right Line and a Circle, found out by the first Geometricians, will easily perceive that Geometry was invented that we might expeditiously avoid, by drawing Lines, the Tediousness of Computation. Therefore these two Sciences ought not to be confounded. The Antients did so

118. Newton, appendix to *Universal Arithmetick*, 225 (118).
119. Ibid., 230–257 (121–134).
120. Newton, *Universal Arithmetick*, 189 (100).

industriously distinguish them from one another, that they never introduced Arithmetical Terms into Geometry. And the Moderns, by confounding both, have lost the Simplicity in which all the Elegancy of Geometry consists.[121]

Interpreted in the context of *Universal Arithmetick,* with its strong component of geometric algebra, this statement is ironic. Moreover, Newton's concern that arithmetic and geometry not be "confounded" was all one-sided, that is, for the preservation of geometric purity. "While Newton apparently had no hesitation in introducing geometry into algebra," Boyer summarized, "he seems to have been reluctant to introduce algebra into geometry."[122]

But just what message did Newton wish to convey in the appendix's second paragraph? Perhaps he intended a narrower interpretation than it has historically received. In the sentence following the previously quoted passage, he wrote: "Wherefore that is *Arithmetically* more simple which is determined by the more simple Equations, but that is *Geometrically* more simple which is determined by the more simple drawing of Lines; and in Geometry, that ought to be reckoned best which is Geometrically most simple." He then appealed to readers to allow him to use the conchoid instead of a conic section in the constructions that followed, since the conchoid, although algebraically more complicated than a conic section, is easier to draw.[123] In short, he perhaps intended primarily to assert the obligation to use geometric, and not algebraic, standards in the construction of equations. More specifically, he seemed not so much to reject analytic geometry as to scoff at Descartes's dictum that the ranking of geometric curves in terms of simplicity be based on the dimensions of their corresponding equations.

Although his intentions remain an open question, his initial disapproval of the inclusion in *Universal Arithmetick* of the appendix's second paragraph is a fact. "The Linear Construction of Equations" was the sixth section of the "Lectures" of 1683–1684. In editing *Universal Arithmetick,* however, Whiston left a blank page between the fifth and sixth sections, leaving the impression that the latter was a freestanding appendix to the preceding textbook. When presented with Whiston's edition, Newton indicated his intent to close the space between the fifth and sixth sections. More substantively, he marked for deletion all but the first paragraph of the introductory remarks to the sixth section.[124] If

121. Newton, appendix to *Universal Arithmetick,* 227–228 (119–120).
122. Boyer, "Cartesian and Newtonian Algebra," 199.
123. Newton, appendix to *Universal Arithmetick,* 228 (120).
124. *Mathematical Papers of Newton,* 5:421–422 n. 615.

Newton had so corrected *Universal Arithmetick* when he published his edition of the work, which soon became the standard, his and all subsequent editions would have been free of the most problematic passages against the mixing of algebra and geometry. In 1722, however, Newton made only a few revisions in Whiston's edition, and let the appendix's second paragraph stand.

As a result, *Universal Arithmetick* bristled with intellectual tension: the work was destined to become a standard algebra textbook in eighteenth-century England, but its author did not endorse an independent algebra and seemed unable to restrain himself from expressing his preference for classical over analytic geometry. Students who read through the second paragraph of the appendix learned that arithmetic and algebra were tedious, whereas classical geometry was simple and elegant.

9

In addition to *Universal Arithmetick,* other sources told eighteenth-century mathematical students of the increased appreciation of classical geometry that had marked Newton's later years.[125] His *Principia* of 1687 was written "in the dress of classical geometry" or "in the language of geometrical figures."[126] Moreover, in his popular *View of Newton's Philosophy* of 1728, Henry Pemberton claimed that Newton's mathematical tastes had changed substantially over time:

> Of their [the ancient geometers'] taste, and form of demonstration Sir Isaac always professed himself a great admirer: I have heard him even censure himself for not following them yet more closely than he did; and speak with regret of his mistake at the beginning of his mathematical studies, in applying himself to the works of Des Cartes and other algebraic writers, before he had considered the elements of Euclide with that attention, which so excellent a writer deserves.[127]

Historians have differed in their estimate of the extent of Newton's eventual disenchantment with Descartes's analytic geometry and attendant attraction toward classical geometry, and Newton himself seems

---

125. Westfall estimates that Newton "became a serious student of classical geometry" around 1680 (Westfall, *Never at Rest,* 377–378).

126. Westfall, *Never at Rest,* 423–425. Westfall writes clearly and well of Newton's use of the facade of classical geometry, which masked his "thoroughly modern" first and last ratios (ibid., 425).

127. Henry Pemberton, *View of Sir Isaac Newton's Philosophy* (London, 1728), preface, 4. Pagination of preface is mine.

never to have explained exactly why his mathematical tastes began to change around 1680.[128] Perhaps the change set in gradually during the course of reading required for his Lucasian lectures. Whatever the initial stimulus, by the late 1670s he was engrossed in a study of the seventh and eighth books of the *Mathematical Collection* of Pappus.[129] Furthermore, according to Westfall, already from the late 1660s he had been very concerned about the atheistic tendencies in Descartes's mechanical philosophy. Within a decade he was inclined to reject the basic tenet of that philosophy, that bodies act on one another solely by direct contact. "[O]nly a wholesale repudiation of his Cartesian heritage would allow him to take that step."[130] Thus it is possible that concerns of natural philosophy were reflected in his evolving mathematical tastes.

Whether from a desire to rationalize his evolving attachment to classical geometry or an honest desire to explain it, Newton occasionally hinted at reasons that were linked more directly to mathematics than to natural philosophy. In *Universal Arithmetick* and elsewhere, he stressed the simplicity and elegance of geometry versus the tediousness of arithmetic and algebra. In an unpublished manuscript, tentatively described as coming from the late 1670s, he claimed that both the ancient geometers and Descartes had solved the problem of demonstrating when a locus problem is a solid locus. "To be sure," he declared,

> their [the ancient geometers'] method is more elegant by far than the Cartesian one. For he [Descartes] achieved the result by an algebraic calculus which, when transposed into words (following the practice of the Ancients in their writings), would prove to be so tedious and entangled as to provoke nausea, nor might it be understood. But they accomplished it by certain simple proportions, judging that nothing written in a different style was worthy to be read, and in consequence concealing the analysis by which they found their constructions.[131]

Like the troubled Hobbes who, too, had lost his father early in life, Newton was attracted also by the certainty of synthetic geometric arguments. Writing anonymously of his own work during the calculus con-

128. Even as he sketches some of Whiteside's reservations concerning Newton's geometric classicism, Westfall summarizes the evidence for at least a partial change in Newton's mathematical tastes from roughly 1680 (Westfall, *Never at Rest*, 377–381).

129. Whiteside, "Introduction" to part 2, *Mathematical Papers of Newton*, 4:218–224; idem, introduction to *Mathematical Works of Newton*, 2:xviii.

130. Westfall, *Never at Rest*, 381.

131. Newton, "Researches into the Greeks' 'Solid Locus'" [? late 1670s], *Mathematical Papers of Newton*, 4:276–277.

troversy, he claimed that he had found most of the results of the *Principia* by analysis "but because the Antients for making things certain admitted nothing into Geometry before it was demonstrated synthetically, he demonstrated the Propositions synthetically, that the system of the Heavens might be founded upon good Geometry."[132]

Too, Newton privately fostered the study of the geometric analysis of the ancients as an alternative to analytic geometry. During the early 1690s he tried to restore some of the lost ancient analytic works, including Euclid's porisms.[133] No publication came of these efforts, but he encouraged similar work by other mathematicians, including his disciples David Gregory and Edmond Halley. Through these disciples Newton's increased appreciation of classical geometry affected not only Cambridge University but Oxford as well. By 1704 he had placed his disciples in all the English university chairs devoted to science and mathematics: upon his recommendation Gregory became the Savilian professor of astronomy at Oxford in 1691; William Whiston succeeded Newton as the Lucasian professor in 1701; and, also with Newton's help, Halley succeeded Wallis as the Savilian professor of geometry at Oxford in 1704. In the latter year Henry Aldrich, dean of Christ Church, asked Gregory and Halley to prepare a translation of the *Conics* of Apollonius; the Savilian professors agreed and collaborated on the project until Gregory's death in 1708. Halley continued the work alone, and indeed in the 1710s he seems to have been completely absorbed in translating and restoring the Apollonian corpus. "So much did Halley love Greek geometry that he taught himself Arabic in order to translate the *De sectione rationis* of Apollonius, a manuscript of which had been found in the Bodleian."[134] Halley's edition of *De sectione rationis,* along with his tentative restoration of Apollonius's *De sectione spatii,* appeared in 1706, and was followed by his nearly definitive edition of the *Conics* four years later. The edition of *De sectione rationis* – which contained an attack on Wallis's algebraic methods[135] – drew special praise from Newton, who thought it "yet the more valuable in that it is a geometrical analysis."[136]

132. [Newton], "An Account of the Book Entitled *Commercium Epistolicum,*" *Philosophical Transactions* 29 (1714–1716): 206. There is, however, no published manuscript evidence that Newton proceeded as outlined here (Westfall, *Never at Rest,* 424).

133. Westfall, *Never at Rest,* 512.

134. G. L. Huxley, "The Mathematical Work of Edmund Halley," *Scripta Mathematica* 24 (1959): 265–273 (quotation from p. 269).

135. Ibid., 271.

136. Quoted in Westfall, *Never at Rest,* 512.

Newton's mathematical legacy was mixed. Positively, he bestowed his name and authority on the algebraic cause. He devoted a decade of Lucasian lectures to algebra, the very subject that Barrow, as his Lucasian predecessor, had dismissed as "yet no Science." Then, albeit reluctantly, he agreed to the publication of *Universal Arithmetick*. The anonymous first edition of the book was immediately attributed to him; the second edition proclaimed: "*If any thing could add to the Esteem every Body has for the* Analytick *Art, it must be, that Sir* Isaac *has condescended to handle it.*"[137] Thus, throughout his career Newton managed to remain above the algebraic fray that engulfed Wallis, Hobbes, and Barrow, and yet his very publication of *Universal Arithmetick* seemed to mark the end of any serious English threat to algebra's right to be considered an academic subject and a mathematical science.

Even as he thus bolstered the arithmetico-algebraic tradition in England, he lent support for the continued cultivation of the strong geometric tradition that Barrow had pushed. Newton's *Universal Arithmetick,* his *Principia,* accounts of his life, his own explanation of the reasons for his adoption of a synthetic style in the *Principia,* and the research tastes of some of his disciples suggested a deep respect for classical geometry (both synthetic and analytic), if not a preference for geometry over algebra. His mathematical legacy, then, was neither exclusively algebraic nor exclusively geometric; it made room for both traditions. Whereas Barrow had argued for a mathematics centered on geometry and Wallis, for a mathematics centered on arithmetic and algebra, Newton seemed to sanction a more elliptical model of the subject – with two focal points, the arithmetico-algebraic and the geometric. Of the eighteenth-century British mathematicians, some would stay rigidly fixed to one or another of these foci; but some of the best of their number, especially Colin MacLaurin, would follow research trajectories that took them, like Newton, alternately closer and farther from each of the two competing foci.

As Newton's elliptical model of mathematics was less extreme than Wallis's arithmetico-algebraic version of the subject, so his views on the foundational problems facing algebra at the turn of the eighteenth century were more conservative and ultimately less fertile than Wallis's. He, like the authors of the other major English algebra textbooks of the period, recognized negative numbers but confusingly defined them as "quantities less than nothing." On complex numbers, he publicly came down somewhere between Kersey and Wallis. He was clearly more progressive than Kersey, who explained "impossible roots" only to dis-

137. Anon., "To the Reader," in Newton, *Universal Arithmetick,* i (4).

miss them, refusing to deal with concrete examples of such roots. But, even as his extension of Descartes's rule of signs implicitly argued for roots involving $\sqrt{-1}$ as essential parts of equation theory, Newton continued to refer to and think of such roots as basically "impossible." Finally, whereas Wallis pushed for "pure algebra," Newton's *Universal Arithmetick* muddied algebra with intrusions from Descartes's geometric algebra and even classical geometry.

Given the strengths and limits of *Universal Arithmetick*, British mathematical thinkers of the early eighteenth century recognized the textbook as the preeminent but not the exclusive algebra of their times. There were at least five printings (in Latin and English) of *Universal Arithmetick* through 1769; but Kersey's *Algebra* enjoyed as many printings, with the last edition published in 1741. Daniel Waterland's *Student's Guide*, appearing toward the end of the first decade of the century, specifically endorsed the study of *Universal Arithmetick* by Cambridge students proceeding to the master's degree. But Robert Green's guide, which was published at roughly the same time and recommended that all second-year undergraduates pursue algebra, listed as possible textbooks those of Pell, Wallis, Harriot, Kersey, Newton, Descartes, Harris, Oughtred, Ward, and Jones! At Cambridge in 1730 there was still no consensus on an algebra textbook, "the usual text-books" being those of Oughtred, Harriot, Wallis, and Newton.[138]

Yet, by the 1720s Newton's algebra had become the textbook of choice of some key British mathematicians. Almost from its initial publication Nicholas Saunderson lectured on *Universal Arithmetick* at Cambridge University, and certainly by the 1720s Colin MacLaurin was lecturing on the book at the University of Edinburgh. These lectures and the students' reactions convinced Saunderson and MacLaurin to write their own textbooks, both of which – at least partially out of deference for Newton – were billed as commentaries on *Universal Arithmetick*.

138. W. W. Rouse Ball, *A History of the Study of Mathematics at Cambridge* (Cambridge: Cambridge University Press, 1889), 92–95.

8

George Berkeley at the Intersection of Algebra and Philosophy

The British intellectual tradition produced three major thinkers, in addition to Newton, who published on algebra in the first half of the eighteenth century: George Berkeley, Colin MacLaurin, and Nicholas Saunderson. Berkeley's writings on algebra were the most philosophical of the works produced by the three men and, in their time, the most neglected. Berkeley was also the only of the three to draw significant inspiration from Wallis, for MacLaurin and Saunderson began their algebras as commentaries on *Universal Arithmetick*. Berkeley wove Wallis's algebraic reflections with the arithmetic insights of Barrow and possibly Hobbes into a coherent philosophy of arithmetic and algebra as sciences of signs, in contrast to his philosophy of geometry as the science of perceptible finite extension.

As Berkeley's general philosophical concerns affected his understanding of early modern mathematics, so mathematics presented him with problems and insights that helped to shape his philosophy. His earliest notebooks referred to Barrow's view of number as a "note" or "sign," and number, so understood, became one of his stock examples against abstraction. Berkeley, moreover, was the first major British (and perhaps European) philosopher to come to terms with the symbolical style of early modern algebra. His early acceptance of symbolical reasoning and Barrow's view of number helped to raise in him "semiotic consciousness" – or "the explicit awareness of the role of the sign as that role is played in a given respect"[1] – and thus contributed toward his innovative theory of language. Algebra provided his prime example of reasoning on signs without concern for significates, and the imaginary numbers, his prime example of reasoning on idealess signs. In short, Berkeley devel-

---

1. For this definition, but not a link to Berkeley, see John Deely, *Basics of Semiotics* (Bloomington: Indiana University Press, 1990), 107.

oped a general philosophy that subsumed, and thereby sanctioned, the philosophy of early modern algebra as a science of signs.[2]

Still, his general philosophy and his philosophy of arithmetic and algebra seem to have exerted but little immediate influence on algebra. Rather, it can be argued, he did more to inhibit than encourage the development of algebra in Great Britain during the late 1730s and 1740s. He inhibited the subject's development not by criticizing it, since he remained an enthusiast through the end of his life, but by publishing *The Analyst*, in which he demanded that mathematicians either geometrically explicate the foundations of calculus or admit that mathematics, like theology, accepted mysteries. By questioning the rigor of calculus as well as the religion of its practitioners, whom he supposed much readier to accept mathematical than theological mysteries, he almost compelled early-eighteenth-century British mathematicians to work on the geometric foundations of calculus. In the elliptical universe of eighteenth-century British mathematics, he helped orient mathematical research toward the geometric focus of mathematics at the expense of the arithmetico-algebraic focus.

I

Berkeley (1685–1753) was the son of an Anglo-Irish family.[3] In England the Berkeleys had suffered during the Civil War for their royalist views

2. Recent works discussing algebra and Berkeley's philosophy include Helena M. Pycior, "Internalism, Externalism, and Beyond: 19th-Century British Algebra," *Historia Mathematica* 11 (1984): 424–441 (Berkeley and the late-eighteenth- and early-nineteenth-century symbolical algebraists); idem, "Mathematics and Philosophy: Wallis, Hobbes, Barrow, and Berkeley," *Journal of the History of Ideas* 48 (1987), 265–286; David Sherry, "The Logic of Impossible Quantities," *Studies in History and Philosophy of Science* 22 (1991): 37–62; and Douglas M. Jesseph, *Berkeley's Philosophy of Mathematics* (Chicago: University of Chicago Press, 1993). The latter is a revised version of Jesseph, "Berkeley's Philosophy of Mathematics" (Ph.D. diss., Princeton University, 1987), which I have not seen. The core of the present chapter is adapted from Pycior, "Mathematics and Philosophy."

3. Biographies of Berkeley include Gerd Buchdahl, "George Berkeley," *Dictionary of Scientific Biography*, ed. Charles C. Gillispie, 18 vols. (New York: Scribner's Sons, 1970–1990), 2:16–18; Augustus De Morgan, "George Berkeley," *Penny Cyclopaedia* (London, 1851), 279–281; and A. A. Luce, *The Life of George Berkeley, Bishop of Cloyne* (London: Thomas Nelson and Sons, 1949).

and, following the Restoration, Berkeley's grandfather had obtained the collectorship of Belfast, Ireland. Berkeley was subsequently born in County Kilkenny, Ireland. In 1696 he enrolled in Kilkenny College, "the Eton of Ireland," and four years later he entered Trinity College, Dublin, from which he obtained a B.A. in 1704 and an M.A. in 1707. A fellow of Trinity College by the latter year and an ordained priest by 1710, he published two major early works, his *Essay towards a New Theory of Vision* of 1709 and *Treatise Concerning the Principles of Human Knowledge* of 1710. In 1713 he used a leave of absence from Trinity to travel to London to facilitate the publication of another book, *The Three Dialogues between Hylas and Philonous*. He then divided his time between London, where he wrote for the popular press, and the continent, which he toured as chaplain and tutor to more prosperous men. While in London in 1720, he published *An Essay towards Preventing the Ruin of Great Britain,* which won him the friendship of Richard Boyle, the earl of Burlington and Cork, who introduced him to George Fitzroy, the duke of Grafton. When the duke was appointed lord lieutenant of Ireland, he promised Berkeley preferment. Berkeley resumed residence in Dublin in 1721, and within months he was appointed divinity lecturer at Trinity. In 1724 he secured the lucrative deanery of Derry.

Sincere if colonialist in his religious convictions, he returned to London in late 1724 to circulate a proposal for a college that he planned to establish in Bermuda to improve the spiritual and moral welfare of the American colonies.[4] According to his plan, the college would accept American children of English and American Indian ancestry; would train some of the English boys for the Anglican ministry; and would convert the American Indian children to Anglicanism. Having secured a charter and parliamentary grant for the college, he and his wife moved in 1728 to Newport, Rhode Island, to establish an estate to supply food for the proposed college. When promised funding for the college did not materialize, he returned to London in 1731.

He now continued to proselytize but in a different fashion, as he published two of his major later works, *Alciphron* of 1732 and *The Analyst* of 1734, in which he applied his mathematical learning to the defense of religion. In 1734 he was appointed bishop of Cloyne, and for the rest of his life he combined pastoral work with occasional writings on philosophical and social issues. In 1752 he moved to Oxford, England, where he died suddenly the next year.

---

4. For details on the "Bermuda Project," see Luce, *Life of George Berkeley,* 94–152.

2

Many of the themes running through Berkeley's writings – from his earliest mathematical publications of 1707 to the *New Theory of Vision* and *Principles* of 1709–1710 and even to *Alciphron* and *The Analyst* of his American and post-American periods – can be traced back to his undergraduate and postgraduate years at Trinity College, Dublin. Although still recovering from the occupation and Civil War, turn-of-the-century Trinity offered an up-to-date undergraduate curriculum, which included John Locke's *Essay Concerning Human Understanding* and Newton's *Principia*.[5] Moreover, by 1706 Berkeley was a leading member – most likely, an officer – of two college societies that met weekly to discuss the new philosophy and science.[6] By 1707–1708 he had begun to respond critically to Locke's philosophy, Newton's science and calculus, and other major mathematico-philosophical reflections that had come out of seventeenth-century England.

He, like Newton, seems to have read widely during his late undergraduate and immediate postgraduate years. Clearly attracted to the new mathematics and science, he read Barrow, Wallis, and Newton and perhaps Hobbes as well. He referred to the ideas of all four thinkers (and frequently to some of their specific writings) in his mathematical publications of 1707 and his *Philosophical Commentaries* – which consist of notes written most likely between the summer of 1707 and the fall of 1708 in preparation for the *New Theory of Vision* and the *Principles*.[7] In his mathematical publications, he cited Wallis's *Mathesis universalis*, *Treatise of Algebra*, and *Arithmetica infinitorum*; in the *Commentaries* he acknowledged the controversy between Wallis and Hobbes and wrote a note to himself to inquire into that dispute when he "treat[ed] of Mathematiques."[8] Other entries of the *Commentaries*

5. G. A. Johnston, *The Development of Berkeley's Philosophy* (New York: Russell & Russell, 1965), 16–17.
6. Luce, *Life of George Berkeley*, 34–36.
7. On the purpose, dating, and ordering of the *Philosophical Commentaries*, see "Editor's Introduction" to George Berkeley, *Philosophical Commentaries*, ed. George H. Thomas, with explanatory notes by A. A. Luce (New York: Garland, 1989), i–xv.
8. *The Works of George Berkeley, Bishop of Cloyne*, ed. A. A. Luce and T. E. Jessop, 9 vols. (London: Thomas Nelson and Sons, 1948–1957), 4:171, 213, 236, and 1:99–100. My references to the *Philosophical Commentaries* will be to vol. 1 of the *Works of Berkeley*; recent improvements in the transcription from Berkeley's manuscript (in Berkeley, *Philosophical Commentaries*, ed. Thomas) will be noted.

displayed his careful and critical study of Barrow's *Mathematical Lectures.*[9] The young Berkeley, however, knew considerably less of Newton's mathematical views. He seems to have known Newton as the mathematician behind the *Principia* and the "Tractatus de quadratura curvarum" of 1704 rather than *Universal Arithmetick,* which appeared only in 1707.[10]

Once exposed to the mathematics of the seventeenth century, Berkeley became temporarily preoccupied with the subject. His earliest publications were the *Arithmetica* and *Miscellanea mathematica,* two tracts that he wrote between 1705 and 1707 when he published them together as a single ninety-two-page volume. Although offering no new mathematical results, these tracts testified to his early enthusiasm for arithmetic and algebra. In the slim *Arithmetica* he aimed to demonstrate the rules of arithmetic without appealing to algebra or geometry. "De radicibus surdis" of the *Miscellanea* showed, in its own way, that he had fallen under the symbolical spell. Here, as he gave rules for operations on surds, he proposed a new notation for square roots: if Latin letters stood for squares, then corresponding Greek letters could stand for their square roots, or $\sqrt{b} = \beta$, $\sqrt{d} = \delta$, and so on. In "De ludo algebraico" of the *Miscellanea,* he presented an algebraic game and, in his conclusion, defended the study of mathematics and algebra in particular.[11]

Berkeley's research interests in mathematics, however, seem to have ceased by the publication of his *Arithmetica* and *Miscellanea.* Of major importance to Western thought, he retained a philosophical interest in early modern mathematics long afterward. In publications stretching over the next quarter-century he would return to such problems as the

9. Scholars long ignored the influence on Berkeley of Barrow's *Mathematical Lectures* (as opposed to his geometric and optical lectures). On Barrow's influence on Berkeley, see Pycior, "Mathematics and Philosophy," 274–283. For evidence of the growing appreciation of this influence, compare the notes on entry 334 in *Works of Berkeley* (1:121) and in *Philosophical Commentaries,* ed. Thomas (212).

10. The "Tractatus de quadratura curvarum" was published in Latin in 1704 as an appendix to Newton's *Opticks.* For details of its publication, see the introduction to *The Mathematical Works of Isaac Newton,* ed. Derek T. Whiteside, 2 vols. (New York: Johnson Reprint, 1964–1967), 1:xviii–xix. The English translation (1710) of the "Quadratura curvarum" is given in *Mathematical Works of Newton,* 1:141–160.

11. *Works of Berkeley,* 4:163–199, 205, 214–220. For an original discussion of the *Arithmetica absque algebra aut Euclide demonstrata* and its "formalistic element," see Jesseph, *Berkeley's Philosophy,* 93–95.

nature of numbers, the legitimacy of reasoning on symbols, the imaginary numbers, and the status of the new calculus.

3

It is quite possible that, despite Berkeley's early familiarity with the major English mathematical thinkers of the seventeenth century, none of these sparked his interest in algebra more than Locke. At the turn of the century Trinity College, Dublin, buzzed of Locke's philosophy. Locke's *Essay Concerning Human Understanding* had been added to the Trinity curriculum almost immediately upon its publication in 1690, at the instigation of William Molyneux, member for the university and a correspondent of Locke.[12] Berkeley would eventually disagree with Locke but Locke's *Essay* nevertheless exercised a formative influence on him. From the perspective of Berkeley's intersecting reflections on mathematics and philosophy, Locke's thoughts on ideas, number, and algebra were extremely important.

In his *Essay,* Locke had endorsed a theory of abstraction according to which the human mind forms general ideas through abstraction from particulars, and general terms are the names of these general ideas.[13] "The *Ideas first* in the Mind, 'tis evident," he had written, "are those of particular Things, from whence, by slow degrees, the Understanding proceeds to some few general ones; which being taken from the ordinary and familiar Objects of Sense, are settled in the Mind, with general Names to them." Displaying the mathematical proclivities of the English philosophers of the Scientific Revolution, he had then given a mathematical example of a general idea, the famous one of the triangle. He had furthermore argued that "*to get and settle in our Minds* determined *Ideas* of those Things, whereof we have general or specific Names" was one of the two "*ways to enlarge our Knowledge.*" In short, clear ideas were essential for knowledge – a point that was borne out by the consideration of mathematics: "he, that has not a perfect, and clear *Idea* of those Angles, or Figures of which he desires to know any thing, is utterly thereby uncapable of any Knowledge about them. Suppose but a Man, not to have a perfect exact *Idea* of a *right Angle,* a *Scalenum,* or

---

12. On the teaching of Locke at Trinity, see Luce, *Life of George Berkeley,* 31.
13. On Locke's abstractionism, often called conceptualism, see A. D. Woozley, "Universals," *The Encyclopedia of Philosophy,* ed. Paul Edwards, 8 vols. (New York: Macmillan, 1967), 1:199–201.

Trapezium; and there is nothing more certain than, that he will in vain seek any Demonstration about them."[14]

Not only were clear ideas essential for knowledge, but one mathematical science, arithmetic, dealt with the simplest and most universal of all human ideas, "*Unity, or One*." From unity, according to Locke, humans formed ideas of greater numbers: "By the repeating . . . of the *Idea* of an Unite, and joining it to another Unite, we make thereof one collective *Idea,* marked by the Name *Two*," and so on. Numbers in general, then, were among the clearest and most distinct of ideas, "two being as distinct from one, as Two hundred; and the *Idea* of Two, as distinct from the *Idea* of Three, as the Magnitude of the whole Earth, is from that of a Mite." Locke had gone on from here to argue for the practical superiority of arithmetic over geometry: "Demonstrations in Numbers, if they are not more evident and exact, than in Extension, yet they are more general in their use, and more determinate in their Application. Because the *Ideas* of Numbers are more precise, and distinguishable than in Extension." Whereas men and women had no trouble distinguishing ninety-one from ninety, "whatsoever is more than just a Foot, or an Inch, is not distinguishable from the Standard of a Foot, or an Inch."[15]

If such reflections fostered Berkeley's interest in the nature of ideas and numbers, Locke's comments on algebra also caught the young Irish philosopher's attention. In his *Essay* Locke had said little about algebra, but what he had said was tantalizing. The second of the two "*ways to enlarge our Knowledge*" was "the Art of *finding out* those *Intermediate Ideas,* which may shew us the Agreement, or Repugnancy of other Ideas, which cannot be immediately compared." Algebra, he had implied, would serve as a model for this art. It was evident, he had written, that mathematicians had not originally reached the Pythagorean and other theorems by mere reflection on mathematical axioms. Perhaps here thinking of geometric analysis, he had explained no more than that these theorems had been "discovered by the Thoughts other ways applied." The final sentence of his paragraph, however, had concerned algebra: "And who knows what Methods, to enlarge our Knowledge in other parts of Science may hereafter be invented, answering that of *Algebra* in Mathematicks, which so readily finds out *Ideas* of Quantities to measure others by, whose Equality or Proportion we could otherwise very hardly,

---

14. John Locke, *An Essay Concerning Human Understanding,* ed. Peter H. Nidditch (Oxford: Clarendon Press, 1975), 595–596 (IV.vii.9) and 648–649 (IV.xii.14–15).

15. Ibid., 205–206 (II.xvi.1–5).

or, perhaps, never come to know?" Even earlier he had declared: "They that are ignorant of *Algebra* cannot imagine the Wonders in this kind are to be done by it."[16] Algebra had thus been touted by Locke as a paradigm of the crucial art of finding ideas intermediate between ideas that were otherwise incomparable.

The young Berkeley not only read Locke's *Essay*; he probably discussed it at meetings of the college society that studied the "new philosophy" and he certainly took notes on it. In entry 354 of the *Commentaries,* he recorded Locke's position that there is "No reasoning about things whereof we have no idea."[17] In entry 697 he reminded himself: "N.B. To Consider well w^t is meant by that wch Locke saith Concerning Algebra that it supplys intermediate Ideas. Also to think of a Method affording the same use in Morals etc that this doth in Mathematiques."[18] In "De ludo algebraico" of the *Miscellanea,* where he mentioned Locke's endorsement of algebra, he looked forward to algebra's application "to the whole extent of mathematics, and every art and science, military, civil, and philosophical." Perhaps thinking as much of Locke as of any early modern algebraist, he exaggerated, "By all it [algebra] is regarded as a great and wonderful art, the topmost pinnacle of human knowledge, and the kernel and key of all mathematical science."[19]

Sometime in 1707–1708 Berkeley moved beyond Locke's reflections on unity, numbers, and ideas. A key element in his intellectual evolution was the inconsistency that he found in Locke's and Barrow's views on unity and number. If for Locke unity and number in general had exem-

16. Ibid., 648–649 (IV.xii.14–15), 549 (IV.iii.18). Oughtred's *Key* seems to have been a major algebraic source for Locke, who in 1681 praised the latter work as the "best Algebra yet extant" (Peter King, *The Life and Letters of John Locke, with Extracts from His Journals and Commonplace Books* [London, 1829; reprint, New York: Burt Franklin, 1972], 122). On the role of analysis in connecting "so many things, so far asunder distant," see William Oughtred, *The Key of the Mathematicks, New Forged and Filed . . .* (London: Thomas Harper, 1647), preface, 3–5.

17. *Works of Berkeley,* 1:42. Luce cites this and other entries to show that Berkeley originally held "Locke's two principles, that (a) no knowledge without ideas, and (b) all significant words stand for ideas," although "he subsequently abandoned both" (see Luce's notes on entries 312 and 354–356, in Berkeley, *Philosophical Commentaries,* ed. Thomas, 205–206, 215–216).

18. *Works of Berkeley,* 1:85.

19. Ibid., 4:219–220. This English translation is taken from Johnston, *Development of Berkeley's Philosophy,* 215.

plified clear and distinct (abstract) ideas, for Barrow: "No Number of itself signifies any thing distinctly, or agrees to any determinate Subject, or certainly denominates anything."[20] In early consecutive entries in the *Commentaries,* Berkeley juxtaposed Barrow's view of the unit with Locke's view of primary qualities. In entry 75 he wrote: "Unite in abstracto not at all divisible it being as it were a point or wth Barrow nothing at all." Referring to primary qualities (which included number), he recorded in the next entry: "Any subject can have of each sort of primary qualities but one particular at once. Locke b.4. c 3.S.15."[21] The young Berkeley was thus left to decide between Barrow's claim that unity was "nothing at all," or a creation of the mind, and Locke's contention that for any object "each particular Extension, Figure, number of Parts, Motion, excludes all other of each kind."[22]

4

Certainly by 1709 Berkeley came to agree with Barrow's view of number and to reject primary qualities. In that year, without giving credit to Barrow, he wrote in his *New Theory of Vision:* "number (however some may reckon it amongst the primary qualities) is nothing fixed and settled, really existing in things themselves. It is intirely the creature of the mind."[23] At the same time as he adopted Barrow's view of number and tentatively extended it to algebra, he seemed to lose interest in the pursuit of mathematics in and of itself and began to emphasize, at least temporarily, the practical rather than theoretical value of arithmetic and algebra. In the *Commentaries* he had moved rapidly from the thesis that "Numbers are nothing but Names, meer Words"[24] to the conclusion that both arithmetic and algebra "are sciences purely Verbal, & entirely

20. Isaac Barrow, *The Usefulness of Mathematical Learning Explained and Demonstrated: Being Mathematical Lectures Read in the Publick Schools at the University of Cambridge,* trans. John Kirkby (London, 1734), 34.
21. *Works of Berkeley,* 1:14–15.
22. Locke, *Essay Concerning Human Understanding,* 546–547 (IV.iii.15).
23. *Works of Berkeley,* 1:214. On Berkeley's debt to Barrow, see Pycior, "Mathematics and Philosophy," esp. 279–281; on the connection between Berkeley's theory of number and his attack on primary qualities, see also Jesseph, *Berkeley's Philosophy,* 97–98.
24. Berkeley, *Philosophical Commentaries,* ed. Thomas, 100, 307. Thomas has corrected the original transcription of this entry: "Numbers are nothing but Names, never Words" (*Works of Berkeley,* 1:92).

useless but for the Practise in Societys of Men. No speculative knowledge, no comparing of Ideas in them."[25] Perhaps significantly, his *Arithmetica* and *Miscellanea mathematica* – which had lauded algebra as "the topmost pinnacle of knowledge" – were not only his first but also his last strictly mathematical publications.

Still, Berkeley's vanishing interest in mathematics per se can be attributed to his having found his forte in more strictly philosophical thought, the firstfruits of which included the *Principles* of 1710 as well as the *New Theory of Vision*. The *Principles* was Berkeley's major philosophical treatise in which he developed his distinctive doctrines of antiabstractionism and immaterialism. But even in these works, especially the latter, mathematical reflections played a somewhat central role, and in these works Berkeley outlined his philosophy of mathematics.

Berkeley's general philosophy began with rejection of abstract ideas and rejection of matter with an existence independent of mind.[26] Humans deceived themselves, he argued, when they imagined that the mind could frame abstract general ideas from particular objects. Locke had, of course, written at some length in favor of abstractionism, at enough length that philosophical scholars have observed that, when his various statements are scrutinized and compared one with another, his theory of ideas seems illusive and, at the extreme, absurd.[27] A masterly debater,

---

25. *Works of Berkeley,* 1:93.
26. The following sketch of Berkeley's antiabstractionism aims only to provide essential background for a discussion of the intersection of the mathematical and philosophical strands of his thought. For his general philosophy, see J. O. Urmson, *Berkeley* (Oxford: Oxford University Press, 1982), and G. J. Warnock, *Berkeley* (Baltimore: Penguin, 1969). On antiabstractionism, see Willis Doney, ed., *Berkeley on Abstraction and Abstract Ideas* (New York and London: Garland, 1989); on immaterialism, Urmson, *Berkeley,* 32–46. Berkeley's maxim was "esse est percipi." He argued that "to exist is either to perceive . . . , which constitutes the existence of spirits, or to be perceived . . . , which constitutes the existence of the inanimate, of ideas. These perceived ideas are objects of the mind which have no existence independent of the mind" (Urmson, *Berkeley,* 33).
27. E.g., "the passages in the *Essay* in which he [Locke] discussed general ideas . . . are neither so clearly thought out and expressed nor perhaps even so consistent as to save him from varying interpretations" (Woozley, "Universals," 199). Noting that "it is usual to hold that Locke's theory of universals is absurd," Aaron tries to refute this charge while acknowledging that Locke's theory is not "wholly satisfactory" (R. I. Aaron, "Locke's Theory of Universals," *Proceedings of the Aristotelian Society* 33 [1933]: 173–202; reprinted in *Berkeley on Abstraction,* ed. Doney, 1–30).

Berkeley turned Locke's own words against abstract general ideas. In the *Commentaries* he determined to use "Lockes general triangle" for "the killing blow ... in the Matter of Abstraction."[28] In the *Principles* he quoted and ridiculed Locke's enigmatic description of the idea of a triangle. "For when we nicely reflect upon them," Locke had begun,

> we shall find, that general *Ideas* are Fictions and Contrivances of the Mind, that carry difficulty with them, and do not so easily offer themselves, as we are apt to imagine. For example, Does it not require some pains and skill to form the *general Idea* of a *Triangle*, (which is yet none of the most abstract, comprehensive, and difficult,) for it must be neither Oblique, nor Rectangle, neither Equilateral, Equicrural, nor Scalenon; but all and none of these at once. In effect, it is something imperfect, that cannot exist; an *Idea* wherein some parts of several different and inconsistent *Ideas* are put together.[29]

Implying his own inability to imagine Locke's triangle, Berkeley then challenged the reader to "fully and certainly inform himself whether he has such an idea or no." The idea that was required, he summarized, was "the general idea of a triangle, which is, *neither oblique, nor rectangle, equilateral, equicrural, nor scalenon, but all and none of these at once.*"[30]

Arguing so against abstraction as well as against the existence of matter independent of mind, he eventually turned to correcting what he saw as the major errors flowing from abstractionism and materialism. As he stated in the *Principles,* he "suspect[ed] that mathematicians are, as well as other men, concerned in ... [these] errors."[31] Seeking therefore to elaborate a new philosophy of mathematics based on neither premise, he eventually argued that, on the one hand, geometry was a science of perceptible finite extension and, on the other, arithmetic and algebra were sciences of signs.

5

As finally elaborated by Berkeley, the view of geometry as a science of perceptible finite extension was revolutionary in its antiabstractionist and immaterialist underpinnings, but in other ways it was compatible

---

28. *Works of Berkeley,* 1:84.
29. Locke, *Essay Concerning Human Understanding,* 596 (IV.vii.9); Berkeley reproduces the passage in the *Principles* (*Works of Berkeley,* 2:32).
30. *Works of Berkeley,* 2:33.   31. Ibid., 2:95.

with the English geometric tradition that had solidified in the seventeenth and early eighteenth centuries. As Hobbes and Barrow took a sense-based approach to geometry, Berkeley argued in the *Principles* that geometry concerned "*extension* . . . considered as relative," and, in the later *Analyst,* that geometry's "object" was "the proportions of assignable extensions" and its "end" was "to measure assignable finite extension."[32] Because he rejected matter existing independently of mind, however, Berkeley's position on geometry differed in a philosophically fundamental way from that of his predecessors. For him, the finite extension of geometry existed only insofar as it was perceivable. As he explained in the *Principles:* "Every particular finite extension, which may possibly be the object of our thought, is an *idea* existing only in the mind, and consequently each part thereof must be perceived."[33]

Echoing Hobbes in rejecting abstract general ideas and emphasizing extension as geometry's object, and compounding his alienation from his contemporaries by his rejection of independent matter, Berkeley was at least as much a geometric maverick as Hobbes had been. For example, both men opposed the Euclidean definition of a straight line. In the *Commentaries,* Berkeley observed, "We can no more have an idea of length without breadth or visibility than of a General figure." The Irish philosopher also rejected the infinite divisibility of extension. In the *Principles,* following his restriction of the material to the perceptible, e declared, "If therefore I cannot perceive innumerable parts in any inite extension that I consider, it is certain they are not contained in it."[34]

Also like Hobbes, Berkeley, despite his unconventional views, followed the mathematical majority in maintaining the generality of geometric arguments. His resolution of the problem of general arguments without abstract general ideas emerged in the *Principles* in response to Locke's statement on the triangle. Explicitly rejecting Locke's version of abstractionism, he resolved the problem of general arguments in a manner reminiscent of Barrow's suggestion that "the intelligible Sphere . . . [was] one understood universally." For his part, Berkeley contended that the "universal" triangle was no more than one triangle taken as a sign of all others. Thus, even without an abstract general idea of a triangle, the geometer could reason generally about triangles. In Berkeley's own words:

> [W]hen I demonstrate any proposition concerning triangles, it is to
> be supposed that I have in view the universal idea of a triangle;

32. Ibid., 2:97, 4:96.   33. Ibid., 2:98.
34. Ibid., 1:61, 2:98.

which ought not to be understood as if I could frame an idea of a triangle which was neither equilateral nor scalenon nor equicrural. But only that the particular triangle I consider, whether of this or that sort it matters not, doth equally stand for and represent all rectilinear triangles whatsoever, and is in that sense *universal.*[35]

Berkeley's philosophy of geometry seems to have remained relatively constant for the rest of his life. However, as noted by G. J. Warnock, in *De motu* of 1721 he observed that "geometers for the sake of their art make use of many devices which they themselves cannot describe nor find in the nature of things." As an example, he pointed to geometers' considering a curve "as consisting of an infinite number of straight lines, though in fact it does not consist of them."[36] Even these remarks do not show that "Berkeley in later years . . . [came] to adopt . . . [an] alternative view" of "geometry . . . as an abstract calculus, *applicable* (more or less roughly) to the physical world but not descriptive of its properties."[37] The later *Analyst,* in fact, clarified Berkeley's position on the geometric devices cited in *De motu:* the devices were part of the art and not the science of geometry. "And whether it would not be righter," query 10 of *The Analyst* asked, "to measure large polygons having finite sides, instead of curves, than to suppose curves are polygons of infinitesimal sides, a supposition neither true nor conceivable?" Furthermore, continuing in *The Analyst* to describe finite extension as geometry's object, Berkeley explained that geometry was at its best when its "objects . . . [were] kept in view . . . [with] the attention ever fixed upon them."[38]

The latter phrases appeared in *The Analyst* as part of a lengthy statement that affirmed Berkeley's adherence to the then current thesis of geometry as a logical paradigm. "It hath been an old remark," he began, "that Geometry is an excellent Logic."

> And it must be owned that when the definitions are clear; when the postulata cannot be refused, nor the axioms denied; when from the distinct contemplation and comparison of figures, their properties are derived, by a perpetual well-connected chain of consequences, the objects being still kept in view, and the attention ever fixed upon them; there is acquired an habit of reasoning, close and exact and methodical: which habit strengthens and sharpens the mind, and

35. Ibid., 2:33–34 (quotation); Pycior, "Mathematics and Philosophy," 278–279. For a detailed study of the evolution of Berkeley's philosophy of geometry to this point, see Jesseph, *Berkeley's Philosophy,* 44–87.
36. *Works of Berkeley,* 4:41, 48.   37. Warnock, *Berkeley,* 210.
38. *Works of Berkeley,* 4:96, 66.

being transferred to other subjects is of general use in the inquiry after truth.[39]

Thus, for Berkeley, as for Hobbes, Barrow, and perhaps the older Newton, geometry was a paradigm of human thought. According to Berkeley, geometry owed its priority not to Platonic or abstract general ideas; rather geometry was sound reasoning applied to extension through the use of particulars as general signs.

6

It was in his philosophy of the sciences of arithmetic and algebra – and his formulation of a new theory of language, which seemed to develop hand in hand with his mathematical philosophy – that Berkeley's synthetic genius shone. He originally shared Hobbes's and Barrow's concern for the lack of an immediate empirical basis for the numbers of arithmetic, and, following Barrow's lead, he developed arithmetic as a science of signs. Unlike Hobbes and Barrow and like Wallis, however, he was enthusiastic about symbolical reasoning and algebra, and presented algebra as an extension of arithmetic and therefore also as a science of signs.

The young Berkeley claimed no originality for the theory of numbers as names. In entry 881 of the *Commentaries* he admitted: "It has already been observ'd by others that names are no where of more necessary use than in Numbering." Moreover, the ideas and terms incorporated into Berkeley's early accounts of numbers and numbering revealed the definite influence of Barrow. As entry 75 gave Barrow's version of the unit, entries 104 and 110 expressed Barrow's theory of numbers. Entry 104 noted: "Number not without the mind in any thing, because tis the mind by considering things as one that makes complex ideas of 'em tis the mind combines into one." Entry 110 reiterated Barrow's thesis that numbers did not stand for any particular bodies but rather were assigned by the mind according to its pleasure. As Berkeley expressed it: "Number not in bodies it being the creature of the mind depending entirely on its' consideration & being more or less as the mind pleases."[40]

Yet, Berkeley's strong appreciation of arithmetic and algebra – acquired at least partially from Wallis and Locke[41] – inclined him away from Barrow's relegation of arithmetic to a part of geometry and away

---

39. Ibid., 4:65–66.   40. Ibid., 1:104, 14, 17–18.
41. Malebranche's *Recherche de la vérité* possibly also contributed to the young Berkeley's favorable attitude toward algebra. See A. A. Luce, *Berkeley and Malebranche: A Study in the Origins of Berkeley's Thought* (Oxford: Clarendon Press, 1934), 15.

from Barrow's rejection of algebra. Entries 834 and 837 of the *Commentaries* referred to the controversy between Wallis and Hobbes, and the former entry proved that Berkeley understood Hobbes's main objection to symbolical reasoning.[42] But, as is well known, Berkeley not only accepted symbolical reasoning but elaborated a view of arithmetic and algebra that was, in modern terms, abstract and formal.[43] He rejected both the contention that arithmetic was based on Platonic or other abstract ideas and that it was immediately conversant about sensible objects. Instead he argued that numbers were "creatures of the mind," that arithmetic and algebra were sciences immediately conversant about signs rather than objects, and that their practitioners could reason without concern for the significance of the signs. Furthermore, he used algebra to support his general contention that reasoning could take place without concern for meaning and that some useful signs stood for no particular or general ideas.

The germs of these views were scattered throughout the *Commentaries*. Along with the entries that restated Barrow's claims about number were others that pointed toward the general view of arithmetic and algebra as sciences of signs, while yet expressing concern about their want of speculative content. As we have seen, entry 763 declared: "Numbers are nothing but Names, meer Words." The next entry highlighted the imaginary numbers as a stimulus to Berkeley's formulation of a new theory of arithmetic and algebra: "Mem: Imaginary roots to unravel that Mystery." Entry 767 asked: "Take away the signs from Arithmetic & Algebra, & pray w^t remains?" The response came in the following entry, where Berkeley admitted that arithmetic and algebra were "sciences purely Verbal, & entirely useless but for Practise in Societys of Men. No speculative knowledge, no comparing of Ideas in them." Then, entry 803 offered support for the philosophy of arithmetic as a science of signs. It stated that "the great use of the Indian figures above the Roman shews Arithmetic to be about Signs not Ideas, or not Ideas different from the Characters themselves."[44] If arithmetic dealt with ideas or physical objects, notational systems (or signs) would be a relatively insignificant aspect of the subject. The power of the Hindu–Arabic numerals over the Roman, Berkeley reasoned, showed that the latter was not the case.

42. *Works of Berkeley*, 1:99–100.
43. See, e.g., Robert J. Baum, "The Instrumentalist and Formalist Elements of Berkeley's Philosophy of Mathematics," *Studies in History and Philosophy of Science* 3 (1972): 119–134.
44. Berkeley, *Philosophical Commentaries*, ed. Thomas, 100; *Works of Berkeley*, 1:92–93, 96.

Direct evidence that the arithmetic and algebraic speculations of the *Commentaries* were key elements in Berkeley's reformulation of philosophy, especially relating to the theory of language, came moreover in a letter of late 1709 to Samuel Molyneux. Here he appealed to algebra and the theory of numbers as names in a crucial discussion of the mind's ability to reason on signs without concern for significates. "We may very well, and in my Opinion often do," he wrote,

> reason without Ideas but only the Words us'd, being usd for the most parts as Letters in Algebra, which tho they denote particular Quantities, Yet every step do not suggest them to our Thoughts, and for all that We may reason or perform Operations intirly about them. Numbers We can frame no Notion of beyond a certain Degree, and yet We can reason as well about a Thousand as about five, the Truth on't is Numbers are nothing but Names.[45]

By 1709, then, Berkeley was crafting a new philosophy of arithmetic and algebra as well as a general philosophy that took early modern algebra into account. In doing so, he was raising fertile questions about language and the role of signs, questions that had however long been implicit in algebra.

7

Although he had explained number as "the creature of the mind" in the *New Theory of Vision*, it was in the *Principles of Human Knowledge* that Berkeley elaborated a philosophy of arithmetic. In part 1 of the latter, he discussed number twice: first in sections 12–13, in the context of proving that primary qualities did not exist independently of mind, and second in sections 118–122, in the context of defining arithmetic as a science of signs. The postulation of primary qualities, he noted, was associated with materialism: some philosophers "will have our ideas of the primary qualities to be patterns or images of things which exist without the mind, in an unthinking substance which they call *matter*." Berkeley of course would have nothing to do with matter existing independently of mind, and aimed to refute the possibility of independent primary qualities. Arguing first that the so-called primary qualities of "extension, figure and motion are only ideas existing in the mind," he

45. Berkeley to Molyneux, 8 Dec. 1709, *Works of Berkeley,* 8:25. This Molyneux was Berkeley's friend from Trinity College, Dublin, to whom the *Miscellanea mathematica* had been dedicated, and the son of William Molyneux, Locke's correspondent.

turned to number as his most convincing example. "That number is entirely the creature of the mind, even though the other qualities be allowed to exist without," he began section 12, "will be evident to whoever considers, that the same thing bears a different denomination of number, as the mind views it with different respects."[46]

He then argued for the relativity of number and unity and attacked the doctrine of abstract general ideas. As an example, he considered a yard-long extension, to which (he explained) the mind could assign variously the numbers 1, 3, and 36, according as it measured the extension in yards, feet, or inches. "Number is so visibly relative, and dependent on men's understanding," he exclaimed, "that it is strange to think how any one should give it an absolute existence without the mind." Having made this point through what he saw as a compelling example, he tackled the relativity of unity in section 13. In a clear reference to Locke, he wrote that "some will have [unity] to be a simple or uncompounded idea, accompanying all others into the mind." But, as he had implied of Locke's triangle, so he said here of unity that he could not find within himself "any such idea answering the word *unity*." "Methinks I could not miss finding it," he added, ". . . it should be the most familiar to my understanding, since it is said to accompany all other ideas, and to be perceived by all the ways of sensation and reflexion." Alas, he concluded, "To say no more, it is an *abstract idea*."[47] And, of course, earlier in this work he had begun to elaborate his antiabstractionist stance.

He returned in sections 118–122 to the question of number as he tried to explain what mathematics was really about. No matter "how celebrated . . . [the mathematical sciences] may be, for their clearness and certainty of demonstration, which is hardly any where else to be found," he began, "[they] cannot nevertheless be supposed altogether free from mistakes." Participating in the common errors of abstractionism and materialism, mathematicians supposed, particularly, that arithmetic had "for its object abstract ideas of *number.*" Now appealing to section 13, Berkeley reminded readers that "we may conclude that, if there be no such thing as unity or unit in abstract, there are no ideas of number in abstract denoted by the numerical names and figures." This being so, arithmetic theories,

> if they are abstracted from the names and figures, as likewise from all use and practice, as well as from the particular things numbered, can be supposed to have nothing at all for their object. Hence we

46. *Works of Berkeley,* 2:44–46.   47. Ibid., 2:46.

may see, how entirely the science of numbers is subordinate to practice, and how jejune and trifling it becomes, when considered as a matter of mere speculation.[48]

As this statement suggests, Berkeley allowed for the practical, if not the speculative, value of arithmetic. Indeed, he used the history of arithmetic to emphasize its practical roots. Briefly sketching the move from counters to numbers and then to the Hindu–Arabic numeral system, he concluded that humans had simply invented "signs . . . to represent aptly, whatever particular things . . . [they] had need to compute." Arithmetic, then, "regard[s] not the *things* but the *signs,* which nevertheless are not regarded for their own sake, but because they direct us how to act with relation to things, and dispose rightly of them."[49]

Although in the *Principles* Berkeley did not extend the science of signs to include algebra, neither did he ignore the subject. In the work's theme-setting introduction, as in his letter of 1709 to Molyneux, he implied a traditional view of algebra as part of the science of quantity but, more important, he used algebra to illustrate the major point that strict reasoning could be conducted on words without concern for their significates. The appeal to algebra came as he gave his version of the origins of the doctrine of abstract ideas. "It is a received opinion," he observed, "that language has no other end but the communicating our ideas, and that every significant name stands for an idea." As men and women encountered significant names that did "not always mark out particular conceivable ideas," they supposed such names to stand for abstract ideas. Objecting fundamentally to this supposition, he countered:

> [A] little attention will discover, that it is not necessary (even in the strictest reasonings) significant names which stand for ideas should, every time they are used, excite in the understanding the ideas they are made to stand for: in reading and discoursing, names being for the most part used as letters are in *algebra,* in which though a particular quantity be marked by each letter, yet to proceed right it is not requisite that in every step each letter suggest to your thoughts, that particular quantity it was appointed to stand for.[50]

Berkeley here suggested algebra as a paradigm of reasoning that proceeded without consideration of particular or general ideas. He used algebra subtly to refute the traditional claim, which Locke had reiterated, that there was no reasoning without ideas. After all, in algebra – which, as mathematics, adhered to the "strictest reasonings" – mathema-

48. Ibid., 2:94–96.   49. Ibid., 2:96–97.
50. Ibid., 2:37.

ticians reasoned first and foremost on symbols. Symbolical reasoning, which had so transformed algebra, was now threatening to transform Western philosophy.

8

In *Alciphron, or the Minute Philosopher* of 1732, Berkeley elaborated his mature theory of arithmetic and algebra as "sciences of great clearness, certainty, and extent, which are immediately conversant about signs." Although between the *Principles* and *Alciphron* his basic view of number remained unchanged, his attitude toward arithmetic seemed to improve. He continued to stress the practical nature of arithmetic; but he now wrote of the subject more positively. Of numbers, he proclaimed: "They direct us in the disposition and management of our affairs, and are of such necessary use that we should not know how to do without them."[51] Furthermore, he built around arithmetic and algebra a larger doctrine of signs, which he sketched confidently in *Alciphron,* and which, in turn, became one of his major contributions to Western philosophical discourse.

In the seventh dialogue of *Alciphron,* he argued forcefully against the traditional contention that: "Words are signs: they do or should stand for ideas, which so far as they suggest they are significant. But words that suggest no ideas are insignificant."[52] He was here ostensibly concerned with religion and words used in religious discourse, and, in particular, with the charge that religion could be ignored because it depended on "empty notions, or, to speak more properly, . . . mere forms of speech, which mean nothing, and are of no use to mankind." The first word tackled in the dialogue was "grace," of which Alciphron said to Euphranor (the spokesman for Berkeley's ideas), "I profess myself altogether unable to understand it, or frame any distinct idea of it." Now, the lack of an idea of grace had potentially serious religious consequences, as Alciphron added, "therefore I cannot assent to any proposition concerning it [grace], nor, consequently have any faith about it." One could, of course, simply assume that murky or mysterious terms

51. Ibid., 3:305, 293.
52. Ibid., 3:287. Note that from the beginning of this dialogue, even as he restated the common doctrine of the association of words with ideas, Berkeley wrote of "signs," which was of course the main subject of the dialogue.

were acceptable in religious discourse. But here as in the later *Analyst* Berkeley required that "the same rules of reason and good sense which obtain in all other subjects ought to take place in religion."[53] In short, Berkeley assigned Euphranor the problem of explaining how a word could be bereft of any clear idea and yet significant.

Euphranor's initial response was one that Berkeley had already sketched in the *Principles,* that is, that men and women could and did reason on words without concern for their significates, not only in religion but in other subjects as well. "Words, it is agreed, are signs," Euphranor began. But counters at a card table are also signs. "Say now, Alciphron," he asked, "is it necessary every time these counters are used throughout the progress of a game, to frame an idea of the distinct sum or value that each represents?" Getting the appropriate (negative) response, Euphranor suggested that it seemed "to follow, that words may not be insignificant, although they should not, every time they are used, excite the ideas they signify in our minds." He continued: "there may be another use of words besides that of marking and suggesting distinct ideas, to wit, the influencing our conduct and actions, which may be done either by forming rules for us to act by, or by raising certain passions, dispositions, and emotions in our minds."[54]

As in the *Principles,* Berkeley now claimed through Euphranor that the false belief that words ought always to stand for ideas had led to the equally false doctrine of abstract general ideas. Alciphron asked: "Will you not allow then that the mind can abstract?" Euphranor gave Berkeley's antiabstractionist response, and a still-unconvinced Alciphron observed: "And yet it is a current opinion that every substantive name marks out and exhibits to the mind one distinct idea separate from all others." At this point, Euphranor engaged Alciphron in a discussion of number, which concluded that numbers were signs that "in their use imply relations or proportions of things," and yet did not stand directly for perceptible objects, abstract ideas of such objects, or abstract ideas of their relations. More germane to the main subject of the seventh dialogue, Euphranor cautioned that "to attain a precise simple abstract idea of number is as difficult as to comprehend any mystery in religion."[55] If mathematicians could have their numbers (and physicists, their force),[56]

---

53. Ibid., 3:287–290.   54. Ibid., 3:291–292.
55. Ibid., 3:292–293, 305, 293.
56. Berkeley also argued that he could frame no idea of force, although there were "very evident propositions or theorems relating to force" (ibid., 3:293–296).

religious men could discourse about grace. Numbers, force, grace, and the like were significant signs of which the mind could form no precise distinct ideas.

Although a religious polemic, *Alciphron* was a brilliant explication of an early modern theory of language or, more specifically, a theory of signs, that was heavily dependent on insights drawn from early modern arithmetic and algebra. "Words are signs," Berkeley repeated throughout the seventh dialogue. As his examples – counters, then numbers, and finally imaginary numbers – became more abstruse, so his theory of signs became more sophisticated. Whereas he used counters to illustrate the possibility of reasoning on signs without concern for their significates, he used numbers to illustrate reasoning on signs that admitted no "precise simple abstract idea." And, at the high point of his theory, he turned to imaginary numbers to clinch his argument that men and women could engage in sound and useful reasoning on signs that stood for absolutely no ideas, either particular or general. He elaborated this philosophical lesson of the imaginary numbers as he developed the point that language does not aim solely at a comparison of ideas, but sometimes at "something of an active operative nature, tending to a conceived good." Such a good, he explained,

> may sometimes be obtained, not only although the ideas marked are not offered to the mind, but even although there should be no possibility of offering or exhibiting any such idea to the mind: for instance, the algebraic mark, which denotes the root of a negative square, hath its use in logistic operations, although it be impossible to form an idea of any such quantity. And what is true of algebraic signs is also true of words or language, modern algebra being in fact a more short, apposite, and artificial sort of language, and it being possible to express by words at length, though less conveniently, all the steps of an algebraical process.[57]

Arithmetic and algebra, Euphranor concluded, "are sciences of great clearness, certainty, and extent, which are immediately conversant about signs, upon the skilful use and management whereof they entirely depend." But the seventh dialogue was not strictly about mathematics, and he soon added: "If I mistake not, all sciences, so far as they are universal and demonstrable by human reason, will be found conversant about signs as their immediate object, though these in the application are referred to things."[58] With *Alciphron,* then, early modern algebra (including the doctrine of imaginary numbers, as developed primarily by

57. Ibid., 3:307.   58. Ibid., 3:305.

Wallis),[59] found a place within Western philosophy, but the philosophy that sanctioned the new algebra was one that was being transformed by it. As Berkeley suggested, no traditional theory of language could prevail in the wake of the arithmetic and algebraic developments of the seventeenth century.

9

By the 1730s, then, Newton's pragmatic algebra (with its limited use of roots involving $\sqrt{-1}$) aside, British thinkers had outlined two alternative and yet somewhat comprehensive theories of algebra. From the mathematical side, Wallis's theory developed algebra as an extension of arithmetic in which symbols stood for quantities; from the philosophical, Berkeley's theory defined arithmetic and algebra as sciences of signs. In their time, both theories were daring and not totally unobjectionable. The negative and imaginary numbers were anomalies in Wallis's theory; they were explainable only by appeal to nonarithmetic considerations such as precedent, geometric interpretation, and usefulness. On the other hand, the negative and imaginary numbers were no problem for Berkeley's very abstract view of algebra. But there was a different rub: Berkeley's philosophy of arithmetic and algebra made sense in the context of his new general philosophy but not in the context of more traditional philosophies. In order to save the negative and imaginary numbers, Wallis and even Harriot had begun rewriting the rules of mathematics; at least partially to make sense of early modern algebra, with its imaginary numbers, Berkeley had begun rewriting the rules of Western philosophy.

Through the early eighteenth century, neither Wallis's nor Berkeley's theory dominated British reflection on algebra. As will be explained, many British mathematical thinkers seemed to know little of, and care not at all for, Berkeley's radical views on arithmetic and algebra. In Great Britain, Newton's *Universal Arithmetick* largely remained the preeminent algebra textbook. Not as daring as Wallis and Berkeley, most British mathematical thinkers probably balked at any conscious tamper-

---

59. Berkeley shared Wallis's view of imaginary numbers, stressing their usefulness over their impossibility. It is likely that he had read Wallis's account of these numbers in chapter 66 of the *Treatise of Algebra*, since in "De cono aequilatero et cylindro" of his *Miscellanea* he referred to chapter 81 of the *Treatise* (*Works of Berkeley*, 4:213).

ing with the standard rules of mathematics or philosophy. It was a major step, whether in mathematics or philosophy, to move beyond the criterion of "clear and distinct ideas" as the basis of knowledge.

Even in Berkeley's *Analyst,* a few of the "queries," on which the author asked mathematicians to "meditate," seemed to leave open the possibility that algebra ranked somewhere below geometry – pedagogically, if not scientifically. Query 38, for example, questioned algebra's role in training the mind. Whereas (as we have seen) the text of *The Analyst* extolled geometry as a subject that sharpened the mind, the latter query asked: "Whether tedious calculations in algebra and fluxions be the likeliest method to improve the mind?" Perhaps tellingly, query 45 described geometry as a science and algebra as "allowed to be a science." Other queries alluded to concern about the applicability of the general rules of symbolical algebra to particular cases, including those of calculus. Thus query 40 considered the "rule, that one and the same coefficient dividing equal products gives equal quotients" – if $a \times b = a \times c$, then $b = c$. Berkeley asked if this rule was indeed general, and "whether such coefficient can be interpreted by o [Newton's increment] or nothing? Or whether any one will say that if the equation $2 \times o = 5 \times o$, be divided by o, the quotients on both sides are equal?"[60]

But query 41 directed mathematicians to meditate on the scientific character of algebra. It suggested that mathematicians "demonstrate as well" in algebra as in geometry; and that they are "obliged to the same strict reasoning" in both subjects. In conclusion, it asked: "And whether such their reasonings are not deduced from the same axioms with those in geometry? Whether therefore algebra be not as truly a science as geometry?"[61] Contained in *The Analyst,* then, were hints of the tension between the old emphasis on content, which led Hobbes and Barrow to rank geometry above arithmetic and algebra, and the newer emphases on method, applicability, and generality, stressed by Wallis. Although by the writing of *Alciphron* he seems to have taken the final step toward accepting a purely symbolical algebra and embedding it into philosophy, Berkeley acknowledged here that as of 1734 mathematical thinkers (did he include himself?) still needed time and meditation to come to terms pedagogically, if not scientifically, with arithmetic and algebra as sciences of signs.

60. Ibid., 4:99–100; see also query 46 (4:100–101). Interpretation of queries is, of course, difficult. Compare my analysis here with Jesseph, *Berkeley's Philosophy,* 93 n. 5.
61. *Works of Berkeley,* 4:100.

10

The Analyst of 1734 and its sequel, *A Defence of Free-thinking in Mathematics,* of 1735 concerned calculus much more than algebra. As in *Alciphron,* so in these later works Berkeley used his mathematical knowledge to defend religion. According to the standard accounts, he was moved to write *The Analyst* when he heard that Sir Samuel Garth had refused to take the last rites because he had lost his faith under Halley's influence.[62] With the subtitle *A Discourse Addressed to an Infidel Mathematician, The Analyst* attacked "infidel mathematicians" who were thought by men like Garth to "apprehend more distinctly, consider more closely, infer more justly, conclude more accurately than other men, and . . . therefore [be] less religious because more judicious." In a nutshell, Berkeley argued that mathematicians were not preeminently judicious; calculus was riddled with meaningless terms and faulty logic; the study of calculus therefore did "not habituate and qualify the mind to apprehend clearly and infer justly"; and mathematicians ought to be thought no more able to discern religious truths and errors than any "other men."[63]

In *The Analyst* Berkeley focused his attention on the fundamental entities of early-eighteenth-century calculus, including Newton's fluxions and Leibniz's infinitesimals.[64] As he made clear (in an argument and terms reminiscent of Hobbes's attack on Wallis's indivisibles), he was skeptical of the foundations of calculus although appreciative of the subject's usefulness. His skepticism stemmed from his view of calculus

62. Many biographies of Berkeley repeat this story (e.g., Luce, *Life of George Berkeley,* 164). But Cantor has argued that there is insufficient evidence to support the claim that *The Analyst* was directed against Halley. Despite Berkeley's reference to "an infidel mathematician," the work may have been directed at deists and freethinkers in general (Geoffrey Cantor, "Berkeley's *The Analyst* Revisited," *Isis* 75 [1984]: 668–683, esp. 673).

63. *Works of Berkeley,* 4:65, 95 (Berkeley summarizes his argument here).

64. The literature on the mathematical aspects of *The Analyst* and the ensuing debate includes Florian Cajori, *A History of the Conceptions of Limits and Fluxions in Great Britain* (Chicago: Open Court, 1919), 57–148; J. O. Wisdom, "The *Analyst* Controversy: Berkeley As a Mathematician," *Hermathena* 59 (1942): 111–128; Carl B. Boyer, *The Concepts of the Calculus: A Critical and Historical Discussion of the Derivative and the Integral* (New York: Hafner, 1949; reprint, as *The History of the Calculus and Its Conceptual Development,* New York: Dover, 1959), 224–236; and Ivor Grattan-Guinness, "Berkeley's Criticism of the Calculus as a Study in the Theory of Limits," *Janus* 56 (1969): 215–227.

as a refined part of geometry. Since he saw geometry as a science of perceptible extension, he expected the fundamental entities of calculus to stand for extension or particular ideas. Already in the early *Commentaries* he had alluded to the core of his problem with calculus. Entry 337 raised the issue of the infinitesimals' "being nothings." Entry 354a explained what prevented his developing calculus as a science of signs similar to arithmetic and algebra: "[N]or can it be objected that we reason about Numbers wch are only words & not ideas, for these Infinitesimals are words, of no use, if not suppos'd to stand for Ideas."[65]

A quarter-century later *The Analyst* developed the hints of the *Commentaries*. In this later work Berkeley described fluxions as "the general key by help whereof the modern mathematicians unlock the secrets of Geometry" and analysis as "a most excellent method"; he characterized calculus, which he saw as analysis applied to geometry, as "the abstruse and fine geometry."[66] Seeing calculus as geometry and clearly not a science of signs, he challenged mathematical thinkers to explain fluxions and infinitesimals. For example, in an oft-quoted passage of *The Analyst,* he asked: "And what are these fluxions? The velocities of evanescent increments? And what are these same evanescent increments? They are neither finite quantities, nor quantities infinitely small, nor yet nothing. May we not call them the ghosts of departed quantities?" Again and again, he argued that fluxions and infinitesimals were "inconceivable."[67] Trying to account for this metaphysical gap in calculus, he suggested that mathematicians were "wonderfully deceived and deluded by their own peculiar signs, symbols, or species." Of fluxions and infinitesimals, he wrote: "if we remove the veil and look underneath, if, laying aside the expressions, we set ourselves attentively to consider the things themselves which are supposed to be expressed or marked thereby, we shall discover much emptiness, darkness, and confusion; nay, if I mistake not, direct impossibilities and contradictions."[68]

---

65. *Works of Berkeley,* 1:41–42. The thesis that Berkeley saw calculus as part of geometry, and therefore demanded ideal backing for its fundamental entities, is developed in Pycior, "Mathematics and Philosophy," 284–285. See also Jesseph, *Berkeley's Philosophy,* 158–159.
66. *Works of Berkeley,* 4:66, 100, 88.
67. Ibid., 4:89; on the inconceivability of fluxions and infinitesimals, see, e.g., ibid., 4:67–68, 95.
68. Ibid., 4:69. In *The Analyst* Berkeley briefly pursued a symbolical approach to calculus: "possibly some men may hope to operate by symbols and suppositions, in such sort as to avoid the use of fluxions, momentums, and infinitesimals." Reaching an inconsistency, he suggested that "all attempts

Compounding his skepticism toward calculus were the logical lapses of its proponents. As he told the mathematical analysts, "your inferences are no more just than your conceptions are clear, and . . . your logics are as exceptionable as your metaphysics."[69] As is well known, his attack on the logic of early modern calculus cited Newton's work on fluxions, including his derivation in the *Principia* of the moment of a product (his product rule for differentiation). Here, in calculating the moment of a rectangle with sides A and B, Newton had considered momentaneous increments and decrements $\frac{1}{2}a$ and $\frac{1}{2}b$ of the sides, subtracted the area of the diminished rectangle ($[A - \frac{1}{2}a] \times [B - \frac{1}{2}b]$) from that of the augmented rectangle ($[A + \frac{1}{2}a] \times [B + \frac{1}{2}b]$), and found the moment to be $aB + bA$. Berkeley corrected Newton:

> it is plain that the direct and true method to obtain the moment or increment of the rectangle AB, is to take the sides as increased by their whole increments, and so multiply them together, $A + a$ by $B + b$, the product whereof $AB + aB + bA + ab$ is the augmented rectangle; whence, if we subduct AB the remainder $aB + bA + ab$ will be the true increment of the rectangle, exceeding that which was obtained by the former illegitimate and indirect method by the quantity ab.

Berkeley's point was that Newton had conveniently ignored the product of the two momentaneous increments (the term ab) to obtain the desired fluxion, and in doing so had violated the norms of mathematical reasoning. Newton himself had defended mathematical strictness in his "Tractatus de quadratura curvarum," and Berkeley concluded his discussion of the moment of a rectangle by quoting (in Latin) Newton's relevant sentence: "For Errours, tho' never so small, are not to be neglected in Mathematicks."[70]

Despite faulty principles and methods, calculus led to true and useful

for setting the abstruse and fine geometry [calculus] on a right foundation . . . will be found impracticable, till such time as the object and end of geometry are better understood" (ibid., 4:88). See also query 27 (ibid., 4:98).

69. Ibid., 4:95.

70. Ibid., 4:70; *Mathematical Works of Newton*, 1:141. On Newton's argument here and Berkeley's criticisms, see, e.g., Cajori, *Conceptions of Limits*, 58–59, and Jesseph, *Berkeley's Philosophy*, 189–192. Berkeley also exposed a flaw in Newton's derivation of the fluxion of any power in "De quadratura curvarum," where to find the fluxion of x^n Newton had introduced the increment o, such that x by flowing became $x + o$. As Hobbes had earlier told Wallis that he was not free to regard an indivisible

conclusions, which, according to Berkeley, deceived mathematicians into thinking that they were engaged in science. The true results, he argued, were products of the "compensation of errors." Offering detailed examples, he claimed to show that in some of their proofs analysts had made not one, but two errors, which canceled each other. "If you had committed only one error," he observed, "you would not have come at a true solution of the problem. But by virtue of a twofold mistake you arrive, not at science, yet at truth."[71]

The thesis that calculus, although leading to truth, was unscientific, was essential to *The Analyst*. No simple mathematical tract, the work was, after all, a religious polemic formally directed against "infidels" who used their mathematical reputations to lead others into disbelief. Berkeley's main point was that mathematicians had no special claim as sound reasoners, since the metaphysical and logical gaps of calculus could be tolerated only by someone "pass[ing] for an artist, computist, or analyst, yet . . . not . . . justly esteemed a man of science and demonstration." Exploiting the shortcomings of calculus to full religious advantage, he gibed that "he who can digest a second or third fluxion . . . need not, methinks, be squeamish about any point in divinity," and that Newton's derivation of the fluxion of any power was "a most inconsistent way of arguing, and such as would not be allowed of in Divinity."[72]

II

Mathematical reaction to *The Analyst* was intense because, while attacking the principles and methods of calculus, its author was seen to have impugned not only the professional standards of mathematicians but their religious beliefs as well. The religious challenge alone may have been sufficient to evoke an angry response from mathematical thinkers, since at least the appearances of religious conformity still remained a prerequisite for British academic appointments. Newton had concealed his Arian or unitarian inclinations, but in 1691 Halley had lost a bid for

alternately as "something" and as "nothing," Berkeley now charged Newton with treating *o* as an increment and as nothing (*Works of Berkeley,* 4:74).

71. *Works of Berkeley,* 4:71, 76, 78 (quotation). See J. O. Wisdom, "The Compensation of Errors in the Method of Fluxions," *Hermathena* 57 (1941): 49–81.
72. *Works of Berkeley,* 4:86, 68, 73.

the Savilian professorship of astronomy at Oxford because of his public irreligion. In 1709 and again in 1714, Samuel Clarke, another disciple of Newton and a suspected Arian, had been pressed to renounce the heresy publicly. Most strikingly, William Whiston's revelation of Arian beliefs had led in 1710 to his expulsion from Cambridge.[73]

A long line of British mathematical thinkers and other Newtonians, then, took up pens to defend themselves against infidelity as well as to defend the great Sir Isaac and his calculus. The first and "the most consequential" of "the above twenty tracts and dissertations" on *The Analyst* that appeared within a decade of its publication[74] was James Jurin's anonymous work, *Geometry No Friend to Infidelity: Or, A Defence of Sir Isaac Newton and the British Mathematicians* of 1734. Jurin, a physician with strong Cambridge ties,[75] discussed what he saw as the three main arguments of *The Analyst*: mathematicians were infidels; they used their standing as the supposed "greatest masters of reason" to convert others to infidelity; and there was "error and false reasoning" in mathematics.[76] Before Jurin was done with Berkeley, he had tried to demolish all three points, defended Newton as promised in his subtitle, portrayed Berkeley as an egomaniac, and even attacked his antiabstractionism. *Geometry No Friend to Infidelity* thus set the stage for the public rejection of Berkeley and his mathematical views by English mathematical thinkers of the late 1730s and 1740s.[77]

73. Richard S. Westfall, *Never at Rest: A Biography of Isaac Newton* (Cambridge: Cambridge University Press, 1980), esp. 315, 500, 649–652. Westfall describes Newton as "an Arian in the original sense of the term. He recognized Christ as a divine mediator between God and man, who was subordinate to the Father Who created him" (ibid., 315). See John Gascoigne, *Cambridge in the Age of the Enlightenment: Science, Religion and Politics from the Restoration to the French Revolution* (Cambridge: Cambridge University Press, 1989), 117–122.

74. David Berman, introduction to *George Berkeley: Eighteenth-Century Responses*, ed. David Berman, 2 vols. (New York: Garland, 1989), 1:vi. This count comes from Berkeley.

75. Jurin had been a student at Trinity College, Cambridge, and "had imbibed Newtonian teachings from Newton himself" (Cajori, *Conceptions of Limits*, 64).

76. Philalethes Cantabrigiensis [James Jurin], *Geometry No Friend to Infidelity: Or, A Defence of Sir Isaac Newton and the British Mathematicians, in a Letter to the Author of the "Analyst"* (1734; reprint, *George Berkeley*, ed. Berman, 1:325–412), 334.

77. Since my intent is merely to suggest how *The Analyst* and ensuing controversy affected British algebra, the following discussion will be restricted to

As Jurin's tract showed, mathematical thinkers and other Newtonians were deeply concerned about Berkeley's charge of infidelity. It was the first issue that Jurin discussed. Mathematicians were not infidels, he began, and, even if they were, their infidelity should not be exposed to public scrutiny. "Though the evidence and certainty of our holy Religion is so firmly established, as not to be shaken . . . by the reputation and authority of any unbelievers whatsoever; yet, I am afraid, it would be a great stumbling block to men of weak heads, if they were made to believe, that the justest and closest reasoners were generally Infidels." But more than concern for the corruption of "weak heads" lay behind Jurin's desire to squelch the charge of infidelity. Pushing the charge to the limit and exaggerating its possible consequences, he exclaimed: "Let us burn or hang all the Mathematicians in *Great Britain,* or halloo the mob upon them to tear them to pieces every mother's son of them. . . . Let us dig up the bodies of Dr. *Barrow* and Sir *Isaac Newton,* and burn them under the gallows."[78] These remarks were dramatic flourish, but underneath them there assuredly lurked the memory of Whiston's ouster from Cambridge.

Jurin and, most likely, other participants in the campaign against *The Analyst* feared also the destruction of the careful case for the study of mathematics that Oughtred, Barrow, and other English thinkers had crafted during the seventeenth century. When Cambridge had a mission to educate future clergymen, God was portrayed as a geometer and mathematics was touted as the main way of reading his "expresse prints" on the world[79] and as the best logic. Even if no infidel mathematician were to suffer academic ostracism because of Berkeley's barbs, there remained the larger intellectual threat that mathematics, once associated with infidelity, would be pushed back to the periphery of British academia.

Jurin, then, carefully refuted Berkeley's first charge that mathematicians were infidels and his second that they corrupted others. He stressed that Berkeley's only evidence of infidelity was hearsay: he had been told by an "informer" of an unnamed mathematician who had corrupted Sir Garth. In reality, Jurin declared, the best of the British mathematicians had been religious. Barrow was "esteemed one of the greatest luminaries of the Christian Church," and Newton – whose Arianism had been

Jurin's work. On other responses to *The Analyst,* see Cajori, *Conceptions of Limits,* 69–148, and Jesseph, *Berkeley's Philosophy,* 231–295.

78. [Jurin], *Geometry No Friend to Infidelity,* 338–339, 354.
79. William Oughtred, *To the English Gentrie, and All Others Studious of the Mathematicks* (1632?), 8.

largely concealed even beyond his death – was "acknowledged to have been a true believer, and to have given some of the strongest and clearest proofs, that have ever been produced, of the goodness, wisdom, and power of the Supreme Being." Against Berkeley's second charge, Jurin argued that the religious beliefs of individual mathematicians did not generally influence the beliefs of their societies. The example of Barrow, who was "a learned, sound and Orthodox Divine," had not swayed "the Arians, the Socinians, the Quakers, and . . . other sects of Dissenters" to Anglican orthodoxy. Similarly, Jurin wrote, "Sir *Isaac Newton* was a greater Mathematician than any of his contemporaries in *France*; . . . and yet I have not heard that the French Mathematicians are converted to the Protestant Religion by his authority."[80] As devout mathematicians had not converted others to their religious beliefs, no one should fear that infidel mathematicians (if there were any) would sow rampant infidelity.

This argument notwithstanding, Jurin was clearly unprepared to abandon the close relationship between religion and mathematics, and sound thinking and mathematics, that had become so crucial a part of the defense of mathematics as an academic subject. He deftly emphasized that mathematics, on the one hand, and logic and religion, on the other, were connected on the individual level, if not on the social. Geometry was "an excellent Logic," and sound reasoners were generally religious men. Quoting Berkeley's lengthy passage on geometry as a logic from *The Analyst*, he asked: "is it probable, or even possible, . . . that a great number of persons, who have *acquired this habit of reasoning,* should generally not see and comprehend the clear, the certain and undeniable evidence of the Christian Religion?" If there were mathematical infidels who were "dangerous adversaries to Christianity," Jurin proposed that they be answered by mathematical clergy. "Let no man enter into orders, unless he be an able Mathematician," he dictated. "When this is done, Sir, let us see what Mathematical Infidel will dare to beard a Christian Priest, as *great a master of reason* as himself, and armed besides with his Theological accomplishments."[81]

As these statements imply, Jurin rejected completely Berkeley's argument that, in practice, calculus and religion did not deal solely with clear ideas. He rather reiterated the thesis that the study of mathematics would train men to appreciate "the clear, the certain and undeniable evidence" of Christianity. In adhering to the norm of clarity of ideas, he and many of his contemporaries refused to appreciate and, in fact,

---

80. [Jurin], *Geometry No Friend to Infidelity,* 354–355, 357–358.
81. Ibid., 356, 343–344; see also 348.

distorted Berkeley's thesis. At the root of the misinterpretation was the incompatibility of Berkeley's special brand of philosophy, religion, and mathematics, on the one hand, and the shared philosophical and religious tenets of many contemporary English mathematicians and scientists, on the other. Locke's philosophy had found a congenial reception at Cambridge University;[82] the tradition of English natural theology sought rational proofs of the existence and attributes of God;[83] and even Wallis had used the analogy of the cube with its three dimensions to demonstrate the reasonableness of the doctrine of the Trinity.[84] Clarke and other Cambridge "latitudinarians," moreover, suggested that reason – and not revelation, clergy, or the Church – ought to be the final arbiter of religious doctrine.[85]

Berkeley and Jurin thus talked at each other, but not to each other. Jurin implied as much at the beginning of his tract, where he admitted that his expectations for *The Analyst* differed radically from the realities of the work. "If your design were to be guessed at from your Title-page," he wrote Berkeley,

> wherein you profess to *examine, whether the object, principles and inferences of the modern Analysis are more distinctly conceived, or more evidently deduced, than religious mysteries and points of faith*, one would be apt to conclude . . . that you were about to give us a Mathematical demonstration, or one of equal clearness and certainty, of the truth of the Christian Religion.

Berkeley had done no such thing. Rather, as Jurin saw it, he had argued "that there is no more evidence and certainty in the modern Analysis, than in the Christian Religion." Jurin then appealed to divines to decide how much "honour to Christianity" Berkeley had done thereby.[86]

Their religious fidelity challenged, their defense of mathematics as the special path to God threatened, British mathematicians and their associates could hardly have been expected as a group to respond posi-

82. Gascoigne, *Cambridge in the Age of the Enlightenment*, 7, 140.
83. See Richard S. Westfall, *Science and Religion in Seventeenth-Century England* (New Haven: Yale University Press, 1958).
84. Wallis wrote eight items on the Trinity, the first of which was his *Doctrine of the Blessed Trinity Briefly Explained in a Letter to a Friend*, dated 1609, in which he compared the threeness of the cube and the Trinity (R. C. Archibald, "Wallis on the Trinity," *American Mathematical Monthly* 43 [1936]: 35–37).
85. Gascoigne, *Cambridge in the Age of the Enlightenment*, 4–6 (on the term "latitudinarian"), 115–141; Cantor, "Berkeley's *The Analyst* Revisited," 670–673.
86. [Jurin], *Geometry No Friend to Infidelity*, 335–337.

tively to *The Analyst* or to Berkeley's other writings, which presumably all now suffered by association with the latter. As Jurin elaborated a religious response to *The Analyst*, so he also sketched a mathematical response and attacked Berkeley personally. He defended and tried to clarify Newton's ideas on calculus, while reminding readers that calculus was after all a "very late invention" and that Berkeley had not questioned traditional geometry or algebra.[87] On a personal level, noting that Berkeley had attacked Locke's abstractionism in the *New Theory of Vision* as well as Newton's calculus in *The Analyst*, he imputed a combination of ignorance and egomania to the Irish bishop. The attacks on Newton and Locke showed, he argued, that Berkeley cared not so much to defend reason and Christianity as to enhance his own reputation. "[Y]ou have too great an opinion of your self, and too mean a one of all other men," he accused Berkeley.

> Hence, not content with the reputation you have deservedly acquired of being a clear and just reasoner, you can never rest, unless you convince the world that all those, who have hitherto been esteemed the *greatest masters of reason,* are in this respect greatly inferior to Dr. B——y.[88]

12

The Analyst was a complicated, elusive tract: it interwove mathematical, philosophical, and religious themes; it couched some of its points in satire or in queries. Still, it was one of the most influential mathematico-philosophical works ever published, as British mathematicians let it largely set their research program early in the second third of the eighteenth century. Attachment to Locke's theory of ideas, a compelling need to defend themselves against the charge of infidelity, belief in the reasonableness of religion, respect for Newton's memory – that is, philosophical, religious, social, and national factors – as well as concern for foundations mandated that British mathematicians try to resolve the metaphysical and logical gaps in calculus that Berkeley had exposed.

Berkeley's philosophy of arithmetic and algebra was, to a certain extent, a casualty of *The Analyst*. For Berkeley the disputant it had been

87. Ibid., 347. On Jurin's and Berkeley's extended debate, see, e.g., Cajori, *Conceptions of Limits,* 64–69, 72–78, 80–85.
88. [Jurin], *Geometry No Friend to Infidelity,* 350–351. Jurin's criticism of Berkeley's antiabstractionism centered on a defense of Locke's idea of a triangle (ibid., 399–412).

essential that the attack on calculus be as clean as possible: the concepts and methods of calculus had to be starkly contrasted with those of classical geometry; a detailed explanation of a second, more abstract kind of mathematics would have been out of place, if not damaging, to his main argument. Therefore *The Analyst* and its sequel, *Defence of Free-thinking in Mathematics* – Berkeley's two works that enjoyed wide readership among British mathematical thinkers and students throughout the eighteenth century – contained but scattered germs of his philosophy of arithmetic and algebra as sciences of signs. If *The Analyst* affected algebra through the midcentury, then, it did so largely indirectly, by focusing British mathematical research on the foundations of calculus. In the wake of *The Analyst*, the foundational questions of calculus were simply more pressing than those of algebra.

Even if Berkeley had not antagonized his mathematical contemporaries with *The Analyst*, it seems unlikely that many would have been won over to his philosophy of arithmetic and algebra, which was intimately connected with his antiabstractionism. His general philosophy was slow to win converts in England, although somewhat faster in Scotland.[89] Jurin's *Geometry No Friend to Infidelity*, in particular, suggested that by the 1730s English mathematical thinkers and their fellow Newtonians were unprepared for an abstract theory of arithmetic and algebra as sciences of signs. Satirically, Jurin declared himself "one of those who still adhere to the vulgar, or rather universal error of all Mankind, that neither Geometry, nor any other general science can subsist without general Ideas."[90]

89. Berman, introduction to *George Berkeley*, 1:v–viii. In England, serious discussion of Berkeley's philosophy seems to have begun around the mid-eighteenth century (ibid., vi).
90. [Jurin], *Geometry No Friend to Infidelity*, 74.

9

The Scottish Response to Newtonian Algebra

Even if Berkeley's philosophy of the abstract sciences of arithmetic and algebra had little effect on British mathematics through the mid-eighteenth century, Newton's bifocal mathematical legacy helped to assure that algebra as well as geometry continued to be cultivated in Great Britain. Probably most early-eighteenth-century British mathematical thinkers were strongly attracted toward the geometric focus of Newtonian mathematics; some, however, felt at least an equal pull toward the arithmetico-algebraic focus.

In Scotland there were two schools of thought on algebra: the somewhat anti-Newtonian school, led by Robert Simson (1687–1768), and the largely Newtonian school, led by Colin MacLaurin (1698–1746). Simson – the "father-figure" of Scottish mathematicians of the first half of the eighteenth century – was primarily attracted to the geometric focus of Newtonian mathematics. By his later years, if not in his early career as well, he came near abandoning early modern algebra. He preferred geometric analysis to analytic geometry; perhaps questioned the symbolical style; and rejected the negative and imaginary numbers. MacLaurin, on the other hand, published first on geometry, then prepared a manuscript on algebra, and next devoted eight years of his life to answering Berkeley's *Analyst* with his *Treatise of Fluxions*, a largely geometric work with, however, a significant algebraic section.

By all accounts, MacLaurin was a brilliant mathematician; in many respects, he was the greatest of Newton's mathematical disciples. Early meetings with MacLaurin seem to have convinced Newton that he and the young Scotsman were kindred mathematical souls, who appreciated both the new and the old mathematics. MacLaurin, for his part, readily assumed the mathematical mantle of the aging Newton, and never quite came out from under it. Most famous for his defense of Newtonian calculus, he also played a major role in elaborating and defending Newtonian algebra. If his defense of calculus against the critique of Berkeley was important and definitive, his defense of algebra against the critique of Simson was equally important but less definitive. In calculus, Mac-Laurin could vindicate Newton's fluxions; in algebra, he could elaborate a new defense of the negative numbers but he could not explain the imaginaries. When writing about calculus, he felt obliged to admit that one should not even call the imaginaries "quantities." As MacLaurin's *Treatise of Fluxions* marked the culmination of the British geometric development of calculus, his *Treatise of Algebra* and the *Treatise of Fluxions* spoke tellingly, if not always directly, of the need for mathematicians to take the foundations of algebra as seriously as those of calculus.

I

Simson came of age in the immediate pre-Enlightenment period in Scotland. A bifocal, Newtonian style of mathematics had already been brought to the northern country by James Gregory (1638–1675), who served as the first Regius professor of mathematics at St. Andrews from 1668 to 1674 and then briefly as the professor of mathematics at Edinburgh. By 1690 Gregory's successor at Edinburgh, his nephew David Gregory, had – at the very least – introduced a few of his students to Newton's *Principia*.[1]

Simson attended Glasgow University, which had been slower to convert to the Newtonian ideas than Edinburgh[2] but by the 1730s would play its own part in the budding Scottish Enlightenment. His education prepared him well for his later mathematical research, which focused on the restoration of ancient geometric texts. Admitted to the university in

1. Christina M. Eagles, "David Gregory and Newtonian Science," *British Journal for the History of Science* 10 (1977): 216–225, esp. 218; Christine M. Shepherd, "Newtonianism in Scottish Universities in the Seventeenth Century," in *The Origins and Nature of the Scottish Enlightenment*, ed. R. H. Campbell and Andrew S. Skinner (Edinburgh: John Donald Publishers, 1982), 65–85, esp. 67. On James Gregory, whose work involved geometry and algebra, see D. T. Whiteside, "James Gregory (Gregorie)," *Dictionary of Scientific Biography*, ed. Charles C. Gillispie, 18 vols. (New York: Scribner's Sons, 1970–1990), 5:524–530, and Antoni Malet, "Studies on James Gregorie (1638–1675)" (Ph.D. diss., Princeton University, 1989). James Gregory (Gregorie) also appears in the literature as James I Gregory, to distinguish him from James Gregory (1666–1742?), his nephew who became the professor of mathematics at Edinburgh in 1692.
2. Shepherd, "Newtonianism in Scottish Universities," 76.

1701–1702, he studied under his uncle John Simson, the professor of divinity at Glasgow who would later be charged with heresy for an "undue" appeal to "reason and nature" in explaining religious doctrines.[3] The younger Simson distinguished himself not only in theology and philosophy but also in the classics, oriental languages, and botany, while managing by the end of his studies to learn some mathematics. Indeed, his notebook of 1705 reveals a familiarity with Oughtred's *Key* and Kersey's *Algebra*.[4]

Simson is reported to have come to the study of mathematics indirectly and mainly as a source of certitude. According to his student John Robison, he originally studied for the ministry under his uncle but, upset by the "vague speculation" of theology, he sought certitude elsewhere. At first he turned to "oriental philology, in which he found something which he could discover to be true or to be false." At those times when philology brought no relief for his intellectual "fatigue," he tried mathematics, "which never failed to satisfy and refresh him." For a long time, however, he was afraid to immerse himself in mathematics for fear of exhausting the subject. But he eventually realized that mathematics was an inexhaustible "cordial" as well as "a manly and important study," and, finally, that he could make a profession of it.[5]

Upon receiving his master's degree in 1711, he capped his education with a year's study in London. There contact with Edmond Halley sealed the direction of his future mathematical studies. As already noted, through 1710 Halley had been knee-deep in the restoration of geometric classics, and "Halley's influence tended to confirm him [Simson] in his predilection for the works of the Greek geometricians, for the study of

3. On the charge of 1715 and the strictures imposed by the Assembly on John Simson, see James K. Cameron, "Theological Controversy: A Factor in the Origins of the Scottish Enlightenment," in *The Origins and Nature of the Scottish Enlightenment,* ed. Campbell and Skinner, 116–130 (quotation from p. 119).

4. William Trail, *Account of the Life and Writings of Robert Simson, M.D.* (Bath, 1812), 2. On Simson's life, see also E. I. C., "Robert Simson," *Dictionary of National Biography,* ed. Leslie Stephen and Sidney Lee, 22 vols. (Oxford: Oxford University Press, 1921–1922), 18:287–288, and John Robison, "Dr Robert Simson," *Encyclopaedia Britannica,* 8th ed. (Boston: Little, Brown, 1853–1860), 20:298–302.

5. Robison, "Robert Simson," 299. Strictly accurate or not, the similar accounts of Simson's and (later) MacLaurin's turns from theology to mathematics probably helped bolster the standing of mathematics as an undergraduate subject that encouraged sound reasoning rather than dissent and controversy.

which his classical learning fitted him."[6] According to Robison, Simson found in Halley a man of "the most acute penetration, and the most just taste in that [mathematical] science." For his part, Halley gave Simson a copy of Pappus's *Mathematical Collection* (replete with Halley's personal notes), and advised him that the restoration of geometric classics was "the most certain way for him, then a very young man, both to acquire reputation and to improve his own knowledge and taste."[7]

At the end of his London year, Simson returned to Glasgow to assume the professorship of mathematics, which he was to hold for the next half-century. Following Halley's advice, he focused his research on the reconstruction of Euclid's porisms. Porisms, which were considered important examples of early geometric analysis, were known through Pappus's *Mathematical Collection,* which contained an abstract of Euclid's lost three-part treatise on the subject.[8] Simson and his disciples contended that porisms and, more generally, the geometric analysis of the ancients "had been entirely mistaken or despaired of by the best of the modern mathematicians."[9] Whereas Wallis and other seventeenth-century mathematical thinkers had implied that the ancients had used, but concealed, a sort of analysis that was akin to modern algebra, Simson argued that the analysis of the ancients was neither algebraic nor

6. E. I. C., "Robert Simson," 287.
7. Robison, "Robert Simson," 299–300.
8. There is a brief discussion of Simson's work on porisms in Anand C. Chitnis, *The Scottish Enlightenment: A Social History* (London: Croom Helm, 1976), 161–162. A problem of the period was the very definition of porism. Chitnis defines a porism as "a corollary or a proposition intermediate between a problem and a theorem" (ibid., 192 n. 76). See also John Playfair, "On the Origin and Investigation of Porisms," *Transactions of the Royal Society of Edinburgh* 3 (1784): 154–204, and *Pappus of Alexandria: Book 7 of the "Collection,"* ed. with translation and commentary by Alexander Jones, 2 pts. (New York: Springer-Verlag, 1986), 2:547–553 ("Euclid's Porisms").
9. This opinion was captured in the senate minutes of Glasgow University on the occasion of Simson's retirement (quoted in James Coutts, *A History of the University of Glasgow* [Glasgow, 1909], 226). Since Simson published little besides restorations and editions of ancient geometric works, his opinions on larger mathematical issues must be reconstructed from such evidence as senate minutes and students' reports of his lectures and table talk. Despite Simson's impression, Newton had tried to restore Euclid's porisms, but had published nothing on the topic (Richard S. Westfall, *Never at Rest: A Biography of Isaac Newton* [Cambridge: Cambridge University Press, 1980], 512).

purposely concealed; rather it was geometric and "described in the fragmentary treatise of Pappus."[10] Simson therefore engaged in painstaking and successful work toward reconstructing Euclid's porisms, "to which [the mathematician John Playfair later quipped] more genius was perhaps required than to the first discovery of them."[11] The research on porisms set the pattern for the rest of his career, during which he prepared commentaries on and restorations of works by Apollonius and, in 1756, a well-received and influential edition of Euclid's *Elements* and *Data*.

A deep appreciation of classical geometry seems to have led Simson to criticize analytic geometry. According to John Leslie, "from his familiarity with the ancient mode of demonstration . . . [Simson had] inhaled the very spirit of the Greeks,"[12] and with his increasing age that spirit seemed to gain the upper hand on any countervailing modern mathematical sympathies.[13] In lecture notes on l'Hospital's treatise on conic sections, he substituted geometric demonstrations for the original algebraic ones.[14] In a late letter to George Lewis Scott, an amateur mathematician and pupil of Abraham De Moivre, he declared: "You think algebra almost indispensably necessary in difficult enquiries: I have, on the contrary, been obliged to make use of the ancient method in many problems, in which I could not find that algebra was of any use to me."[15] Still, some of Simson's students and colleagues remembered a

10. William Trail, a student of Simson, summarized this opinion in his *Life and Writings of Simson*, 12–13. For Wallis's comments, see John Wallis, *A Treatise of Algebra, Both Historical and Practical, Showing the Original, Progress, and Advancement Thereof, From Time to Time, and by What Steps It Hath Attained to the Height at Which Now It Is* (London, 1685), preface.

11. John Playfair, "Account of Matthew Stewart, D.D.," *Transactions of the Royal Society of Edinburgh* 1 (1788): 57–76, on 59. Alexander Jones commends Simson's study of porisms in *Pappus of Alexandria: Book 7 of the "Collection,"* ed. Jones, 2:548.

12. John Leslie, *Dissertation Fourth; Exhibiting a General View of the Progress of Mathematical and Physical Science, Chiefly during the Eighteenth Century* (Edinburgh, 1824), 582.

13. Simson's "strong bias . . . to the analysis of the ancient geometers . . . increased as he went forward" (Robison, "Robert Simson," 299). See also Leslie, *Dissertation Fourth*, 582.

14. Trail, *Life and Writings of Simson*, 4. L'Hospital's *Traité analytique des sections coniques* appeared in 1720 and Edmund Stone's English translation in 1723.

15. Simson to Scott, 1 May 1764, reproduced in Trail, *Life and Writings of Simson*, 121.

mathematician who appreciated the old and new mathematics – or at least the new mathematics in its proper place. He taught some algebra and analytic geometry at Glasgow, but shared his reservations about these subjects with students. Robison and William Trail, who had also studied under Simson, stressed that his "animadversions . . . on the application of algebra to geometry, chiefly referred to those cases where it was not necessary, and in which the more excellent method of the ancients could be successfully employed."[16]

Newton, of course, had also expressed reservations about the mixing of algebra and geometry and, in particular, about analytic geometry, but – of interest to historians probing the reception of Newton's mathematical corpus in the eighteenth century – Simson apparently studied the *Principia* and *Universal Arithmetick* and concluded that Newton was more of an algebraist than a geometer:

> It would take up too much time, and writing too much for my eyes, to shew the advantage of the ancient method above the algebraic, and how the precepts given in this last method, particularly in the *Arithmetica Universalis,* lead those who observe them, from the right solution of geometrical problems, into such as are quite out of the natural method; many instances of this occur in that book.[17]

If Simson praised Newton, he praised the *Principia.* When illustrating to his students "the superiority of the geometrical over the algebraic analysis," he compared Newton's solution of the inverse problem of centripetal forces, in proposition 42 of the *Principia,* with the more algebraic solution given by Johann (Jean) Bernoulli in 1713.[18] Thus recognizing Newton's two public mathematical faces, he yet took the algebraic face as the truer expression of Newton the mathematician. He told his students that "to his own knowledge Newton frequently investigated his propositions in the symbolical way, and that it was owing chiefly to Dr Halley that they did not finally appear in that dress."[19]

Simson seems to have criticized not only the application of algebra to geometry but also such key components of early modern algebra as the symbolical style and algebra's expanded universe. Robison reported that Simson

16. Trail, *Life and Writings of Simson,* 67; see also Robison, "Robert Simson," 301–302.
17. Simson to Scott, 1 May 1764, 121–122.
18. Robison, "Robert Simson," 299.
19. Ibid. Simson's remarks here must be compared with evidence of Halley's algebraic interests. See, e.g., Edmond Halley, "To the Reader," in William Oughtred, *Key of the Mathematicks* (London, 1694).

felt a dislike to the Cartesian method of substituting symbols for operations of the mind, and still more was he disgusted with the substitution of symbols for the very objects of discussion, for lines, surfaces, solids, and their affections. He was rather disposed, in the solution of an algebraical problem, where quantity alone was considered, to substitute figure and its affections for the algebraical symbols, and to convert the algebraic formula into an analogous geometrical theorem. And he came at last to consider algebraic analysis as little better than a kind of mechanical knack, in which we proceed without ideas of any kind, and obtain a result without meaning, and without being conscious of any process of reasoning, and therefore without any conviction of its truth.[20]

By all accounts, he rejected negative numbers. As Trail explained, algebra had originally concerned whole and fractional numbers but then had become

encumbered with the metaphysical difficulties which arose from the processes and reasonings of the followers of DES CARTES. Some of these were reasonably considered by Dr. SIMSON as defective in that precision of definition, and strictness of argument, which have ever been the boast of pure geometry.... The Doctor, from this cause perhaps, conceived a prejudice against an application of algebra, which was accompanied with such difficulties; and was thence led to treat of that science after the manner of the early writers on it, with the limited definition and use of the negative sign.[21]

Simson's rejection of negative numbers shows just how vulnerable the expanded algebraic universe remained into the eighteenth century. Although one of the two textbooks on which Simson had cut his algebraic teeth was Kersey's, which endorsed the negative numbers, the other was Oughtred's *Key*. As Oughtred had ignored the negative and imaginary

20. Robison, "Robert Simson," 299. I have been unable to locate any other evidence of Simson's reservations about the symbolical style.
21. Trail, *Life and Writings of Simson*, 66–67. Instead of elaborating Simson's conclusions about the negatives, Trail noted that they were similar to those of Francis Maseres, *A Dissertation on the Use of the Negative Sign in Algebra* (London: Samuel Richardson, 1758). On Maseres's views on the negatives, see the Epilogue. Interestingly, James Gregory – like Descartes – described negative roots as "false" and he (at least sometimes) omitted a negative sign in front of a symbol for a geometric magnitude and yet accurately assumed the sign when operating on the symbol. On the latter point, see Herbert Westren Turnbull, ed., *James Gregory Tercentenary Memorial Volume* (London: G. Bell & Sons, 1939), 213–217, 365–367, and, for a more refined analysis, Malet, "Studies on James Gregorie," 187–189.

roots of quadratic equations, Simson wrote "some short [unpublished] essays . . . on cubic equations, in which he endeavour[ed] . . . to explain them without admitting negative or impossible roots."[22]

Further evidence of Simson's rejection of negatives survives in his previously cited letter to Scott. Here he attacked the strong version of the fundamental theorem and claimed to offer a counterexample. This was the equation: $xx - 4x = 12$, for which he immediately gave the root $x = 6$. He then attempted to show that the equation has no other root. Suppose, he began, there is another root, say y. Then y is either greater than or less than $x = 6$. Taking $y > x$, he reasoned: $y - 4 > x - 4$, and then $y (y - 4) > x (x - 4)$, or $yy - 4y > xx - 4x = 12$. "Therefore," he correctly concluded, "$yy - 4y$ is not equal to 12; that is, y is not the root of the equation." Without any further explanation, he added: "The same thing follows, if y be said to be less than 6." Of course, in reaching a contradiction for the case $y < x$ through multiplication of inequalities, Simson had also to assume that y was positive – an assumption that, in his peculiar algebraic framework, required no statement.[23]

Publishing nothing on his rejection of the negative numbers but making "no secret of" it to his students or colleagues,[24] Simson exerted a strong conservative influence on Scottish algebra. Through his students, who included Matthew Stewart and James Williamson as well as MacLaurin, Robison, and Trail, and through his geometric writings, he became (as George Elder Davie put it) "the father-figure and seminal mind of Scottish mathematics."[25] His love of classical geometry, preference for geometric analysis over analytic geometry, rejection of the negative and imaginary numbers, and emphasis on mathematical history and textual scholarship helped to imbue Scottish mathematics with a bias toward geometry and a certain wariness of algebra, or at least certain aspects of the subject, which persisted through the end of the eighteenth century.

But such a philosophical victory was bought at a professional cost. At least partially because of his turning away from Cartesian and Newtonian algebra, Simson so restricted his mathematics that his research fell

22. Trail, *Life and Writings of Simson*, 67.
23. Simson to Scott, 1 May 1764, 123. Simson's assumptions ruled out $y = -2$, which satisfies the given equation.
24. Trail, *Life and Writings of Simson*, 123.
25. George Elder Davie, *The Democratic Intellect: Scotland and Her Universities in the Nineteenth Century* (Edinburgh: Edinburgh University Press, 1961), 109.

short of his innate talents. Even Robison, who was sympathetic to classical geometry, said so. According to him, Simson should have aimed both to exhibit ancient geometric analysis "in its most engaging form, elegant, perspicuous, and comprehensive" and to transform that analysis into a viable research tool for his own century; instead he had merely reconstructed ancient texts. Robison admitted that Simson had "contributed greatly to the entertainment of the speculative mathematician, who is more delighted with the conscious exercise of his own reasoning powers than with the final result of his researches." But Robison's point – the point that kept victory for classical mathematics at bay in the eighteenth century – was that the criteria for good mathematics include applicability and fertility as well as perspicuity and elegance. Thus it was "deeply to be regretted that . . . [Simson] occupied, in this superstitious palaeology, a long and busy life, which might have been employed in original works of infinite advantage to the world, and honour to himself."[26]

2

The criticisms leveled against Simson did not apply to MacLaurin, who was the beneficiary of somewhat inconsistent intellectual traditions stemming not only from Simson but from Newton, Berkeley, and Locke as well. As a teenager, he fell under Simson's influence.[27] Born in Kilmodan, Scotland, he had lost his father, a Presbyterian minister, when he was six weeks old and his mother, at the age of nine. He and his older brother John had subsequently been cared for by an uncle-minister. Sent

26. Robison, "Robert Simson," 301.
27. There is no full-length biography of MacLaurin. An early study is Patrick Murdoch, "An Account of the Life and Writings of the Author," in Colin Maclaurin, *An Account of Sir Isaac Newton's Philosophical Discoveries, in Four Books* (London, 1748; reprint, with an introduction by L. L. Laudan, The Sources of Science, no. 74, New York: Johnson Reprint, 1968). Other sources are Charles Tweedie, "A Study of the Life and Writings of Colin Maclaurin," *Mathematical Gazette* 8 (1915): 133–151; idem, "Notes on the Life and Works of Colin Maclaurin," *Mathematical Gazette* 9 (1919): 303–305; and J. F. Scott, "Colin Maclaurin," *Dictionary of Scientific Biography*, 8:609–612. For newer biographical sketches, see Erik Lars Sageng, "Colin MacLaurin and the Foundations of the Method of Fluxions" (Ph.D. diss., Princeton University, 1979), and Stella Mills, "Historical Introduction" to *The Collected Letters of Colin MacLaurin*, ed. Stella Mills (Nantwich: Shiva Publishing, 1982), xv–xx.

to the University of Glasgow in 1709, he soon discovered his talent for mathematics. According to a story reminiscent of Hobbes's mathematical awakening and clearly intended to emphasize MacLaurin's genius, he "accidentally met with a copy of *Euclid* in a friend's chamber, [and] in a few days he became master of the first six books without any assistance." The dramatic appeal of this standardized account notwithstanding, Simson, then professor of mathematics at Glasgow, certainly encouraged MacLaurin's pursuit of mathematics, if he did not inspire it. He opened his library to the young man and spent time outside the classroom with him.[28] By 1714 MacLaurin earned a master's degree for his thesis, "On the Power of Gravity."

At this point he was supposed, like his brother, to enter the ministry. Turned off by the internal conflicts of the Scottish Presbyterians, he instead returned to his uncle's home in Kilfinan, Scotland, where for the next three years he continued his study of mathematics and did some private tutoring.[29] In 1717 he competed successfully against Walter Bowman for the chair of mathematics at Marischal College, Aberdeen. According to Charles Gregory, professor of mathematics at St. Andrews and one of the examiners, although Bowman "was much readier and distincter" in his answers to questions on Euclidean geometry, MacLaurin "plainly appeared better acquainted with the speculative and higher parts of Mathematicks."[30]

By 1719 MacLaurin had published two papers in the *Philosophical Transactions:* one on the "construction and measure of curves" and another that extended Newton's classification of cubic curves into component species. Encouraged by a letter from Halley, secretary of the Royal Society, MacLaurin traveled to London in May of that year. Here he met many of the society's fellows, including Newton; was admitted to the society; and made arrangements for publication of his first book, *Geometria organica,* which incorporated his earlier papers and appeared in 1720 under Newton's authority as president of the Royal Society.[31] The book and conversations with MacLaurin convinced Newton that the young Scotsman was a mathematician of his own ilk. In the mid-1720s, then, he strongly endorsed MacLaurin for appointment as deputy

28. For the story of MacLaurin's mathematical awakening as well as Simson's encouragement, see Murdoch, "Account," ii.
29. Mills, "Historical Introduction" to *Letters of MacLaurin,* xvii.
30. Quoted in Tweedie, "Study of the Life and Writings," 134.
31. For MacLaurin's account of these events, see MacLaurin to Colin Campbell, 6 July 1720, *Letters of MacLaurin,* 162–165. On MacLaurin's geometry, see Tweedie, "Study of the Life and Writings," 134, 139–142.

to the ailing James Gregory (1666–1742?), who had succeeded his brother David Gregory as professor of mathematics at the University of Edinburgh. In a letter to MacLaurin of 1725, which the recipient was allowed to show the patrons of Edinburgh, Newton praised MacLaurin's mathematical style (and, in the process, exposed his own mathematical leanings): "I am very glad to hear that you have a prospect of being joyned with M$^{r\cdot}$ James Gregory in the Professorship . . . , not only because you are my friend, but principaly because of your abilities, you being acquainted as well with the new improvem^{t}s of Mathematicks as with the former state of those sciences." Newton's enthusiasm was so sincere that he wrote a second letter, this time to the provost of Edinburgh, offering to contribute twenty pounds per year toward MacLaurin's salary as deputy to Gregory.[32]

The appointment at Edinburgh, which evolved into a full professorship upon Gregory's death in the early 1740s, brought MacLaurin into contact with the Rankenian Club. This student society devoted itself to critical analysis of Berkeley's philosophical ideas. MacLaurin's association with the club assured that he knew Berkeley not merely through his later mathematical writings, but also through at least one of his major philosophical works, the *Principles of Human Knowledge*. Familiarity, however, did not guarantee acceptance. MacLaurin joined the Rankenian Club in its declining years, when general disillusionment with Berkeley's philosophy seems to have replaced initial enthusiasm.[33] MacLaurin himself went on to reject Berkeley's immaterialism and antiabstractionism,[34] and to write the most complete of the many mathematical answers to *The Analyst*. Still, Berkeley's writings perhaps influenced MacLaurin's practice and philosophy of mathematics in subtle ways: he thought seriously about the status of numbers although he did not side

32. Newton to MacLaurin, 21 Aug. 1725, M132, Sir Isaac Newton Collection, Manuscripts Division, Department of Special Collections, Stanford University Libraries, Stanford, California; reproduced in *Letters of MacLaurin*, 171–172. The relevant passages from the preceding letter and from Newton's letter to the provost are quoted by Murdoch, "Account," iv–v.

33. G. E. Davie, "Berkeley's Impact on Scottish Philosophers," *Philosophy* 40 (1965): 222–234, on 222–223, 226–227. Davie raises the possibility that prior to coming to Edinburgh MacLaurin had already discussed Berkeley with George Turnbull, a fellow professor at Marischal College who had studied at Edinburgh from 1717 to 1721 (ibid., 225–226). Moreover, Davie asserts that in the 1730s MacLaurin discussed "with his classes . . . the more specialised topic of Berkeley's views on physics and mathematical foundations" (ibid., 222–223).

34. MacLaurin, *Account*, 97–99.

with Berkeley on the issue; despite his training under Simson, he regarded algebra as a science, using language reminiscent of Berkeley to
make this point; more so than Newton, he carefully separated geometry
and algebra; and, displaying a debt to Locke but perhaps Berkeley as
well, he described mathematics as the science of relations.

3

As a mathematical professor in early-eighteenth-century Scotland, Mac-
Laurin could hardly have ignored the philosophical aspects of his major
subject. The sifting and winnowing of the opinions of Berkeley, Locke
(with whom MacLaurin was also familiar), and other philosophical
thinkers on such topics as abstract ideas and number were no easy
tasks for mathematical thinkers of the period. As suggested by scattered
comments in his scientific books, analysis of which is complicated by the
fact that two were left in manuscript form upon his death, MacLaurin
found some of the major philosophical questions raised by Berkeley and
Locke interesting but perhaps irresolvable. At the least, he stated that
natural philosophers did not need to resolve such questions beyond a
certain point.

He discussed the problem of ideas and, less so, mathematical foundations in his *Account of Sir Isaac Newton's Philosophical Discoveries*,
much of which was written in the 1730s but which was published
posthumously in 1748, as well as in the *Treatise of Fluxions* of 1742.[35]
In book I of the *Account*, which focused on "the method of proceeding
in natural philosophy, and the various systems of philosophers," he
criticized those philosophers, including Descartes, who had "pretended
to explain the whole constitution of things by what they call clear ideas,
and by mere abstracted speculations."[36] Instead he supported what he
portrayed as the empiricist methodology of Newton.[37] Conceding that
some of the heavens and the microscopic level were out of the immediate
reach of the senses, he yet argued that:

35. Since I have found no comprehensive, historical analysis of MacLaurin's
    philosophy of mathematics, mine is largely original and intended to gener-
    ate further research. For the thesis that MacLaurin presented mathematics
    as a paradigm of Baconian science, see Sageng, "Colin MacLaurin," 112–
    117.
36. MacLaurin, *Account*, 3, 14 (quotation).
37. As Sageng elaborates, MacLaurin argued that Bacon founded the experi-
    mental philosophy and Newton perfected it (Sageng, "Colin MacLaurin,"
    114).

[I]t is from those [objects] which are proportioned to sense that a philosopher must set out in his enquiries, ascending or descending afterwards as his pursuits may require. He does well indeed to take his views from many points of sight, and supply the defects of sense by a well regulated imagination; . . . but as his knowledge of nature is founded on the observation of sensible things, he must begin with these, and must often return to them, to examine his progress by them.[38]

If his main design had been philosophical, MacLaurin would probably have elaborated on the relationship between sensible objects and ideas. In the *Account* he specifically declined to do so beyond a certain point. In the section in which he rejected Berkeley's immaterialism and antiab-stractionism, he wrote: "As we are certain of our own existence, and that of our ideas, by internal consciousness; so we are satisfied, by the same consciousness, that there are objects, powers, or causes without us, and that act upon us."[39] Cautious in his philosophy as in his mathematics, he then refused to take the discussion of ideas much farther. "How external objects . . . act upon the mind, by producing so great a variety of impressions or ideas," he stated, "is not our business at present to enquire: neither is it necessary for us to determine how exact or perfect the resemblance may be between our ideas and the objects or substances they represent."[40]

Having set this limit to his philosophical inquiries, he penned a few additional sentences to emphasize that he, unlike Hobbes and Berkeley but like Locke, saw ideas (including those of number) as more than mere words or signs. "It is, however, rating our ideas of external objects by much too low, to compare them to words or mere arbitrary signs, serving only to distinguish them from each other." It is from ideas of objects, he continued, that "we learn their properties, relations, and their influences upon each other, and upon our minds and those of

38. MacLaurin, *Account,* 17–18. Although MacLaurin does not mention Locke here, his remarks about the limits of the human senses are close to Locke's (John Locke, *An Essay Concerning Human Understanding,* ed. Peter H. Nidditch [Oxford: Clarendon, 1975], 301–304).

39. MacLaurin, *Account,* 97. On MacLaurin's arguments against Berkeley (as lacking a "real grasp of [some of] the serious questions Berkeley posed"), see M. A. Stewart, "Berkeley and the Rankenian Club," in *George Berkeley: Essays and Replies,* ed. David Berman (Dublin: Irish Academic Press, 1986), 25–45, 38–39. Stewart suggests that the *Account* shows that MacLaurin rejected Hume's emphasis on perceptions even if "it does not indicate a reading of . . . [Hume's *Treatise*] at first hand" (ibid., 38).

40. MacLaurin, *Account,* 98.

others, and acquire useful knowledge concerning them and ourselves." The very usefulness of ideas, he here suggested, implied a higher status than mere words or signs. Furthermore, "by comparing and examining our ideas, we judge of order and confusion, beauty and deformity, fitness and unfitness, in things. The ideas of number and proportion, upon which so useful and extensive sciences are founded, have the same origin." Thus MacLaurin assumed that there are abstract ideas of number; the last sentence perhaps also suggested that the ideas of number and proportion came from a comparison of other ideas.[41]

Whereas the *Account* did not refer specifically to Locke's philosophy, the *Treatise of Fluxions* did. In the introduction to the *Treatise*, MacLaurin criticized earlier analysts for "admitting quantities, of various kinds, that were not assignable" and for applying "every operation in geometry and arithmetic . . . to them with the same freedom as to finite real quantities." He praised Newton for introducing the doctrine of fluxions "that admits of strict demonstration, which requires the supposition of no quantities but such as are finite, and easily conceived." But, all this stated, he declared that geometers and analysts were not limited to the study of finite extension. Indeed, on the issue of the objects of mathematics he claimed to follow Locke's "excellent essay."[42] Chapter 1 of book I of the *Treatise of Fluxions* opened with an explanation that mathematics concerned relations and not the essences of objects. In a creative application of Locke's reflections on relations, MacLaurin argued that:

> The mathematical sciences treat of the relations of quantities to each other, and of all their affections that can be subjected to rule or measure. . . . We enquire into the relations of things, rather than their inner essences, in these sciences. Because we may have a clear conception of that which is the foundation of a relation, without having a perfect or adequate idea of the thing it is attributed to*, our ideas of relations are often clearer and more distinct than of the things to which they belong, and to this we may ascribe in some measure the peculiar evidence of the mathematics.
>
> *Essay concerning human understanding, book 2, chapt. 25, s. 8.

"It is not necessary," he added, "that the objects of the speculative parts [of mathematics] should be actually described, or exist without the mind; but it is essential, that their relations should be clearly conceived, and evidently deduced."[43]

41. Ibid.
42. Colin MacLaurin, *Treatise of Fluxions, in Two Books,* 2 vols. (Edinburgh: T. W. and T. Ruddimans, 1742), 1:38–39, 49, 45.
43. Ibid., 1:51–52. MacLaurin's reference is to Locke, *Essay Concerning Human Understanding,* 322–323 (II.xxv.8), where Locke discusses relations

The view of algebra expressed in the *Treatise of Fluxions* seemed also to have Lockean roots. Near the beginning of book II, where he considered the foundations of algebra, MacLaurin wrote that algebra "is a general kind of arithmetic; and this is what renders its usefulness so universal." The "evidence" of algebra was exemplary, "if we have no ideas more clear or distinct than those of numbers, and often acquire more satisfactory and distinct knowledge from computations than from constructions."[44] In language and spirit, this endorsement of algebra built on Locke's reflections on arithmetic:

> The Clearness and *Distinctness of each mode of Number* from all others, even those that approach nearest, makes me apt to think, that Demonstrations in Numbers, if they are not more evident and exact, than in Extension, yet they are more general in their use, and more determinate in their Application. Because the *Ideas* of Numbers are more precise, and distinguishable than in Extension; where every Equality and Excess are not so easie to be observed, or measured.[45]

MacLaurin thus seems to have made a sort of mathematician's peace with some of the foundational issues highlighted by the British mathematico-philosophical thinkers of the seventeenth and early eighteenth centuries, a peace for which he was most dependent on Locke. He accepted some form of empiricism and rejected the nominalism of Hobbes and the antiabstractionism of Berkeley. Although he sought "finite real quantities" as the basis for geometry and calculus, he argued against the necessity of understanding the essences of all mathematical objects; he argued for at least some mathematical objects that did not "exist without the mind" but whose relations were clearly understood. Significantly but not unexpectedly, his position on the philosophy of mathematics was one that sanctioned large components of the expanded universe of early modern mathematics, as developed principally by Newton.

4

In his major mathematical works he treaded a moderate path between the ancients and the moderns, just as Newton had predicted for him in 1725. As he did so, he served Newton, his geometry, his calculus, and

---

but not mathematics. For an introduction to Locke's reflections of relations, see John L. Kraus, *John Locke: Empiricist, Atomist, Conceptualist and Agnostic* (New York: Philosophical Library, 1968), 108–117.

44. MacLaurin, *Treatise of Fluxions*, 2:576.
45. Locke, *Essay Concerning Human Understanding*, 205–206 (II.xvi.4).

his algebra well. His greatest accomplishment as a Newtonian was the writing of the *Treatise of Fluxions,* a hefty work of over seven hundred pages written in response to Berkeley's attack on calculus. But, even this work, so long taken as an example of the strength of the classical geometric tradition in Great Britain of the eighteenth century, was, by page length, six parts geometry and at least one part algebra.

The very first sentence of the *Treatise* encapsulated MacLaurin's moderate mathematical stance: "Geometry is valued for its extensive usefulness, but has been most admired for its evidence; mathematical demonstration being such as has been always supposed to put an end to dispute, leaving no place for doubt or cavil." Thus MacLaurin, who had fled to mathematics as a refuge from religious controversy, sought a means of preserving the "extensive usefulness" of the new calculus without sacrificing the "true and real excellency [of geometry], which chiefly consists in its perspicuity and perfect evidence." He worried, too, that errors in calculus might taint the areas of natural philosophy to which it was being extensively applied.[46]

If the analysts of the seventeenth and early eighteenth centuries had departed from "the strictness of geometry" anywhere, according to MacLaurin, they had done so in their appeal to infinites and infinitesimals. "These terms," he wrote, "imply something lofty, but mysterious; the contemplation of which may be suspected to amaze and perplex, rather than satisfy or enlighten the understanding." Still, not every early modern analyst had handled the foundations of calculus so cavalierly. The major exception was Newton, "whose caution was almost as distinguishing a part of his character as his invention." Newton had rejected infinites and infinitesimals, and tried to "establish ... [fluxions] in a way more agreeable to the strictness of geometry." Berkeley's failure to understand calculus, MacLaurin speculated, was however due to the "concise Manner" in which Newton had described fluxions.[47]

MacLaurin's explanation of Newton's caution was revealing of why Newton, MacLaurin, and their British disciples tried to hold modern mathematics to "the strictness of geometry," at least as far as possible. While appreciative of the lure and power of the new mathematics, they feared getting so caught up in it that they would exceed what they recognized as the bounds of mathematical reason. An "admirer" of infinites in his early career,[48] MacLaurin now wrote: "We acknowledge

46. MacLaurin, *Treatise of Fluxions,* 1:1–2. Sageng stresses the latter point (see, e.g., Sageng, "Colin MacLaurin," 137).
47. MacLaurin, *Treatise of Fluxions,* 1:2.
48. In preparing the *Treatise of Fluxions,* MacLaurin feared that his earlier enthusiasm for infinites could be used as an argument against the work,

..., that there is something marvellous in the doctrine of infinites, that is apt to please and transport us; and that the method of infinitesimals has been prosecuted of late with an acuteness and subtlety not to be parallelled in any other science." But the crucial question was, Where should mathematicians draw the line in pursuing mathematics? After explaining how Cavalieri had checked conclusions based on his indivisibles against conclusions based on "old methods," MacLaurin added that many later mathematicians were less careful in their use of indivisibles, infinitesimals, and the like, and so "it was difficult for the Geometricians to determine where they should stop." "Suppositions, that were supposed at first diffidently, as of use for discovering new theorems . . . with the greater facility, and were suffered only on that account, have been indulged, till it [calculus] has become crowded with objects of an abstruse nature, which tend to perplex it and the other sciences that have a dependence upon it." As the exception to mathematical indulgence there stood Newton, who became cautious "especially after he saw that this [mathematical] liberty was growing to so great a height."[49]

MacLaurin and (he implied) Newton believed that calculus should be pursued only to the extent that its foundations could be justified according to geometric strictness. That is, calculus should be developed synthetically from self-evident axioms about real quantities – "the most effectual way to set the truth in a full light, and to prevent disputes." Hence the design of the *Treatise of Fluxions* was not

> to alter Sir Isaac Newton's notion of a fluxion, but to explain and demonstrate his method, by deducing it at length from a few self-evident truths, in that strict manner: and, in treating of it, to abstract from all principles and postulates that may require the imagining any other quantities but such as may easily be conceived to have a real existence.[50]

In book I of the *Treatise,* MacLaurin defined a fluxion as "the velocity with which a quantity flows, at any term of the time while it is supposed to be generated." He left velocity undefined, but described motion as "the power by which magnitudes are conceived to be generated in

and for that reason considered publishing it anonymously. In 1734 he recounted that in his later years he had studied classical mathematics more carefully than before, and this study, combined with exposure to Fontenelle's work on infinites, left him "disgust[ed]" with infinites (MacLaurin to James Stirling, 16 Nov. 1734, *Letters of MacLaurin*, 251). MacLaurin also referred to Locke's *Essay Concerning Human Understanding* when arguing against infinites (MacLaurin, *Treatise of Fluxions*, 45–46).

49. MacLaurin, *Treatise of Fluxions*, 1:47, 38–39, 2.
50. Ibid., 1:2–3.

geometry" and wrote of "discover[ing] the relations of these powers and of any quantities that are supposed to be represented by them"[51] – all after appealing to Locke's theory of relations. His subsequent development of the calculus of fluxions from additional definitions and four axioms followed the canons of classical geometry, with indirect geometric proofs being given even for advanced topics.[52] According to Florian Cajori, "Barring some obvious slips that are easily remedied, Maclaurin certainly reached the ideal he had set." However, traditional geometric "strictness" did not come easily to calculus; it took MacLaurin nearly six hundred pages to complete his geometric development of fluxions. As his first biographer put it, "the work cost him infinite pains, but he did not grudge it."[53]

By length, book I on the geometry of fluxions was the core of the *Treatise*. At the same time, book I was preparatory to the shorter book II, where MacLaurin considered "Fluxions . . . abstractly, or as represented by general Characters in Algebra." Defending his earlier concentration on geometry, he wrote in book II that Newton's fluxions "seemed to be more immediately applicable to geometrical magnitudes . . . than to quantities considered abstractly, or as they are expressed by general symbols in algebra." He admitted also that he had hitherto taken most of his "demonstrations from geometry, because these are often preferred, as more satisfactory than algebraic computations." Too, without naming Berkeley, he recalled that some had suggested that symbols "serve[d] to cover defects in the principles and demonstrations" of calculus. But usefulness as well as evidence counts in mathematics, and MacLaurin now argued that algebra combined both qualities and that the geometric reins ought to be lifted from fluxions:

> The improvements that have been made by it [calculus], either in geometry or in philosophy, are in great measure owing to the facility, conciseness, and great extent of the method of computation, or algebraic part. It is for the sake of these advantages that so many symbols are employed in algebra, the number and complication of which (together with the greater care there has been taken in treating of geometry, after the excellent models left us by the ancients,) have contributed more to occasion the preference that is often as-

51. Ibid., 1:57, 52.
52. See here Niccolò Guicciardini, *The Development of Newtonian Calculus in Britain, 1700–1800* (Cambridge: Cambridge University Press, 1989), 47–51, and Sageng, "Colin MacLaurin," 190–223.
53. Florian Cajori, *A History of the Conceptions of Limits and Fluxions in Great Britain from Newton to Woodhouse* (Chicago: Open Court, 1919), 187; Murdoch, "Account," xiv.

cribed to geometry, in respect of perspicuity and evidence, than any essential difference that can be supposed to be between them.[54]

Continuing to defend algebra, he offered next the strong endorsement of the subject that he had built on Locke's reflections on arithmetic. He acknowledged that there had been some "obscurity" in algebra, which he now tried to avoid by "defining clearly the import and use of the symbols." Not surprisingly, the only symbols he discussed were those of negative and imaginary quantities. He began: "The use of the negative sign in algebra is attended with several consequences that at first sight are admitted with difficulty, and has sometimes given occasion to notions that seem to have no real foundations." Anticipating his *Treatise of Algebra,* which existed at this point only as a manuscript (and is discussed subsequently), he legitimated negatives as quantities to be subtracted and hinted at a basic opposition between positive and negative quantities. He conceded, however, that "the $\sqrt{-1}$, or the square-root of a negative, . . . is a mark or character of the impossible cases of a problem." Then supposing "that the common algebra is admitted," he moved to algebraic calculations on fluxions.[55]

5

MacLaurin's strong, long-standing commitment to algebra was evidenced not only by his stunning defense of the subject in the *Treatise of Fluxions* but also by his early algebraic research, defense of algebra in a synopsis of book II of the *Treatise of Fluxions* that he published in the *Philosophical Transactions,* and popular textbook on the subject. The textbook traced its origin to lecture notes. From the late 1720s the notes circulated freely around Edinburgh, and by 1729, before MacLaurin even thought of writing his *Treatise of Fluxions,* he had prepared a manuscript copy of his *Treatise of Algebra.*[56] His decision to turn the

54. MacLaurin, *Treatise of Fluxions,* 2:575–576.
55. Ibid., 2:576–578. A similar discussion of negatives and imaginaries opened his synopsis of book II for the *Philosophical Transactions* ([Colin MacLaurin], "The Continuation of an Account of a Treatise of Fluxions, &c. Book II. by Colin McLaurin," *Philosophical Transactions,* no. 469 [1742–1743]: 403–415, on 403–405).
56. In 1730 MacLaurin stated that he had composed the manuscript in 1726; in 1732 he recalled that in 1727 he had added some material to the manuscript (MacLaurin to an unknown recipient, 6 June 1730, *Letters of MacLaurin,* 240, and MacLaurin to John Machin, 15 Nov. 1732, ibid., 248). Mills reports that there is an extant manuscript dated 1729 (ibid., 248 n. 390).

algebra notes into a textbook was linked to his successful completion of research on Newton's rule for determining the number of imaginary roots of an equation – the complicated rule that Newton had stated without proof near the end of *Universal Arithmetick*.[57] In an article of 1726 he offered a proof of Newton's rule; in an article published three years later he announced his discovery of "a great Variety of new Rules, different from his [Newton's], and from any other hitherto published, for discovering when an Equation has imaginary Roots." In the latter article he wrote also of "the Design ... [he] for some Time had of publishing a Treatise of Algebra, where ... [he] proposed to treat this [the determination of the number of imaginary roots] and several other Subjects in a new Manner."[58]

Despite the existence of the textbook in manuscript form by the late 1720s and the foregoing declaration of intent to publish it, it did not appear in MacLaurin's lifetime. His teaching schedule of five hours a day, social commitments, and (after 1737) service as a secretary of the Edinburgh Philosophical Society left him little time for research and writing.[59] Then, following the publication in 1734 of Berkeley's *Analyst,* he focused all his research on the foundations of calculus. Upon complet-

57. Isaac Newton, *Universal Arithmetick: or, A Treatise of Arithmetical Composition and Resolution,* trans. Mr. Ralphson and Rev. Mr. Cunn, 2d ed. (London, 1728), reprinted in *The Mathematical Works of Isaac Newton,* ed. Derek T. Whiteside, 2 vols. (New York: Johnson Reprint, 1964–1967), 2:195–198 (103–105).

58. Colin MacLaurin, "A Letter from Mr. Colin Mac Laurin, Professor of Mathematicks at Edinburgh, and F.R.S. to Martin Folkes, Esq; V. Pr. R.S. concerning Aequations with Impossible Roots," *Philosophical Transactions,* no. 394 (1726): 104–112; idem, "A Second Letter from Mr. Colin Mc Laurin, Professor of Mathematicks in the University of Edinburgh and F.R.S. to Martin Folkes, Esq; concerning the Roots of Equations, with the Demonstration of Other Rules of Algebra; Being the Continuation of the Letter Published in the *Philosophical Transactions,* No 394," *Philosophical Transactions,* no. 408 (1729): 59–96, on 59–60. MacLaurin's desire to publish an algebra may have been connected to his dispute with George Campbell over priority for extensions and proofs of Newton's rule on impossible roots. On the dispute, as well as MacLaurin's and Campbell's various results, see Stella Mills, "The Controversy between Colin MacLaurin and George Campbell over Complex Roots, 1728–1729," *Archive for History of Exact Sciences* 28 (1983): 149–164; on Campbell, see Dennis Weeks, "The Life and Mathematics of George Campbell, F.R.S.," *Historia Mathematica* 18 (1991): 328–343.

59. For his laments on the activities that kept him away from mathematical research, see, e.g., MacLaurin to Sir Martin Folkes, 2 Apr. 1743, *Letters of MacLaurin,* 384.

ing the *Treatise of Fluxions,* which appeared in 1742, he turned to preparation of an English translation of David Gregory's *Geometria practica*[60] (published in 1745 under the title of *A Treatise of Practical Geometry*) and, finally, to completing his account of Newton's physical discoveries. Only after MacLaurin's premature death in 1746 was his *Treatise of Algebra* published by Martin Folkes and his other literary heirs to alleviate the financial problems of his widow and children.

Religious, professional, and mathematical reasons help to explain why in the 1730s and early 1740s MacLaurin prepared his *Treatise of Fluxions* rather than his *Treatise of Algebra* for publication. The son, nephew, and brother of clergymen and himself a former divinity student, MacLaurin seemed to believe sincerely in an intimate relationship between mathematics and religion. Of his master's thesis on gravity, he wrote that he

> had most in . . . view throughout all the discourse . . . the establishing the universality of the law of gravity & the necessity of referring it to a first cause the first because it is the most pleasant entertaining truth in Natural Philosophy the other because it is of the greatest importance & use seeing it furnishes us with a most clear & mathematical proof of the existence of a god and his providence that he not only made the world but that he rules and governs it.[61]

Imbued with such sentiments, he was affected to the quick by *The Analyst;* he determined to compose an "Ansuer" to the work but then changed his mind and began his *Treatise of Fluxions.*[62] In a manuscript of this period, probably written for either the "Ansuer" or the treatise, he identified himself as "a sincere wellwisher to Religion" and a devotee of mathematics, in short, the very kind of scholar who felt most betrayed by *The Analyst.* He speculated that some mathematicians may have abused their mathematical authority by employing it to criticize religion, but he added quickly: "that can never justify his [Berkeley's] representing Mathematicians as generally of that character, and his attacking the science itself under the pretence of serving the cause of Religion." Citing Wallis, Barrow, and Newton as "ornaments to the practise of Christianity," he declared that he personally was "satisfied that the interests of true Science and true Religion are united, & that they do real prejudice

60. David Gregory's *Geometria practica* seems to have derived from work of the same title by his uncle, James Gregory (Malet, "Studies on James Gregorie," 92).

61. MacLaurin to Colin Campbell, 12 Sept. 1714, National Library of Scotland, MS. 3440 f.32; reproduced in *Letters of MacLaurin,* 159–161.

62. MacLaurin to James Stirling, 16 Nov. 1734, *Letters of MacLaurin,* 250.

to Mankind who endeavour to represent them as opposite in any measure."[63]

Throughout the eight years that he labored on the *Treatise of Fluxions,* there was an air of anticipation that his proposed algebra textbook could never have generated. His contemporaries were interested in seeing that Berkeley's double threat to mathematics and religion be squashed, and quickly so. Around 1737, for example, he was informed of "the Impatience of . . . [the] Virtuosi in Dublin about your Fluxions." "Your Friends," Francis Hutcheson reported, "are angry at the Delay and Bp Berkelys [sic] are triumphing already."[64] At least book I of the work promised to be popular without being controversial, mathematically creative and yet philosophically cautious. In a letter of 1736, thanking MacLaurin for drafts of the introduction and other sections of the *Treatise of Fluxions,* Simson declared: "The Ancient Heroes are much indebted to you for the Justice you have done them in the fine defence of their Accurate method."[65]

MacLaurin had to have different expectations for the *Treatise of Algebra.* It would be introductory, largely derivative, not as "strict" or rigorous as that on fluxions, and perhaps controversial. At least in Scotland, the work would face a critical audience, including Simson. And, importantly, whereas MacLaurin could confidently claim to establish sound foundations for the calculus of fluxions, he could not pretend that roots involving $\sqrt{-1}$ were anything more than signs of impossibility. Moreover, a good sketch of what he had to say on the legitimation of the negatives and imaginaries appeared in the last section, book II, of the *Treatise of Fluxions,* a section that could be ignored by the more geometric mathematicians of the period and yet assured circulation of some of MacLaurin's most important foundational reflections on algebra. For these and perhaps other reasons, he failed to complete his algebra textbook even after the publication of the *Treatise of Fluxions.* For those final four years of his life (1742–1746), he worked on geometry and on Newton's physical sciences but not on algebra.

63. MacLaurin to recipient not stated, n.d., University of Aberdeen Library, MS.206.65; reproduced in *Letters of MacLaurin,* 425–435, on 425–427. This document has been variously identified as a possible draft of the proposed "Ansuer" to Berkeley, dating from late 1734 or early 1735 (ibid., 425 n. 21), and as a preface or introduction to the proposed *Treatise of Fluxions* (Sageng, "Colin MacLaurin," 139).

64. Hutcheson to MacLaurin, 21 Apr. [1737], University of Aberdeen Library, MS.206.11; reproduced in *Letters of MacLaurin,* 274.

65. Simson to MacLaurin, 22 Sept. 1736, University of Aberdeen Library, MS.206.10; reproduced in *Letters of MacLaurin,* 263.

6

He conceived the proposed *Treatise of Algebra* as a textbook that incorporated his results on the enumeration of roots involving $\sqrt{-1}$ and filled the gaps in *Universal Arithmetick* that he and others had noticed as they worked through Newton's book with students. This pedagogical concern reflected the seriousness with which he approached his teaching responsibilities at Edinburgh. With his income partially dependent on student enrollments, he built his following to about a hundred students per year. The students were divided into four or five classes that met for an hour each day. In the lowest class he covered part of Euclid's *Elements,* plain trigonometry, practical geometry, fortification, and introductory algebra; in the second class, algebra, more Euclid, spherical trigonometry, conic sections, and basic astronomy; in the third, astronomy, perspective, some of Newton's *Principia,* and the elements of fluxions; and in the fourth, his system of fluxions, the doctrine of chances, and the rest of the *Principia.* He offered his students not only breadth of mathematical coverage but clarity as well. "All Mr. *Maclaurin*'s lectures on these different subjects were given with such perspicuity of method and language, that his demonstrations seldom stood in need of repetition."[66]

In keeping with MacLaurin's pedagogical intent, the note "To the Reader" prefixed to the posthumous *Treatise of Algebra* presented the work as a commentary on *Universal Arithmetick* and one adapted to the pressing needs of algebra students:

> Sir Isaac Newton's Rules, in his *Arithmetica Universalis,* concerning the Resolution of the higher equations, and the Affections of their roots, being, for the most part, delivered without any demonstration, Mr. Maclaurin had designed, that his Treatise should serve as a Commentary on that Work. For we here find all those difficult passages in Sir Isaac's Book, which have so long perplexed the Students of Algebra, clearly explained and demonstrated.

Since no criticism of Newton was taken lightly in early-eighteenth-century Britain, the note was quick to stress that MacLaurin was not alone in perceiving the need for a new algebra. Illustrating "how much such a Commentary was wanted," the note cited Willem 'sGravesande's

---

66. Murdoch, "Account," v. On the salary arrangements for Edinburgh professors, see Paul Wood, "Science, the Universities, and the Public Sphere in Eighteenth-Century Scotland," *History of Universities* 13 (1994): 99–135, esp. 101–105.

call for such a work on the grounds that it was a "necessity" and would be an additional honor for Newton.[67]

In many but not all respects, the *Treatise of Algebra* followed Newton's *Universal Arithmetick* fairly closely. Its introductory paragraph defined algebra as "a general method of computation by certain signs and symbols which have been contrived for this purpose." It noted that algebra was "called an Universal Arithmetick, and proceeds by operations and rules similar to those in common arithmetick, founded upon the same principles."[68] In this and the next two paragraphs, however, MacLaurin superseded Newton in a spirited defense of algebra, which hinted at Berkeleyan and Lockean influences. He described arithmetic as "allowed to be one of the most clear and evident of the sciences,"[69] and then explained algebra as a kind of language with wide applicability. He began with a comparison between geometry and algebra. The representations of geometry are "natural" (Berkeley's term) – "lines are represented by a line, triangles by a triangle, and other figures by a figure of the same kind." Those of algebra, on the other hand, are "arbitrary" (also Berkeley's term) – "quantities are represented by the same letters of the alphabet; and various signs have been imagined for representing their affections, relations, and dependencies." The representations of geometry, MacLaurin concluded,

> are like the first attempts towards the expression of objects, which was by drawing their resemblances; the latter [those of algebra] correspond more to the present use of languages and writing. Thus the evidence of Geometry is sometimes more simple and obvious; but the use of Algebra more extensive and often more ready.[70]

The usefulness of algebra, then, compensated for its relative complexity and abstruseness. Algebra enjoyed wide, physical applicability because it could handle not only magnitude but also affections and opera-

67. Anon., "To the Reader," in Colin MacLaurin, *Treatise of Algebra in Three Parts* (London, 1748), n.p.

68. MacLaurin, *Treatise of Algebra*, 1.

69. Ibid., 2. Compare this with Berkeley's description of arithmetic and algebra as "sciences of great clearness, certainty, and extent" (*The Works of George Berkeley, Bishop of Cloyne*, ed. A. A. Luce and T. E. Jessop, 9 vols. [London: Thomas Nelson and Sons, 1948–1957], 3:305) and query 45 of *The Analyst* according to which algebra is "allowed to be a science" (ibid., 4:102).

70. MacLaurin, *Treatise of Algebra*, 2. Berkeley distinguished "natural" and "arbitrary" signs, for example, in *Alciphron* (*Works of Berkeley*, 3:157–158). Newton, however, had written of "the Algebraical language" (Newton, *Universal Arithmetick*, 70 [41]).

tions. Here MacLaurin asked his readers to consider the study of a moving object, which would involve its speed, the direction of the object's motion, and the magnitude and direction of any forces acting on the object. Giving no other details, he implied that algebra could be made to handle both magnitude and the affection of direction.[71]

This third paragraph set the stage for the defense and explanation of the negative numbers. MacLaurin, who had, after all, learned his early mathematics from Simson, struggled mightily to justify the negatives. Unlike Simson, he could not abandon the negative numbers; unlike Berkeley, he could not accept numbers as mere signs; and unlike Newton, he could not offer a few examples of their applicability and cavalierly define negatives as "quantities less than nothing." Instead he charted his own cautious path to full acceptance of negative quantities based on the relation of opposition or contrariety.

The opening for the discussion of the negatives came in the last sentence of the third paragraph, which alluded to "other symbols [that must] be admitted into Algebra beside the letters and numbers which represent the magnitude of quantities." A brief fourth paragraph defined three of these symbols: = (is equal to), $\sqsubset$ (is greater than), and $\lrcorner$ (is less than); the fifth through seventh paragraphs covered positive and negative quantities. MacLaurin began the fifth paragraph with a definition of quantity as "what is made up of parts, or is capable of being greater or less." Appealing to this definition, he observed that there were "two primary operations" relating to quantity: addition, by which quantity is increased, and subtraction, by which it is diminished. "Hence it is," he explained, "that any quantity may be supposed to enter into algebraic computations two different ways which have contrary effects; either as an *increment* or as a *decrement;* that is, as a quantity to be added, or as a quantity to be subtracted." Using the intermediate terms "increment" and "decrement," he defined:

> The sign + (*plus*) is the mark of *Addition,* and the sign − (*minus*) of *Subtraction.* Thus the quantity being represented by *a,* +*a* imports that *a* is to be added, or represents an increment; but −*a* imports that *a* is to be subtracted, and represents a decrement.

The fifth paragraph concluded with definitions relating to subtraction: "+*a* − *b* denotes the quantity that arises when from the quantity *a* the quantity *b* is subtracted; and expresses the excess of *a* above *b.* When *a* is greater than *b,* then *a* − *b* is itself an increment; when *a* = *b,* then *a* − *b* = o; and when *a* is less than *b,* then *a* − *b* is itself a decrement."[72]

In the sixth paragraph MacLaurin used examples drawn from every-

71. MacLaurin, *Treatise of Algebra,* 2–3.   72. Ibid., 3–4.

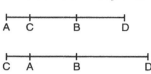

Figure 3. MacLaurin's geometric illustration of subtraction leading to positive and negative quantities (Colin MacLaurin, *Treatise of Algebra* [London, 1748], 5).

day life, geometry, and the physical sciences to illustrate opposition or contrariety, the relation at which he had hinted in the fifth paragraph (recall the reference to "contrary effects") and the relation on which he was ultimately to ground the distinction between positive and negative quantities. "As addition and subtraction are opposite, or an increment is opposite to a decrement, there is an analogous opposition between the affections of quantities that are considered in the mathematical sciences," he began. Examples of opposition included money due to a person versus money owed by the person, a line drawn to the right versus a line drawn to the left, gravity versus levity, and elevation above the horizon versus depression below it. He then explored by illustration the effects of "joining" or "taking" quantities of the same or opposite qualities to or from one another. He noted: "100 *l.* due to a man and 100 *l.* due by him balance each other, and . . . may be both neglected." More generally, when two unequal quantities of opposite qualities are joined, "the greater prevails by their difference." Then he offered an easy example involving unequal quantities of the same quality, where the lesser was to be taken from the greater. Considering the subtraction of the line BD from the line AB (and using figure 3), he noted that "BC = BD is to be taken the contrary way towards A, and the remainder is AC."[73]

He next wrote at more length and with more qualification about taking a greater from a lesser. In illustration, he subtracted the line BD from the line AB, where now "BD, or BC exceeds AB," with the result that AC "becomes a line on the other side of A" (figure 3). This illustration aside, he quickly – and significantly – admitted that the taking of a greater from a lesser was interpretable in some contexts, but not in all:

> When two powers or forces are to be added together, their sum acts upon the body: but when we are to subtract one of them from the other, we conceive that which is to be subtracted to be a power with an opposite direction; and if it be greater than the other, it will

73. Ibid., 4–5.

prevail by the difference. This change of quality however only takes place where the quantity is of such a nature as to admit of such a contrariety or opposition. We know nothing analogous to it in quantity abstractly considered; and cannot subtract a greater quantity of matter from a lesser, or a greater quantity of light from a lesser. And the application of this doctrine to any art or science is to be derived from the known principles of the science.[74]

Specifically, whereas one can subtract a longer line from a shorter line, one "cannot subtract a greater quantity of matter from a lesser" – although, we may add parenthetically, Newton's definition of negative quantities as "quantities less than nothing" seemed to make no such distinction.

In the seventh paragraph MacLaurin tried to vindicate Newton's definition even as he continued to develop the relation of opposition or contrariety as a solid basis for the positive and negative quantities of algebra. First, abandoning the terms "increments" and "decrements," he wrote that "A quantity that is to be added is likewise called a *positive* quantity; and a quantity to be subtracted is said to be *negative:* they are equally real, but opposite to each other, so as to take away each other's effects, in any operation, when they are equal as to quantity." Directly after this statement, he gave the equations of relation:

$$3 - 3 = 0 \text{ and } a - a = 0.$$

Feeling success in hand, he tried now to clarify the problematic definition of a negative quantity as less than nothing: "It is on account of this contrariety that a negative quantity is said to be less than nothing, because it is opposite to the positive, and diminishes it when joined to it, whereas the addition of 0 has no effect. But a negative is to be considered no less as a real quantity than the positive."[75]

Thus went the core of MacLaurin's defense of negative quantities. In one sense, perhaps in response to Simson's critique of the negatives, he had retreated to a conservative definition of positive and negative quantities as those to be added and those to be subtracted. Such had been the tentative definition offered by Oughtred, among others. However, as shown already, MacLaurin subtly elaborated this definition. His defense depended less on the operations of addition and subtraction per se and more on the relation of opposition or contrariety – in the *Treatise of Fluxions* he had after all stated that mathematics had more to do with relations than objects.

But MacLaurin's case for the negatives was even more complicated.

74. Ibid., 5–6.   75. Ibid., 6–7.

As he stressed in the long, previously quoted passage from paragraph 6, quantity did not always admit of opposition. Whereas forces came in "plus" and "minus" (that is, could be directed oppositely), he pointed out that the quality of "minus" made no sense for matter or light, and hence the subtraction of a greater from a lesser made no sense with respect to matter or light. Implicit, then, was his recognition of the right of mathematicians to define and work with a relation that had some physical applications but did not apply universally to "quantity abstractly considered." In paragraph 5 he proclaimed that $a - b$, where b was greater than a, was "itself a decrement," and in paragraph 7, just after introducing the terms "positive" and "negative," he defined $a - a = 0$, all despite his admission in paragraph 6 that such an equation (and the relation of opposition behind the equation) made no sense for abstract quantity. His implicit, certainly not explicit, point was that algebraists could use the equation $a - a = 0$, which captures the relation of opposition, as their working definition of positive and negative quantities, and worry about individual applications only later. Applications to a particular science, he mandated, were "to be derived from the known principles of the science."

Although in the *Treatise of Algebra* he neither highlighted nor elaborated on the new emphasis on relations and equations capturing relations that lay embedded in his defense of negative quantities, MacLaurin's use of the equation $+a - a = 0$ as a "definition" spoke tellingly of the direction in which he was taking the algebra of the negatives. At the beginning of a quite rigorous proof that the product of a positive by a negative was negative, he wrote: "By the Definition, $+a - a = 0$; therefore, if we multiply $+a - a$ by n, the Product must vanish or be 0, because the Factor $a - a = 0$." He continued: the product of $+a$ by $+n$ is $+na$, as he had already proved; but then the second product, $-a$ by $+n$, must be $-na$, the term that "destroys $+na$."[76] He had thus moved from emphasis on negative quantities as supposedly meaningful objects to emphasis on the equations capturing the relation of opposition between the negative and the positive.

7

Whereas the *Treatise of Algebra* testified to MacLaurin's coming to terms with the negative numbers, the work offered little new on the imaginary numbers. MacLaurin used the terms "impossible" and "imag-

76. Ibid., 13.

inary" interchangeably, and, perhaps for the benefit of the Scottish professors as well as students, he emphasized the point that equations with real coefficients sometimes had imaginary roots.[77] He otherwise followed Newton closely in recognizing imaginary roots but still seeing them primarily as impossibilities. In his section on the extraction of roots, he explained that "since there is no Quantity that multiplied into itself an even Number of Times can give a negative Product ... the square Root of $-a^2$ cannot be assigned, and is what we call an '*impossible* or *imaginary* Quantity.' " In a later section on quadratic equations, he stated that "there are some Quadratic Equations that cannot have any Solution," and offered as an example the equation: $y^2 - ay + 3a^2 = 0$, "whence the two Values of y must be imaginary or impossible." Finally: "we shewed that 'the Square of any Quantity positive or negative, is always positive,' and therefore 'the Square Root of a Negative is impossible or imaginary.' "[78]

As other statements in the *Treatise* clarified, imaginaries counted as roots of equations. MacLaurin accepted the fundamental theorem of algebra, of which he gave a moderately strong version: "an Equation of any Dimension may be considered as produced by the Multiplication of as many simple Equations as it has Dimensions."[79] One of the more creative sections of the *Treatise* offered the proof and elaborations of Newton's rules for determining the number of imaginary roots of an equation that MacLaurin had first published in the late 1720s.[80] Moreover, toward the end of the *Treatise,* he argued that imaginary roots were useful. Like Newton, he observed that they indicated when problems were impossible. In addition, he stressed, such roots contributed to the generality of algebra. For quadratic equations, for example, there is "a general Solution, and there is an Expression, in all Cases, of the thing required; only, within certain Bounds, that Expression represents an imaginary Quantity, or rather, '*is the Symbol of an Operation which, in that Case cannot be performed.*' "[81]

77. See, e.g., ibid., 136–137.   78. Ibid., 43, 87, 136.
79. Ibid., 132.
80. The editor of the *Treatise* reported: "The Rules concerning the Impossible Roots of Equations, our Author had very fully considered ... [in] his Manuscript Papers. But as he had no where reduced any thing on that Subject to a better form, than what was long ago published in the Philosophical Transactions ..., we thought it best to take the Substance of Chapt. II. Part II. from thence" (Anon., "To the Reader," in MacLaurin, *Treatise of Algebra,* n.p.).
81. MacLaurin, *Treatise of Algebra,* 299–300. MacLaurin writes generally about equations; the quadratic equation is my example.

Still, MacLaurin may have struggled with the imaginaries more than the posthumous *Treatise of Algebra* let on. In book II of the *Treatise of Fluxions,* he explained as usual that "the $\sqrt{-1}$, or the square-root of a negative, implies an imaginary quantity, and in resolution is a mark or character of the impossible cases of a problem." But, going beyond Newton's *Universal Arithmetick* and expressing a willingness to manipulate imaginary quantities, he noted that these quantities sometimes canceled one another. An imaginary could be "compensated by another imaginary symbol or supposition, when the whole expression may have a real signification." Giving the example of the addition of $1 + \sqrt{-1}$ and $1 - \sqrt{-1}$ which produced the real number 2, he then suggested that other imaginary expressions were compensated "in a manner that is not always so obvious." Additionally, he claimed that theorems arising from reasoning on imaginaries could be demonstrated without them "by the inverse operation, or some other way." But how convincing was this defense? MacLaurin helped to answer this question, when in the very next clause he reassured readers that "tho' such [imaginary] symbols are of some use in the computations in the method of fluxions, its evidence cannot be said to depend upon any arts of this kind."[82]

The abstract of book II of the *Treatise of Fluxions,* which MacLaurin prepared for the *Philosophical Transactions,* gave a much briefer account of the imaginaries and one that highlighted only their impossibility. Having justified positive and negative quantities as increments and decrements, he wrote that the "algebraic expression" $\sqrt{-1}$ should not

> be supposed to represent a certain Quantity; for if the $\sqrt{-1}$ should be said to represent a certain Quantity, it must be allowed to be imaginary, and yet to have a real Square; a way of speaking which it is better to avoid. It denotes only, that an Operation is supposed to be performed on the Quantity that is under the radical Sign. The operation is indeed in this Case imaginary, or cannot succeed; but the Quantity that is under the radical Sign, is not less real on that Account.[83]

MacLaurin – a master of mathematical foundations, who had beaten a reluctant calculus into the geometric mold and constructed an elaborate defense of the negatives based on the relation of opposition – had thus virtually to admit defeat in his quest to legitimate the imaginaries. Imaginaries sometimes compensated one another, but still $\sqrt{-1}$ was

82. MacLaurin, *Treatise of Fluxions,* 2:577–578. Although he did not say so, it is possible that MacLaurin was testing a version of Berkeley's "compensation of errors" (*Works of Berkeley,* 4:78–79) here.

83. [MacLaurin], "The Continuation of an Account of a Treatise of Fluxions," 404–405.

only a "mark," a "character," an "imaginary symbol," a "supposition," an "algebraic expression." Although useful and essential to algebraic generality, imaginaries were not real quantities, should not be called "quantities," and, as he wanted the readers of his work on fluxions to remember, were not a part of the evidence or science of calculus.

8

On another issue, the relationship between algebra and geometry, Mac-Laurin seemed to refine the Newtonian algebraic paradigm. In one sense, he brought it closer to the pure algebraic tradition represented by Wallis and Berkeley and, in another sense, he perhaps made it conform with the ideal set by Newton himself in the appendix to *Universal Arithmetick*. As we have seen, in the latter book Newton had intermingled geometry with algebra, although warning in the much quoted appendix that "these two Sciences ought not to be confounded."[84] Perhaps taking his cue from Newton's remark, but perhaps also writing under the influence of Berkeley's rigid distinction between algebra and geometry, MacLaurin made a real effort to separate the two mathematical sciences.

In his first paper on Newton's rule for enumerating impossible roots, he announced the discovery of a proof of the rule "requiring nothing but the common Algebra, and being founded on some obvious Properties of Quantities . . . , without having recourse to the Consideration of any Curve whatsoever, which does not seem so proper a Method in a Matter purely Algebraical."[85] In his *Treatise of Fluxions*, he segregated geometry to book I and algebra to book II. In his *Treatise of Algebra* he confined geometric problems to a special section on applications and an appendix. In the first two parts of the latter book, its editor explained, MacLaurin had

> omitted the Algebraical solution of particular Geometrical problems, as requiring the knowledge of the Elements of Geometry; from which those of Algebra ought to be kept, as they really are, entirely distinct; reserving to himself to treat of the mutual relation of the two Sciences in his Third Part, and, more generally still, in the Appendix.[86]

MacLaurin's own comments matched the impression conveyed by his editor that he believed in independent sciences of algebra and geometry, the mutual relations of which were nevertheless worthy of study. His

84. Newton, "Appendix," *Universal Arithmetick*, 228 (120).
85. MacLaurin, "A Letter from Mr. Colin Mac Laurin . . . to Martin Folkes," 104.
86. Anon., "To the Reader," in MacLaurin, *Treatise of Algebra*, n.p.

prefatory remarks to the third part of the *Treatise* reiterated a near-Wallisian ideal of an independent algebra: "In the two first Parts we considered Algebra as independent of Geometry; and demonstrated its Operations from its own Principles." Only the third part of the work focused on the "Application of Algebra and Geometry to each other." Here MacLaurin sought to explain "the Use of Algebra in the Resolution of Geometrical Problems; or reasoning about Geometrical Figures" as well as "the Use of Geometrical Lines and Figures in the Resolution of Equations."[87] Of course, segregating geometry from algebra spoke to a respect for the purity of geometry as well as algebra. And, in earlier drafts of the third part of the *Treatise,* even more so than in the published version, MacLaurin cautioned against applying algebra to problems that began in and could more naturally be solved by geometry.[88] His overall intent, then, seems to have been to avoid confounding algebra and geometry – Newton's worry and perhaps Newton's occasional trap – and, more positively, to show the applicability of the two independent sciences to one another.

9

MacLaurin's mathematical work spoke tellingly of the strengths and weaknesses of the Newtonian mathematical legacy. He respected ancient geometry but not at the expense of early modern mathematics; he worked in both the geometric and algebraic traditions. His onerous *Treatise of Fluxions* was possibly the last great achievement of the geometry of calculus. His *Treatise of Algebra* attested to a vibrant algebra, from which new, significant results were forthcoming, witnessed by its sections on enumerating the imaginary roots of an equation and on elaborating Newton's rules for solving simultaneous linear equations. In the latter section MacLaurin solved sets of two and three linear equations involving two and three unknowns, respectively, and in the process offered enough "rudimentary information on determinants" to be given credit for anticipating the general rule enunciated by the Swiss mathematician Gabriel Cramer in 1750.[89]

87. MacLaurin, *Treatise of Algebra,* 297–298.
88. On the drafts, see Sageng, "Colin MacLaurin," 34.
89. MacLaurin, *Treatise of Algebra,* 81–85; Carl Boyer, "Colin Maclaurin and Cramer's Rule," *Scripta Mathematica* 27 (1956): 377–379; and Eberhard Knobloch, "Determinants," in *Companion Encyclopedia of the History and Philosophy of the Mathematical Sciences,* ed. I. Grattan-Guinness, 2 vols. (London: Routledge, 1994), 1:766–774 (quotation from p. 769).

But the *Treatise of Algebra* also attested to an algebra in need of foundational reflection. In a period when mathematicians still insisted on clear definitions of mathematical objects, MacLaurin was unable to offer any definitions of the negatives beyond their being quantities to be subtracted (a definition that, with qualification, Simson could have accepted) and quantities less than nothing (a definition that was just so much ammunition for the attack on the negatives). Also, when mathematics was still largely thought of as the science of quantity, he had to admit that it was better not to talk of $\sqrt{-1}$ in such terms. He had quashed Berkeley's critique of calculus, but not Simson's critique of algebra.

Further frustrating Newton's and MacLaurin's bifocal mathematics in Scotland, Simson outlived MacLaurin by twenty-two years, and Mac-Laurin's successor at Edinburgh was Matthew Stewart, Simson's most able disciple. Stewart had trained at Glasgow under Simson and then at Edinburgh under MacLaurin. When he moved to Edinburgh, Stewart was already a confirmed disciple of Simson, who continued from afar to direct his researches in classical geometry. Stewart established a mathematical reputation with his *General Theorems,* a thoroughly geometric work that was published in 1746, just in time to put him in contention for the professorship left vacant by MacLaurin's early death. Simson encouraged Stewart to apply for the professorship and, according to some of their contemporaries, he permitted Stewart to include (without proper attribution) some of his own porisms in the *General Theorems.*[90]

It was not the case that in the post-MacLaurin period Simson, Stewart, and their disciples rejected Newtonian mathematics. Rather, they picked and chose among that mathematics, and in their regular selection of the geometric over the algebraic they subverted the bifocal Newtonian mathematical legacy. In 1763, for example, Stewart published a book in which he used only geometric methods to calculate the distance of the sun from the earth. His value was far from correct – and within a decade his book came under serious attack by some of England's more algebraic mathematicians, John Dawson and John Landen.[91]

Yet Stewart and his son, Dugald, who was elected joint professor with his father in 1775, reigned over Edinburgh mathematics until 1785 when the elder Stewart died and the younger Stewart was elected professor of moral philosophy at the university. Although somewhat chastened by

90. E. I. C., "Matthew Stewart," *Dictionary of National Biography,* 18:1223–1224; Ian N. Sneddon, "Matthew Stewart," *Dictionary of Scientific Biography,* 13:54–55.

91. E. I. C., "Matthew Stewart," 1223; Sneddon, "Matthew Stewart," 54–55.

Matthew Stewart's encounter with Dawson and Landen, Scotsmen like Dugald Stewart and John Leslie (who was elected the professor of mathematics at Edinburgh in 1805) continued to argue through the early nineteenth century for the pedagogical superiority of geometry over algebra, if not for the absolute superiority of the former over the latter. Within the university, they maintained, the major role of mathematics was to train the human mind, and geometry was the preferred instrument for such training.[92] Of all the professors of mathematics who served at Edinburgh from 1746 through the early decades of the nineteenth century, only John Playfair, who in 1785 succeeded Dugald Stewart as joint professor of mathematics with Adam Ferguson, regularly argued against the geometric bias of Scottish mathematics and for algebra as a major component of the undergraduate curriculum.[93]

---

92. On the geometric biases of Dugald Stewart, John Leslie, and other Scottish mathematicians of the period, see Richard Olson, *Scottish Philosophy and British Physics, 1750–1880: A Study in the Foundations of the Victorian Scientific Style* (Princeton: Princeton University Press, 1975), esp. 20–25, 55–93.
93. See, e.g., [John Playfair], Article VIII, *Edinburgh Review* (1804): 257–272.

10

Algebra "Considered
As the Logical Institutes
of the Mathematician"

Nicholas Saunderson's Elements of Algebra

Whereas MacLaurin worked comfortably in the geometric and algebraic
traditions, some of his mathematical contemporaries at Cambridge University developed the algebraic tradition more so than the geometric. A
bias toward algebra characterized the lectures and writings of Nicholas
Saunderson, who held the Lucasian professorship from 1711 to 1739,
and those of his Lucasian successors, John Colson and Edward Waring.

Published posthumously in 1740, Saunderson's *Elements of Algebra
in Ten Books* was the English counterpart of MacLaurin's *Treatise of
Algebra*. Both textbooks evolved from classroom lectures that began as
commentaries on Newton's *Universal Arithmetick*. Both tried to show
the reasonableness of, if not demonstrate, some of the algebraic rules
that Newton had stated without proof. Both singled out the negative
numbers for lengthy explanation while largely accepting Newton's view
of imaginary numbers as impossible. Still, there were important differences between the textbooks that spoke to Saunderson's relative detachment from the British geometric tradition. Saunderson not only fostered
algebra's independence from geometry, as did MacLaurin, but also
boldly declared the superiority of analysis over synthesis. No earlier
British algebraist had defended algebraic analysis as strongly and explicitly as he now did in his *Elements*. He argued that analysis was a method
of demonstration; as such, analysis was the equal of synthesis; and,
furthermore, the analytic method was actually to be preferred to the
synthetic because "a synthetical demonstration only shews that a proposition is true; whereas an analytical one shews not only that a proposition is true, but why it is so."[1]

This defense of algebraic analysis was essential to his subtle and
significant recrafting of Newtonian mathematics. This recrafting
strengthened the lure of the arithmetico-algebraic focus, and helped

1. Nicholas Saunderson, *The Elements of Algebra in Ten Books*, 2 vols. (Cambridge: Cambridge University Press, 1740), 1:215.

lay the groundwork for the deductive approach to algebra that would characterize the major work of the British symbolical algebraists of the next century.

I

To historians, Saunderson (1682–1739) is better known as a blind mathematician than as a major force in the development of British algebra. He lost his eyesight at the age of twelve months as a result of smallpox. The son of a father with a small estate in Yorkshire and a place in the excise, he received an erratic education, including some training at the Free School at Penniston, instruction in arithmetic from his father, mathematical studies with private tutors, and a brief period at Attercliffe Academy, an English dissenting school. By the age of twenty-five, he had acquired considerable mathematical knowledge and was brought to Cambridge by Joshua Dunn, a fellow commoner and sometime resident of Christ's College.[2]

Early sources state that Saunderson went to Cambridge "with the intention of fixing himself in the university by means of " his mathematics.[3] Whether by design or not, that was exactly what he did. Establishing himself as a resident (but not member) of Christ's College and a specialist in Newtonian mathematics and science, he was granted permission to lecture on the *Principia, Opticks,* and *Arithmetica universalis* by Whiston, then Lucasian professor. He may, indeed, have been the first scholar at Cambridge to lecture on the latter book, the original Latin edition of *Universal Arithmetick* that was published the very year that his private lectures commenced. Benefiting from a general enthusiasm for Newtonian science in the early 1700s, the lectures that he delivered privately through 1710–1711 and then as Whiston's successor attracted large gown and town audiences.[4]

2. Anon., "The Life and Character of the Author," in Saunderson, *Elements of Algebra,* 1:ii–iv. The "Life and Character" was "derived from his friends," Thomas Nettleton, Richard Wilkes, J. Boldero, Gervas Holmes, Granville Wheeler, and Richard Davies (H. F. B., "Nicholas Saunderson," *Dictionary of National Biography,* ed. Leslie Stephen and Sidney Lee, 22 vols. [Oxford: Oxford University Press, 1921–1922], 17:821–822).

3. Quoted in Christopher Wordsworth, *Scholae Academicae* (London: Frank Cass, 1968), 68.

4. Wordsworth, *Scholae Academicae,* 68–69. University lectures by Whiston and Saunderson were attended by students and members of Cambridge and other adults with scientific interests (Niccolò Guicciardini, *The Development*

In obtaining the Lucasian chair when Whiston was stripped of it, Saunderson bested Christopher Hussey, the inside candidate who was a Trinity College man supported by Richard Bentley. Bentley, master of Trinity, had solicited Newton's support for Hussey, but Newton had resolved "not to meddle with this election . . . any further then [*sic*] in answering Letters" and gave but a lukewarm endorsement of Hussey.[5] On November 19, 1711, then, following Queen Anne's recommendation, the university conferred on Saunderson a master of arts degree, a prerequisite for the Lucasian professorship. The next day the electors for the professorship voted six to four in favor of Saunderson, who in early 1712 delivered his inaugural Lucasian address.[6]

As the Lucasian professor, he was a dedicated teacher and a mathematician seemingly called to clarify and proselytize for a new mathematical way – in this case, the modern aspects of Newtonian mathematics. According to his Cambridge friends, in a period when many professors did not lecture, he "gave up his whole time to his Pupils" and "seemed perfectly to know what Difficulties young Minds are apt to be involved in, and how best to obviate or remove them."[7] As MacLaurin was teaching five hours a day at Edinburgh, he taught for seven or eight hours a day at Cambridge.[8] At least partially because of the demands of teaching, he published nothing in his lifetime. Only in 1733, after his brush with death from a fever had apparently started his colleagues worrying that he would leave no publishable legacy, did he begin to transform his algebra lectures into the textbook that appeared posthumously in 1740.[9]

Since it was understood that he "enjoy'd his [Newton's] frequent Conversation concerning the more difficult Parts of his Works,"[10] his explications of Newtonian mathematics could be seen as coming from the master himself. Saunderson's pedagogical skills and the Newtonian connection, then, made *The Elements of Algebra* into a popular text-

    of Newtonian Calculus in Britain, 1700–1800 [Cambridge: Cambridge University Press, 1989], 18).

5. On this election, see Richard S. Westfall, *Never at Rest: A Biography of Isaac Newton* (Cambridge: Cambridge University Press, 1980), 700 (quotation).

6. J. J. Tattersall, "Nicholas Saunderson: The Blind Lucasian Professor," *Historia Mathematica* 19 (1992): 356–370, on 360.

7. Anon., "Life and Character," viii, xiv.

8. H. F. B., "Nicholas Saunderson," 821.

9. Tattersall, "Nicholas Saunderson," 363–364.

10. Anon., "Life and Character," vi.

book, which was reprinted in 1741, condensed and printed four more times between 1756 and 1792, and also translated into French and German.

2

Behind his dedication to mathematics teaching, there lay his deep belief in mathematics as the privileged way to truth and God and as an instrument for improving the mind. This belief seemed to lead him to appreciate mathematics as much as a means to religious and educational ends as in and of itself. For him, the study of mathematics was more important than philosophy, metaphysics, and perhaps even theology. "It is in Mathematics only that truth, that is, rational truth, appears most conspicuous," he wrote near the end of his *Elements of Algebra*. As he saw it, some nonmathematical sciences concern very simple truths, the discovery of which "afford but little pleasure"; in some more complicated sciences, truth "lies deeper and must be dug for, [but] she is seen for the most part through so much dross, obscurity and confusion, that falsehood herself under a plausible disguise often passes for truth." Mathematics alone led to deep and clear truth. In an exceptionally strong endorsement of mathematics as the path to God, he exclaimed: "Whoever then would be thoroughly acquainted with the nature, beauty and harmony of truth; whoever to the utmost of his finite capacity would see truth as it has actually existed in the mind of God from all eternity, he must study Mathematics more than Metaphysics."[11]

What had mathematics finally wrought at Cambridge? Only a century earlier Oughtred had argued for the study of mathematics as one of the two major paths to knowledge of God, the first being sacred scripture. Here Saunderson mentioned nothing about scripture or traditional theology; and he put mathematics ahead of metaphysics. Indeed, he seems to have embraced a sort of neo-Pythagorean philosophy, according to which mathematical truth was eternal and, as such, best revealed the mind of God. That he made his philosophical and related theological views public during his lifetime and that some worried about their incompatibility with Anglicanism are hinted at in earlier sources. In 1824 George Dyer reported that Halley had quipped that "Whiston was dismissed [from the Lucasian professorship] for having too much religion, and Saunderson preferred for having none." According to Dyer,

11. Saunderson, *Elements of Algebra*, 2:740.

Saunderson was "no friend to Divine revelation," although he was said to have requested communion on his deathbed.[12]

Saunderson also developed the more traditional justification of the study of mathematics as a means of improving the reasoning powers. "He considered Mathematics as the Key to Philosophy," his friends wrote, ". . . and thought the Mind was more highly entertained as well as improved in unravelling Her works, than investigating the most subtile Properties of abstract Quantity." Furthermore, they attested, a mathematical "Proposition must have its Uses, in order to engage his Attention. Either the Method of Enquiry must help to form the Mind, and teach new Modes of Reasoning, or the Proposition itself must tend to some Good, to the Improvement of Life or Science."[13] Although his contemporaries thus conveyed the impression of a scholar who sincerely believed in mathematics as logic, Saunderson may also have felt a responsibility as the Lucasian professor to so promote the study of mathematics at Cambridge. Despite the student following that he enjoyed, it was still possible as late as 1730 for a Cambridge undergraduate to study no mathematics and yet pass his final examinations. Some students were so resistant to mathematical studies that Saunderson once reflected that, if he died and went to hell, his eternal torment would be reading lectures to Cambridge fellow commoners.[14] Thus personal conviction and the continuing need to promote mathematics meshed nicely with Lucasian tradition as Saunderson stressed "the Excellence and Advantage of this [mathematics] above every other Method of Reasoning" in his inaugural Lucasian address.[15]

Given his emphasis on mathematics as the way to truth and as a logic, it was fitting that the exposition of the methods of mathematics – more so than any original mathematics – emerged as the hallmark of his professorship. He evaluated geometry and algebra, synthesis and analysis, against one another, and ultimately promoted the analytic method as used in algebra. As his friends summarized, he taught that "Algebra is in it's own Nature an Art of Reasoning, and may be considered as the Logical Institutes of the Mathematician." Yet, as a Newtonian, he respected synthesis as well as analysis and taught his students to do so.

12. George Dyer, *The Privileges of the University of Cambridge*, supplement to vol. 2 (1824; reprint, New York: AMS Press, 1978), 142–143.
13. Anon., "Life and Character," xiv–xv.
14. Wordsworth, *Scholae Academicae*, 66; [Edward Cave], letter to Mr. Urban, *Gentleman's Magazine* 24 (1754): 372–374, at 373.
15. Anon., "Life and Character," vii–viii.

[He] has been every where attentive to improve the Mind, and to furnish it with every Method of Reasoning that may be useful in our Researches into Nature. He has often exposed the same Truths to us in several Lights, as we arrive at them by different Methods of Enquiry: since this served to illustrate the Consistency of those Methods.[16]

Similarly, Richard Davies reported that Saunderson was "justly famous for the display he made of the several methods of Reasoning, for the improvement of the mind."[17]

Saunderson's preference for algebra over geometry and for analysis over synthesis seems to have resulted from an internal consideration of algebra's power as well as predisposing external factors. According to his friends, he praised algebraic analysis as the approach that advanced "us in Science much faster and farther than we could have gone by all the Methods of the Ancients, and . . . [as] the very Art and Principle of Invention."[18] But perhaps he could so strongly endorse the arithmetico-algebraic focus of Newtonian mathematics because he had never become as attached to the geometric focus as most of his British contemporaries. Largely a mathematical autodidact, he escaped early institutional indoctrination in the standing of Euclidean geometry as the paradigm of human knowledge. At Attercliffe Academy, which he attended briefly, mathematics including "geometry, was by no means a leading subject"; students concentrated rather on metaphysics and traditional logic.[19] Furthermore, it is possible that before Newton's *Universal Arithmetick* appeared, Saunderson read Kersey's *Algebra*, the preface of which discussed algebra as an analytic art proceeding by "undeniable consequences," a view of the subject certainly compatible with that developed in Saunderson's *Elements*.[20]

Even if his early exposure to geometry had been considerable, part of the traditional argument in favor of the superiority of geometry as a mathematical science would not have meant exactly the same to Saunderson as to his sighted contemporaries. Mathematical thinkers from Barrow to Berkeley associated the special perspicuity of geometry with

16. Ibid., xvii.   17.   Quoted in Wordsworth, *Scholae Academicae*, 70.
18. Anon., "Life and Character," xv.
19. Giles Hester, *Attercliffe as a Seat of Learning and Ministerial Education* (London: Elliot Stock, 1893), 31.
20. John Kersey, *The Elements of That Mathematical Art Commonly Called Algebra, Expounded in Four Books*, 2 vols. (London: William Godbid, 1673–1674), 1: preface (ii). Saunderson referred to Kersey's *Algebra* in his *Elements* (Saunderson, *Elements of Algebra*, 2:411, e.g.).

the sense of sight. In his defense of geometry Barrow had asked, "[W]ho does not view with the Eye and feel with the Hand all the particular dimensions of Bodies?" In *The Analyst* Berkeley declared, "when from the distinct contemplation and comparison of figures, their properties are derived, by a perpetual well-connected chain of consequences, the objects being still kept in view, and the attention ever fixed upon them; there is acquired an habit of reasoning, close and exact and methodical."[21] Saunderson's inability to keep figures "in view" may have helped free him from the powerful attraction that geometry exercised on so many of his British contemporaries.[22] Of course, if he could not see physical objects and geometric figures, he could still feel them. "He had a Board made with Holes bored at the equal Distance of half an Inch from each other: Pins were fixed in them, and by drawing a Piece of Twine round their Heads, he could more readily delineate all rectilinear Figures used in Geometry, than any Man could with a Pen."[23] But then, by another device he invented – a sort of "*Abacus,* or Calculating Table," described in John Colson's essay on "Palpable Arithmetic" – he could also (in a way) feel the numbers of arithmetic.[24]

Despite his geometric and arithmetic boards, he tended to deny the advantages of using diagrams in mathematics. His friends reported that he frequently observed that "Diagrams which are intended only as helps to the Imagination, are often the means of misleading the Judgment. It is certain, however useful they may be to the Learner, yet the Inventer

21. Isaac Barrow, *The Usefulness of Mathematical Learning Explained and Demonstrated: Being Mathematical Lectures Read in the Publick Schools at the University of Cambridge,* trans. John Kirkby (London, 1734), 19; *The Works of George Berkeley, Bishop of Cloyne,* ed. A. A. Luce and T. E. Jessop, 9 vols. (London: Thomas Nelson and Sons, 1948–1957), 4:65–66.

22. For the learning of mathematics the blind or partially sighted pupil's "deprivation of the clear visual experience of form and shape may put him at some disadvantage compared with the fully sighted pupil" (Elizabeth K. Chapman, *Visually Handicapped Children and Young People* [London: Routledge & Kegan Paul, 1978], 91). Still, the quest for understanding the relationship between blindness and spatial awareness is complicated by the need to consider many factors, including possible differences between persons who are blind from birth and persons who, like Saunderson, have early vision (David H. Warren, *Blindness and Early Childhood Development* [New York: American Foundation for the Blind, 1977], 31–38).

23. Anon., "Life and Character," xii.

24. John Colson, "Dr. Saunderson's Palpable Arithmetic Decypher'd," in Saunderson, *Elements of Algebra,* 1:xx–xxvi.

must in all Cases proceed without them."[25] With geometry stripped of its claim to some sort of guidance from sensible figures, the gap between geometry, on the one hand, and arithmetic and algebra, on the other, narrowed – a point on which Saunderson capitalized in his *Elements of Algebra*. He would argue (if somewhat indirectly) that, if geometry was not guided in any special way by sensible figures and if synthesis was no more demonstrative than analysis, the beloved synthetic geometry was not superior to algebra.

3

It was with the view of algebra as an "Art of Reasoning" and as the "Logical Institutes of the Mathematician" that Saunderson prepared his *Elements*. He explained troublesome algebraic concepts, gave new proofs of difficult algebraic rules, discussed the methods of mathematics in general, and frequently gave both analytic and synthetic proofs of the same proposition. In short, although billed as a textbook for beginners, *The Elements* had a subtext related to mathematical methodology, which was as relevant to mathematical initiates as to novices struggling to come to terms with the new mathematics embedded in Newton's bifocal legacy.

Its important subtext notwithstanding, the work was first and foremost an introductory algebra textbook that had emerged from lectures on Newton's *Universal Arithmetick*. And, in publishing the work, Saunderson's heirs, like MacLaurin's, felt obliged to explain why any commentary was needed for one of Sir Isaac's works. Newton had written his *Universal Arithmetick* in a "masterly Style," they stressed, but a style distinguished for its "brevity" and hence one "leaving the Mathematical Reader to furnish himself with every thing before known, and often to take large Steps alone."[26] By implication, Newton's algebra textbook was not quite adequate for "young beginners, and for . . . those who have such under their care," the precise audience to whom Saunderson addressed his *Elements*.[27]

Claiming to write for beginners, then, Saunderson seems to have aimed at a textbook that was to be the algebraic counterpart of Euclid's

25. Anon., "Life and Character," x. Kersey had said much the same in his *Algebra* (Kersey, *Algebra,* 1: preface [i]).
26. Anon., "Life and Character," v.
27. Anon., "Advertisement," in Saunderson, *Elements of Algebra,* 1:n.p.

geometric compendium. Divided into ten books, the new algebraic *Elements* was to explain all that the beginner needed to know about algebra: its definition, language, concepts, axioms, methods, and results. What Saunderson wrote on each of these points, and what he did not write, spoke tellingly of algebra at mid-eighteenth-century Cambridge. Hardly uneasy about algebra, this Lucasian professor was confident, if not exultant.

Following Newton's lead as an author, Saunderson seemed almost to begrudge the time and space that he devoted to the metaphysical issues of algebra. His definition of algebra was terse. His discussion of the troublesome concept of negative numbers was lengthy but seemed designed more to appease his readers than to answer what he saw as pressing questions about the real nature of such numbers; he broached the topic of imaginary quantities only when absolutely forced to do so. In fact, as we shall see, he occasionally suggested that metaphysical questions were the preoccupations of "narrow minds." Neither did he see any need to cover the history and etymology of algebra. For students with such interests, he recommended Wallis's *Treatise of Algebra;* but he discouraged such interests, for the algebraic "art, like many others, . . . [has] considerably outgrown it's name, and . . . [is] often employed in arithmetical operations very different from what it's name imports."[28] In short, he argued that students needed to know only what algebra was at their particular point in time.

Having thus dismissed historical and etymological approaches to his subject, he concluded the first paragraph of the first section of book I of *The Elements,* which was entitled "The Definition of Algebra," by advancing "by way of definition . . . that *Algebra,* in the modern sense of the word, *is the art of computing by symbols,* that is, generally speaking, by letters of the alphabet; which for the simplicity and distinctness both of their sounds and characters, are much more commodious for this purpose than any other symbols or marks whatever."[29] This was clearly an adaptation of Newton's definition of arithmetic and algebra as "one perfect *Science of Computing.*"[30] But, unlike Newton, Saunderson "advanced" his definition without any discussion of the relationship between algebra and arithmetic and without the barest expla-

---

28. Saunderson, *Elements of Algebra,* 1:49.   29. Ibid.
30. Isaac Newton, *Universal Arithmetick: or, A Treatise of Arithmetical Composition and Resolution,* trans. Mr. Ralphson and Rev. Mr. Cunn, 2d ed. (London, 1728), reprinted in *The Mathematical Works of Isaac Newton,* ed. Derek T. Whiteside, 2 vols. (New York: Johnson Reprint, 1964–1967), 2:2 (7).

nation of number. Perhaps he spoke indirectly to a growing sense of the independence of algebra from arithmetic as well as geometry. On the other hand, he may simply have assumed that his readers could consult *Universal Arithmetick* for Newton's sparse treatment of these points.

The second and longest paragraph of the section on "The Definition of Algebra" highlighted symbolical language. Here Saunderson, ever the careful teacher, considered two problems back to back in an attempt to impress students with the necessity and power of algebraic symbolism. Problem 1 involved the substitution of letters for "unknown quantities"; problem 2, the substitution of letters for "known quantities" as well. The first required finding two numbers the sum of which is 48 and the difference, 14. "[H]ere," he explained, "if I only put x, or some other letter for one of the unknown quantities, and use the known ones 48 and 14 as I find them in the problem, I shall only come to this particular conclusion, to wit, that the greater number is 31, and the less 17, which numbers will answer both the conditions of the problem." Algebra, however, permitted solutions of more general problems, a point stressed by mathematicians from Viète on, and the point of Saunderson's second example, which required finding two numbers the sum of which is a and the difference, b.

> I shall then come to this general conclusion, *viz.* that *Half the sum of* a *and* b *will be the greater number, and half their difference will be the less:* which general theorem will suit not only the particular case abovementioned [problem 1], but also all other cases of this problem that can possibly be proposed.[31]

He concluded the section with a third paragraph, which noted that the symbols of algebra stood not so much for "particular quantities" as for "the relation they have to one another in any problem or computation." Although the latter statement is tantalizing for its appeal to mathematical relations over objects and its echoing of MacLaurin's remarks on the subject, Saunderson pursued the matter no farther than examples. He wrote that "letters represent quantities in Algebra just in the same manner as they do persons in common life, when two or more persons are distinctly to be considered with regard to any compact, lawsuit, or in any other relation whatever."[32] Since a contract or compact, which incorporates certain specific relationships, can be standardized so that any pair of names can be entered into its blank spaces, the relationships incorporated into the document take precedence over the identities of the individual signees. And so, similarly, Saunderson suggested (almost in passing), algebraic relations took precedence over objects.

31. Saunderson, *Elements of Algebra,* 1:49–50. 32. Ibid., 1:50.

4

Following the format of *Universal Arithmetick,* the second section of book I of *The Elements* concerned "affirmative and negative quantities." The length and content of the section witnessed to lingering uneasiness about the negative numbers, even at Cambridge University after Newton had sanctioned such numbers. Never himself agreeing that there was any substantial basis for the uneasiness, Saunderson constructed a rather elaborate defense of the negatives around Newton's problematic definition of a negative number as a quantity less than nothing. However, the elaborateness of the defense and his berating those who questioned such numbers seem to have exacerbated the problem rather than alleviating it.

He opened the second section with Newton's terms and definitions: "Algebraic quantities are of two sorts, affirmative and negative; an affirmative quantity is a quantity greater than nothing, and is known by this sign +; a negative quantity is a quantity less than nothing, and is known by this sign −." Then he immediately conceded: "The possibility of any quantitie's being less than nothing is to some a very great paradox, if not a downright absurdity; and truly so it would be, if we should suppose it possible for a body or substance to be less than nothing." Thus he implied that the definition of a negative as a quantity less than nothing was a truncated definition, one bereft of the nuances that surrounded the concept. It was an appropriate definition only so long as one realized that "quantities, whereby the different degrees of qualities are estimated, may be easily conceived to pass from affirmation through nothing into negation." Here followed some of the (by this point) standard illustrations: for example, "a person in his fortunes *may be said* to be worth 2000 pounds, or 1000, or nothing, or − 1000, or − 2000, in which two last cases he *is said* to be 1000 or 2000 pounds worse than nothing." As we have seen, Newton had similarly explained that "Possessions or Stock *may be call'd* affirmative Goods, and Debts negative ones."[33]

In short, Newton and Saunderson seem to have largely conceptualized the affirmative and the negative within the mathematical paradigm that defined mathematics as the science of quantity. If algebra was the study of quantity, then the negative numbers had to be quantities. "Quantities less than nothing," both men succinctly defined. But, as close readings of their textbooks show, both knew as well as their critics that a quantity

33. Ibid., 1:50–51; Newton, *Universal Arithmetick,* 2:3 (7). The emphases in the latter quotations are mine.

less than nothing was paradoxical. Both said so: Newton in so many words and Saunderson directly. Quantities could be "called," "reckoned," and "conceived of" as less than nothing but they were never actually so. Moreover, according to Saunderson, who here toyed with transcending the definition of mathematics as the science of quantity, the concepts of affirmative and negative involved qualities as well as quantities.

Since Saunderson knew from experience that Newton's implicit strategy of terseness on the negatives had proved ineffective, he next tried to explain at length why the concept of a negative was so difficult. "Certain it is," he continued, "that all contrary quantities do necessarily admit of an intermediate state, which alike partakes of both extremes, and is best represented by a cypher or o." Using the example of heat – of which a body "may be said to have" two, one, no, − one, or − two degrees – he wrote, "if it is proper to say, that the degrees on either side this common limit [o] are greater than nothing; I do not see why it should not be as proper to say of the other side, that the degrees are less than nothing; at least in comparison to the former." Why, then, he asked, did some find this concept of the negative so problematic? In a sentence that addressed this question even as it insulted those who criticized the negatives, he explained:

> That which most perplexes narrow minds in this way of thinking, is, that in common life, most quantities lose their names when they cease to be affirmative, and acquire new ones so soon as they begin to be negative: thus we call negative goods, debts; negative gain, loss; negative heat, cold; negative descent, ascent, &c.: and in this sense indeed, it may not be so easy to conceive, how a quantity can be less than nothing, that is, how a quantity under any particular denomination, can be said to be less than nothing, so long as it retains that denomination.

In this interesting twist, he tried to reduce the difficulty of the negatives to one of prose language. The English language obfuscated the relationship between affirmative and negative. "Difficulties that arise from the imposition of scanty and limited names, upon quantities which in themselves are actually unlimited, ought to be charged upon those names," he declared, "and not upon the things themselves."[34] And so (for another dozen lines) went Saunderson's explication of the negatives.

The defense of Newton's definition behind him, he offered algebra students some general advice for understanding the effects of algebraic operations on negative quantities. Like MacLaurin but without any

34. Saunderson, *Elements of Algebra,* 1:51.

elaboration, he emphasized that affirmative and negative quantities were contrary to one another

> in their own natures, . . . [and] likewise . . . in their effects, a consideration which if duly attended to, would remove all difficulties concerning the signs of quantities arising from addition, subtraction, multiplication, division, &c: for the result of working by affirmative quantities in all these operations is known; and therefore like operations in negative quantities may be known by the rule of contraries.[35]

If Newton had been "brief" on the negatives, Saunderson was diffuse: he filled nearly two pages of book I with reflections on the problem and took a variety of different approaches. The negatives were "quantities less than nothing," but not quite so; an understanding of the negatives had something to do with qualities; "quantities . . . may be easily conceived to pass from affirmation through nothing into negation"; and affirmative and negative quantities were contraries. On top of this, Saunderson introduced yet another twist to the concept of the negatives in book VIII (on the application of algebra to geometry), where he discussed "the transition from an infinitely great affirmative to an infinitely great negative." As v approaches o, he observed, "the fraction $1/v$ will increase by degrees till it becomes infinite" and then, as v crosses from o to very small negative values, the fraction becomes "an infinitely great negative." Insisting that he was clarifying a seeming "mystery," he concluded "that to be less than nothing, or to be greater than infinity, are but two different appellations of the same thing, that is, of a negative quantity; or that if there be any difference, it lies only in our manner of conception."[36] In short, as his critics were to note, the space and ingenuity devoted to negative quantities in books I and VIII of *The Elements* belied Saunderson's dismissal of the problem as one plaguing "narrow minds."

5

His introduction of the imaginaries followed the general scheme used by Newton and subsequently MacLaurin. Since not one of the three men

35.   Ibid., 1:51–52.
36.   Ibid., 2:521, 524. In his *Arithmetica infinitorum* Wallis had similarly discussed quantities passing from positive to negative via infinity. On the latter point, see J. F. Scott, *The Mathematical Work of John Wallis, D.D., F.R.S. (1616–1703)* (London: Taylor and Francis, 1938), 44–45.

could quite bring himself to accept the imaginaries unconditionally as algebraic objects, they all discussed "negative quantities" at the beginning of their textbooks among the fundamental entities of algebra and imaginaries only when there was a need for roots involving $\sqrt{-1}$. Such placement distinguished between a conceptual coming to terms with the negative numbers and a pragmatic détente with the imaginaries.

The imaginaries surfaced first in Saunderson's section on quadratic equations in book III. "The roots of quadratic equations," he began, "are not only very often inexpressible [irrational], but sometimes even impossible, as will appear by the following example." He then solved the equation $xx - 4x + 6 = 0$, getting the roots $x = 2 + \sqrt{-2}$ and $x = 2 - \sqrt{-2}$. Following Newton, he stressed the "impossibility" of such roots despite their "justness" (Newton's terms). "[A]s no quantity whatever, either affirmative or negative, being multiplied into itself will produce a negative," the by-now standard argument ran, "it follows, that $\sqrt{-2}$ is not only an inexpressible quantity, but also an impossible one; and consequently, that the two values of x in this equation . . . will both be impossible." But still – and here in an "N.B." Saunderson got to the heart of the dilemma of the imaginaries – "Though the roots of this last equation be impossible in their own natures, yet they may be abstractedly demonstrated to be just" (that is, to satisfy the given equation).[37]

Saunderson's remaining comments on the imaginaries bear witness to a strong temptation to treat them as full-blown algebraic entities. In his section on "how to generate a quadratic equation that shall have any two given numbers whatever for it's roots," he made some space for impossible roots. By his earlier admission, such roots were not numbers, not real quantities of any kind, and did not deserve to be included in the rule under discussion. Still, he fit them in, albeit at the very end of the section and with the introductory remark: "I think I ought not to omit here." Tentatively, he continued, "if any one has a mind to form a quadratic equation, with any two given impossible roots whatever, (if I may be allowed the expression,) it may be done by the foregoing rule."[38]

In the preceding and later sections, he seemed to inch ever closer to acceptance of the imaginaries as full-blown algebraic entities, or quantities. Indirectly and directly he occasionally applied the term "quantity" to such roots. When generating a quadratic equation from two impossible roots, he explained that "though no possible quantity multiplied into itself can produce a negative, yet an impossible one may." Later, in his

37. Saunderson, *Elements of Algebra*, 1:183–184; see also 1:191.
38. Ibid., 1:186.

discussion of the rule for finding the number of positive and negative roots of equations, he complained, "such is the capriciousness of these [impossible] quantities, (if I may call them so,) that the very self-same roots often appear in both shapes,"[39] that is, as positive and as negative.

Thus the more Saunderson wrote about the imaginaries, the more he revealed about the pragmatic détente with these "quantities" that some of the progressive British mathematicians had reached by the mid-eighteenth century. No one argued that these entities were really quantities. If $\sqrt{-2}$, for example, were a quantity, then $\sqrt{-2}$ was either affirmative or negative. But both affirmative and negative numbers when squared gave positive numbers. Still, Newton, MacLaurin, and Saunderson reminded their readers that impossible roots were "just" – that is, satisfied definite equations with real coefficients. The justness of these roots made them essential components of a general theory of equations, and thus there was a good practical reason for writing of impossible roots. But could mathematicians simply record such roots and then, so to speak, cordon them off from the rest of algebra, which dealt with real quantities? Saunderson's experience shows the difficulty of sustaining such a rigid distinction, once impossible roots were recognized. Algebraists did more than solve specific equations: for example, as algebraists went from given equations to their roots, so they went from given roots to their equations. Ought impossible roots to be included in rules for the latter procedures? Generality dictated so, and thus (as we have seen) Saunderson quickly but tentatively began operating on impossible roots as if they were quantities. By the end of *The Elements* he was actually writing of "impossible quantities," although he continued to indicate that there was some question of whether he ought to do this.

Clearly, time and use had helped to narrow the gap between the real and the imaginary in algebra. Saunderson was more at ease with imaginaries than Newton. Not only did he refer to them as quantities more often than his mentor, but he avoided the murkiness that surrounded the imaginaries in some sections of *Universal Arithmetick*. Whereas Newton had implied that impossible roots were affirmative or negative, Saunderson stated very clearly that "Impossible quantities, properly speaking, belong to no class, either of affirmatives or negatives, and yet they always appear under one form or the other." Similarly, whereas in *Universal Arithmetick* Newton had reverted to the weak version of the fundamental theorem of algebra, Saunderson used a strong version that clued students into the fact that the theorem's gener-

39. Ibid., 1:186, 2:684.

ality was possible only because of the assumption of impossible roots. Accordingly, "Every equation hath as many roots, possible and impossible, as there are dimensions in the highest power of the unknown quantity." And in his book X Saunderson solved one irreducible cubic equation by stating, but not showing how to derive, the cube roots of the complex binomials resulting from application of Cardano's formula. Admitting that he knew neither "a way of extracting the cube root of an impossible binomial" nor if any mathematician had found such a general method, he explained that he offered the solution "to shew the irresistible force and immutable nature of truth, which is able to penetrate even through impossibilities, and can never be so distressed or severely tried, as to be found inconsistent with herself." If concern for mathematical purity and consistency led the older Newton to back away from manipulating "impossible" roots and leave his work on extracting cube roots of complex binomials unpublished, a strong commitment to generality and a belief in mathematics as a divine and inherently consistent "gift" to humans caused Saunderson to look forward to the publication of just such a method. Indeed, appended to *The Elements* by its editor(s) was Abraham De Moivre's trigonometric "Rule for Extracting the Cubic, or Any Other Root of an Impossible Binomial."[40]

6

Although he wrote so extensively on affirmative and negative quantities, and wondered in print if he ought to call roots involving $\sqrt{-1}$ "quantities," Saunderson expressed the opinion that, for mathematicians, excursions into the metaphysics of their subject were risky. Toward the end of *The Elements* and following a discussion of the mathematical infinite, he wrote, "For my own part, it was not without a great deal of reluctance that I prevailed upon myself to say what I have done on this head." It was only because he knew, "by daily experience, how strong a curiosity young Gentlemen have to pry into these notions, (perhaps more that they ought, unless they were better acquainted with all the other parts of mathematical knowledge,)" that he discussed the infinite at all.[41]

Behind his reluctance to venture into the more metaphysical aspects of mathematics was his threefold belief that mathematicians knew what

40. Ibid., 2:684, 677, 706–708, 741, 743–748. The irreducible cubic that Saunderson solved was Bombelli's $x^3 = 15x + 4$.
41. Ibid., 2:560–561.

they were doing without elaborating the metaphysics of their subject; such discussions exposed mathematicians to the futile cavils of those who argued the fine points of mathematics without knowing much of the subject; and the uninitiated should trust established mathematicians. "Mathematicians, and more especially those who best understand this subject," he began, "are, generally speaking, reserved enough upon it, and chuse rather to be deficient than redundant in their expressions upon these occasions," for example, discussions of the infinite. Mathematicians were tight-lipped on such issues not from ignorance or confusion, "not from any diffidence in their own principles, but knowing very well how liable matters of this nature are to be drawn into disputes by such as lie upon the catch, and make it their chief business to oppose those truths which they themselves could never have discovered, nor perhaps will ever be able to understand."[42]

These remarks – coupled with his dislike of public controversy,[43] apparent disengagement from traditional religious issues, and poor health from 1733 on – help to explain why Saunderson had taken no formal part in *The Analyst* debate. Assuredly, he taught and wrote about calculus, enough so that his friends were able to publish his *Method of Fluxions* posthumously in 1756.[44] But the latter work was not written in response to *The Analyst* and some of it betrayed Saunderson's analytical bias. In keeping out of *The Analyst* debate, he had avoided the near necessity of answering Berkeley's critique in its own terms, that is, elaborating geometric foundations for calculus. Instead in the late 1730s he had been free to concentrate on the arithmetico-algebraic focus of Newtonian mathematics, to which he was naturally drawn.

But what if, in avoiding unproductive disputes, all mathematicians were to write sparsely? What was to be expected of their students? Saunderson argued basically that students ought to persevere despite unanswered questions. Students learned mathematics by closely following the reasoning of established mathematicians, and not by questioning that reasoning at every turn. As he put it in his discussion of "infinite quantities," the reader-student should

> suffer himself to be perswaded, as well upon this, as upon almost all other occasions, when he is reading the words and sense of another, to apply himself wholly to them, to view every thing in the light it is placed in, and not to form a judgment of things from his own narrow conceits and little prejudices on one hand, or from high

42. Ibid., 2:560.   43. Anon., "Life and Character," xv.
44. For a brief description of the *Method of Fluxions*, see Guicciardini, *Development of Newtonian Calculus*, 24–25.

flown, metaphysical, and perhaps chimerical notions on the other, which serve but to bewilder his understanding, and to draw off his thoughts from the main subject, which ought to take up his whole attention.[45]

His advice that students follow established mathematicians closely, without asking a lot of questions, seemed to stem not so much from a respect for authority in mathematics as from his deep belief in the fundamental simplicity and consistency of the subject. Mathematical truths were, after all, truths that "existed in the mind of God from all eternity." As Newton had assumed that the physical world was simple, Saunderson assumed that mathematical truths were simple. Having (in the previously quoted passage) advised students to "view every thing in the light it is placed in," he then explained that "Geometrical truths are plain simple truths, and whoever would see them in the clearest light, ought to view them in the simplest: for it is here, as in Optics, where all light let in upon an object besides what is proper to give a distinct view of it, is not only superfluous, but tends rather to obscure than to illustrate the object."[46]

7

Although it was thus with some misgivings that Saunderson discussed the concepts and principles of mathematics, he enthusiastically probed the subject's methods. "Narrow minds" might require some help with the metaphysics of algebra and calculus, but the methods of mathematics were central to all mathematicians' work. An early example of his emphasis on proof came in book I of his *Elements* as he, having just given the rule of signs for multiplication, hypothesized that "the reader expects a demonstration of this rule." The demonstration depended on two special assumptions or, as he described them, "advertisements," which were "in a manner self-evident":

> *first,* that numbers are said to be in arithmetical progression, when they increase or decrease with equal differences, as 0, 2, 4, 6; or 6, 4, 2, 0; also as 3, 0, -3; 4, 0, -4; 12, 0, -12; or -12, 0, $+12$:
> ... 2*dly,* If a set of numbers in arithmetical progression, as 3, 2 and 1, be successively multiplied into one common multiplicator, as 4, or if a single number, as 4, be successively multiplied into a set of numbers in arithmetical progression, as 3, 2 and 1, the products 12, 8 and 4, in either case, will be in arithmetical progression.

45. Saunderson, *Elements of Algebra*, 2:518–519.   46. Ibid., 2:740, 519.

Having stated this, he then turned to the proofs of:

$$+4 \times +3 = +12 \qquad (1)$$
$$-4 \times +3 = -12 \qquad (2)$$
$$+4 \times -3 = -12 \qquad (3)$$
$$-4 \times -3 = +12 \qquad (4)$$

Viewing the first case as $+4$ taken 3 times, he observed that rule (1) "is self-evident, and needs no demonstration." But for the proof of rule (2) he appealed to an arithmetic progression 4, 0, -4. "[M]ultiply the terms of this arithmetical progression . . . into $+3$," he directed, "and the products will be in arithmetical progression, as above; but the two first products are 12 and 0; therefore the third will be -12; therefore -4 multiplied into $+3$, produces -12." Proof of rule (3) was similar, starting with the progression $+3$, 0, -3. Proof of rule (4) depended on the earlier rules and the progression 3, 0, -3. The reader was directed to multiply the progression by -4. As Saunderson explained, rule (2) now gave $-4 \times 3 = -12$; $-4 \times 0 = 0$; and "therefore the third product [-4×-3] will be $+12$." QED.[47] Thus his proof was a sound "demonstration," that is, it was a deductive proof that involved careful reasoning from "advertisements" or axioms.

On first glance, however, this proof seems to be a lot to do about relatively little, since Saunderson had proved only a special case of the rule of signs – that involving $+3$, -3, $+4$, and -4. He soon, however, claimed larger significance for the result, as he defined the multiplication of simple algebraic quantities: "These things premised, the multiplication of simple algebraic quantities is performed, first by multiplying the numeral coefficients together, and then putting down, after the product, all the letters in both factors, the sign (when occasion requires) being prefixed as above directed." For the example $-5ab \times 6bc$, one found, first, $-5 \times 6 = -30$, presumably by recognizing that -5 and $+6$ could be substituted for -4 and $+3$ in the preceding proof with no change of resulting sign, and, second, $ab \times bc = abbc$, by using the definition just given, with the result that $-5ab \times 6bc = -30abbc$.[48]

Less than ten pages after offering his specific demonstration of the rule of signs and seeming to apply the rule liberally in his definition of algebraic multiplication, Saunderson stressed the insufficiency of specific numerical arguments as proofs of general algebraic theorems: "I will give notice once for all, that instances in numbers serve well enough to illustrate a general theorem, but they must not by any means be looked upon as a proof of it; because a proposition may be true in some

47. Ibid., 1:56–58.   48. Ibid., 1:58–59.

particular cases instanced in, and yet fail in others." He continued: algebraic theorems demanded general "demonstrations," and these were invariably "in speciebus" demonstrations, that is, symbolical proofs. As he put it: "whenever a proposition is found to be true *in speciebus,* that is, in letters or symbols, it is a sufficient demonstration of it, because these are universal representations." As an example, he offered the following:

$$a + b$$
$$a + b$$
$$aa + ab + bb$$
$$+ ab$$
$$aa + 2ab + bb.^{49}$$

What is worthy of remark here is not simply Saunderson's emphasis on algebraic symbols as essential elements in general algebraic proofs but also the extent to which he wrote about and set the goal of "sufficient demonstrations" for general algebraic theorems, even if he could not always meet that goal.

As implied in his careful proof of the special case of the rule of signs and in the preceding discussion of *in speciebus* demonstrations, he began to try to make of algebra a demonstrative science. His *Elements of Algebra* was not as tightly and successfully deductive as Euclid's *Elements,* but nevertheless it represented a new peak of concern for proof and, in particular, logical demonstration in British algebra (and perhaps in algebra in general). Saunderson did not open the work with a list of axioms or postulates, but rather stated some axioms as he needed them in the work's course. Thus he gave the two "advertisements" for the early proof of the rule of signs, and at the beginning of his section "Of the resolution of simple equations," he stated four "Axioms," including that of transposition, while reserving for himself the right to "take notice of [the rest of the required axioms] occasionally, as they offer themselves." "If a quantity be taken from either side of an equation," axiom 4 began, "and placed on the other with a contrary sign, which is commonly called transposition, the two sides will still be equal to each other."[50] Although some of these axioms, including transposition, were mere restatements of some of Newton's seven "Rules" for "ordering a single Equation,"[51] it was perhaps significant that Saunderson, like Kersey before him, wrote specifically of "axioms," the term mathematicians traditionally used for self-evident propositions.

49. Ibid., 1:63–64.   50. Ibid., 1:95–96.
51. Newton, *Universal Arithmetick,* 2:56–60 (34–36).

It was definitely significant that in his *Elements* he explicitly declared that analysis was a method of demonstration on par with synthesis. Many British mathematicians, like their French counterparts, had cited analysis as the special method of algebra. However, Hobbes, Barrow, Newton (on occasion), and others had extolled the method of synthesis at the expense of analysis, with Hobbes explicitly arguing that synthesis was the only demonstrative method. Saunderson now boldly declared that algebra was a science that employed two demonstrative methods, synthesis and the more "satisfactory" analysis.

Appealing to the traditional categories of analysis and synthesis, he explained that algebraic results could be proved either analytically or synthetically, but that analytic proofs were preferable. A revealing discussion of "a general theorem for resolving all quadratic equations," a theorem that he proved both ways, came in book III of his *Elements*. Here he first demonstrated the result analytically, that is, he began with the general equation $Axx = Bx + C$ and worked step-by-step to the conclusion that $x = (B + s)/2A$ and $x = (B - s)/2A$, where $BB + 4AC = ss$. The careful analytic proof began with transposition, whereby the original equation was transformed into $Axx - Bx = C$; then each term was divided by A; next, the square was completed, as Saunderson referred back to an earlier proposition sanctioning this step; and so on, until the roots were identified. This proof was *in speciebus*, analytic, and demonstrative. Clearly, it was *in speciebus* and therefore general; it was analytic in the classical sense, that is, the proof began with the taking of something that was unknown, the root or *x,* as if it were known. In addition, Saunderson stressed that the proof was demonstrative or mathematically compelling. He wrote that he had "demonstrated analytically, that if Axx be equal to $Bx + C$, then x must necessarily be equal both to $B + s / 2A$, and to $B - s / 2A$." When he later gave the synthetic proof of the same theorem, he noted that the second proof was not required: "Now it may not be improbable but that the learner, especially if he has any taste or genius, may have a curiosity to see the same demonstrated again synthetically. . . . it is therefore to gratify the learner in this particular, that I have added the following demonstration." Saunderson then showed that, upon substitution, the roots $x = (B + s)/2A$ and $x = (B - s)/2A$ satisfied the original equation.[52] In summary, the primary proof of the quadratic formula was the analytic, with the synthetic supplied as a concession to mathematical curiosity and taste.

In book IV, he elaborated on *in speciebus* proofs and, in the process,

52. Saunderson, *Elements of Algebra*, 1:172–175.

returned to the question of the ways in which algebraic theorems were demonstrated. Unlike proofs that used symbols for unknown quantities only, he explained, *in speciebus* proofs employed symbols also for known quantities. "By this means in the first place he [the algebraist] will obtain indefinite answers, which ... suit and solve all particular cases to which they are applicable; and in the next place he will be able to prove his work synthetically." Synthetic proofs, he continued to suggest, had a place in algebra, if only to "confirm" analytic proofs and "further enure and reconcile ... [the student] to the operations of symbolical or specious Arithmetic; and so render him entire master of this sort of computation."[53]

At this point he referred his readers back to the two proofs of the quadratic formula, and then turned to analytic and synthetic treatments of the general form of the problem with which he had opened *The Elements*: "What two numbers are those, whose difference is 14, and whose sum when added together, is 48?" More generally, he now asked: "What two numbers are those, whose sum is *a,* and difference *b?*" Solving the problem *in speciebus,* he let x stand for the smaller number; then $x + b$ is the greater, and their sum is $2x + b = a$. Thus $2x = a - b$, and x (the smaller number) $= (a - b)/2$; and $x + b$ (the greater number) follows from:

$$\frac{a - b}{2} + \frac{b}{1} = \frac{a - b + 2b}{2} = \frac{a + b}{2}.$$

He next gave the synthetical demonstration of the same result, which centered on showing that the two expressions obtained here for the smaller and greater numbers met the stated conditions of the problem. Or,

$$\frac{a - b}{2} + \frac{a + b}{2} = \frac{2a}{2} = a$$

$$\frac{a + b}{2} - \frac{a - b}{2} = \frac{2b}{2} = b.\text{[54]}$$

As the preceding problem exemplified the power of symbolical reasoning, it was also used by Saunderson as a basis for a reaffirmation of the demonstrative nature of analysis and, indeed, the further claim that analytic proofs were preferable to synthetic proofs. The latter proof, he explained, "is what is called a synthetical demonstration, and doubtless shews the truth of the theorem to which it belongs, as well as the

53. Ibid., 1:213. 54. Ibid., 1:107, 214–215.

analysis whereby that theorem was investigated." He continued: synthesis, indeed, shows the truth of a theorem "but not so much to the satisfaction of the mind: for a synthetical demonstration only shews that a proposition is true; whereas an analytical one shews not only that a proposition is true, but why it is so, places you in the condition of the inventer himself, and unveils the whole mystery." If analysis was the superior mathematical method, why had ancient mathematicians, some early modern mathematicians, and, on occasion, even Newton seemed to privilege synthesis? According to Saunderson, the main appeal of synthetical demonstrations had rested not with their greater certainty but with their requiring "fewer principles than analytical ones . . . and those too, such as were commonly known."[55]

Saunderson thereby cast a new light on the analytic method of algebra. According to the classical distinction, in synthesis mathematicians assumed "what is [already] admitted [and worked] through the consequences [of that assumption]"; in analysis mathematicians assumed "that which is sought as if it were admitted [and worked] through the consequences [of that assumption]."[56] With some exceptions, including Viète, mathematical thinkers had viewed reasoning from the known, or that which was already admitted, as superior to reasoning from the unknown, or that which was merely taken as admitted. For Saunderson, however, the nature of the terms with which reasoning began – what was admitted versus what was only taken as if admitted – did not really matter, or at least mattered a lot less than the reasoning process itself. Whether applied to what was admitted or just taken as admitted, deductive reasoning led to undeniable conclusions. Analysis as well as synthesis, he stressed, was deductive reasoning. In addition, reasoning on what was only taken as if admitted was more satisfactory than reasoning on what was admitted. The equal of synthesis in demonstrative force, analysis was its superior in inventiveness.

Despite Saunderson's clear commitment to demonstrative algebra, there was little apparent consistency in his development of the various rules and theorems of the subject. Some, such as the rule of signs and the quadratic "theorem," were carefully demonstrated; others received scanty justification. The commutative property of multiplication fell into the latter group. "It is a matter of no great consequence in what order the letters are placed in a product," he simply wrote, "for *ab* and *ba*

55. Ibid., 1:215.
56. See, e.g., François Viète, *The Analytic Art: Nine Studies in Algebra, Geometry and Trigonometry* . . . , trans. T. Richard Witmer (Kent, Ohio: Kent State University Press, 1983), 11.

differ no more from one another than 3 times 4, and 4 times 3." Other rules were singled out as needing proofs. Saunderson wrote, for example, that, although algebra textbooks routinely stated the rule for determining the number of affirmative and negative roots of an equation and offered many examples of it, he had never encountered a "demonstration" of the rule.[57]

The justification offered for the rule of signs and even the brief defense of the commutative property suggest that, despite his omission of a discussion of the foundations of algebra, Saunderson agreed with Newton in seeing the subject as some sort of extension of arithmetic. Nevertheless, in *The Elements,* the grounds for deciding which rules were self-evident and which required proof – and even the bases for such proofs – remained murky. Not ready to proclaim the mathematician's right to frame axioms at will, Saunderson had described the "advertisements" for his proof of the rule of signs as "in a manner self-evident." In short, a major message of his *Elements* was that most algebraic rules and theorems should be proved, analytically if not also synthetically, with Saunderson's successors left to work out the details of the proofs.

8

Although, in some respects including the methodological, Saunderson's algebra was more innovative than MacLaurin's, it was not as tidy or well organized as a textbook. Put together during Saunderson's declining years from earlier notes, *The Elements* was a sometimes rigorous, sometimes chatty algebra textbook, bloated to over seven hundred pages by the inclusion of material relating particularly to number theory, geometry, and calculus. Much of the work was devoted to standard algebraic topics, for example, the solutions of quadratic through biquadratic equations and of simultaneous linear equations. Saunderson improved on Newton's solutions of the latter through a "criss-cross scheme of elimination," which foreshadowed determinants but compared unfavorably to MacLaurin's related work of the same period.[58] *The Elements,* like MacLaurin's *Treatise of Algebra,* featured a separate section (book VIII) on applying algebra to geometry.

Less typically, *The Elements* included books on Diophantine equations

57. Saunderson, *Elements of Algebra,* 1:59, 2:683.
58. Ibid., 1:237–238; Carl B. Boyer, "Cartesian and Newtonian Algebra in the Mid-Eighteenth Century," *Actes du XIe congrès internationales d'histoire des sciences* (Warsaw, 1968), 3, 195–202, on 201.

(book VI), Euclid's theory of proportion (book VII), and "miscellaneous topics" (book IX) as well as sections on Archimedes' geometry (in book VIII) – accompanied by the author's rationale or apology for the material. Saunderson, for example, admitted that Diophantus had not dealt with the first principles and rules of algebra but defended a whole book of Diophantine equations since every "particular problem puts us upon a new way of thinking, and furnishes a fresh vein of analytical treasure." He justified book VII as a necessary preliminary to book VIII. In book VII he considered proportion "in it's full extent as it is laid down in the fifth book of the elements of Geometry," and he briefly discussed his opposition to "representing ratios by fractions, or even fraction-wise, as is done by Barrow and others." He confessed that he had "been prevailed upon to insert" into book VIII sections on the circle, sphere, and cylinder, drawn from Archimedes. Although the sections "have no immediate relation to Algebra," he added, they were included "for the sake of those who cannot read *Archimedes,* and for the ease of those who can."[59]

The Elements of Algebra thus fit into the Newtonian bifocal mathematical tradition. But a comparison of the mature Newton with Saunderson highlights a shift in emphasis within that tradition over the course of the early eighteenth century. The anonymous "Note to the Reader" prefacing *Universal Arithmetick* explained that Newton had "condescended" to write on algebra whereas the "Advertisement" to *The Elements* stated that Saunderson "intended this treatise, not as a course of Algebra only, but also to promote, as far as possible, the study of Geometry."[60] Whereas Newton's friends felt the need to stress that he sanctioned algebra, Saunderson's friends felt the need to stress that he sanctioned geometry.

Clearly known in his lifetime for his commitment to the arithmetico-algebraic focus of mathematics, Saunderson was perhaps redeemed posthumously as a bifocal mathematician. Even mathematicians who opposed Newtonian algebra, such as the English mathematician Francis Maseres, praised sections of *The Elements of Algebra,* most especially the book on Euclid's theory of proportion.[61] Moreover, Saunderson's call for a demonstrative system of algebra, replete with axioms and

59.   Saunderson, *Elements of Algebra,* 2:363, 437, 468, 569.
60.   Anon., "Advertisement," in Saunderson, *Elements of Algebra,* n.p.
61.   On "Dr. Saunderson's excellent discourse" on ratios, see Francis Maseres, *A Dissertation on the Use of the Negative Sign in Algebra* (London: Samuel Richardson, 1758), 6.

deductive proofs and in this way akin to geometry, hit a responsive chord. Largely (not exclusively) an algebraist in a bifocal mathematical community, Saunderson had pointed the way toward finally establishing algebra as geometry's equal and thus helped set the stage for the evolution of algebra from its post-Newtonian form to the much touted British symbolical algebra of the early nineteenth century.

9

Saunderson's bias toward the arithmetico-algebraic focus of mathematics took hold at mid-eighteenth-century Cambridge. At the instigation of "some of the principal Tutors" at Cambridge, his *Elements* was trimmed of those sections and chapters that were "rather of Advantage and Amusement to Proficients in the general Science of Mathematics, than of necessary Use to Students in Algebra," and reappeared as the more popular *Select Parts of Saunderson's Elements of Algebra for the Use of Students at the Universities.*[62] His Lucasian successor was another algebraically inclined mathematician, John Colson (1680–1760), whom Saunderson had praised as a man of "great genius and known abilities in these [mathematical] sciences."[63]

Colson was educated at Christ Church, Oxford, without taking a degree. Prior to assuming the Lucasian professorship in 1739 at the late age of sixty, he served as the first upper master of the Mathematics School at Rochester, which had opened in 1709 to train young men for the maritime occupations,[64] and as vicar of Chalk, near Gravesend. As befitted the master of the Rochester School, he worked on some practical mathematics, including the construction and use of spherical maps, on which he published in 1736. His pure research centered on algebra, although in the 1730s he, like many of his British mathematical contemporaries, became involved in *The Analyst* controversy. His major mathematical paper, "The Universal Resolution of Cubic and Biquadratic Equations, as well Analytical as Geometrical and Mechanical," was

62. *Select Parts of Saunderson's Elements of Algebra for the Use of Students at the Universities* (London, 1756), iii.
63. Saunderson, *Elements of Algebra*, 2:720.
64. D. E. L. Flower, ed., *A Short History of Sir Joseph Williamson's Mathematical School: Rochester, 1701–1951* (printed privately, Rochester, Kent, 1951), 23, 28. On Colson, see also T. C., "John Colson," *Dictionary of National Biography*, 4:861–862.

originally published in the *Philosophical Transactions* of 1707 and later appended to 'sGravesande's Latin edition of Newton's *Universal Arithmetick* of 1732. In 1713 this paper and perhaps his connection to Thomas Sprat, then the bishop of Rochester, gained him entry into the Royal Society of London.[65] His second arithmetico-algebraic paper, "Account of Negative-Affirmative Arithmetic," followed in the *Philosophical Transactions* of 1726.

With the exception of his essay on Saunderson's "palpable arithmetic," all Colson's remaining publications involved translations of the works of other scholars, the most famous of which was his translation into English of Newton's manuscript *De methodis*. This was the early manuscript in which Newton had expanded algebraic expressions into infinite series, used the rules of algebra on such series, and then applied the series in developing his calculus of fluxions. Written just after Newton had completed his "Observations" on Kinckhuysen's *Algebra* and for a while intended as the companion piece to the "Observations" (see Chapter 7), the manuscript was not published during Newton's lifetime. It was only in 1736 that the work appeared in Colson's translation, *The Method of Fluxions and Infinite Series*. Colson claimed to publish the work to give both the defenders and critics of calculus the benefit of firsthand knowledge of Newton's views on the subject.[66]

The Method of Fluxions testified to the bifocal Newtonian mathematical legacy. It offered one of Newton's more algebraically elaborated versions of the calculus supplemented by about two hundred pages of commentary in which Colson worked on the sensible, geometric realization of fluxions. Unlike the Jurins of *The Analyst* controversy, Colson respected Berkeley and was "desirous to make" his commentary "as satisfactory as possible, especially to the very learned and ingenious Author of the Discourse call'd *The Analyst*."[67] In "a naive anticipation of Maclaurin's kinematic definitions of the basic concepts of the calculus," then, Colson tried to exhibit fluxions and fluents geometrically and mechanically.[68] As he put it, Newton's "Principles of the Doctrine of Fluxions" were "chiefly abstracted and Analytical," but he now "endeavour[ed] . . . , to shew something analogous to them in Geometry

---

65. On Colson and Sprat, see *Williamson's Mathematical School*, ed. Flower, 30.

66. John Colson, preface to *The Method of Fluxions and Infinite Series; with Its Application to the Geometry of Curve-Lines. By the Inventor, Sir Isaac Newton . . .* , trans. John Colson (London, 1736), ix–x.

67. Ibid., xii.   68. Guicciardini, *Development of Newtonian Calculus*, 57.

and Mechanicks; by which they may become not only the objects of the Understanding, and of the Imagination ... but even of Sense too, by making them actually to exist in a visible and sensible form."[69]

Still, it was not the case that in *The Method of Fluxions* Colson showed himself a geometric partisan. Besides outlining the kind of geometric and mechanical vision of the fluxions that he believed a response to *The Analyst* required, he made space in his commentary to promote algebra as well as a somewhat instrumentalist philosophy of mathematics. Writing of "the Analytical Art" and the "Synthetical Method of Demonstration," Colson stressed that Newton "has very much improved both Methods, and particularly in this Treatise he wholly applies himself to cultivate Analyticks, in which he has succeeded to universal applause and admiration." Possibly drawing on Saunderson's reflections on the methods of mathematics, he noted that analysis was a demonstrative method, "the several steps [of which] have a necessary connexion with each other ... [and lead to] the knowledge of the proposition required."[70]

Lest any reader remain unclear about the type of analysis being promoted, he quickly distinguished between geometric and algebraic analysis. As he reminded readers, some "modern Geometricians" had concluded that ancient mathematicians had derived their major results by geometric analysis but had then "studiously conceal'd" that analysis. Showing no sympathy for the reconstructive work of British mathematicians from the seventeenth century on, Colson declared that it did not really matter if there was a lost, ancient analysis because geometric analysis had been so far surpassed by algebraic analysis as to make its loss mathematically insignificant. In his own words, "however this may be, the loss of that [geometric] Analysis, if any such there were, is amply compensated, I think, by our present Arithmetical or Algebraical Analysis, especially as it is now improved, I might say perfected, by our own sagacious author in the Method before us." Appearing then to accept an expanded view of the new analysis coming at least partially from Newton, he concluded that algebraic analysis was "more universal

69. Colson, "A Perpetual Comment upon the Foregoing Treatise," in Newton, *The Method of Fluxions,* 266. For an explanation of how Colson tried to "prove the reality of fluxions by making them sensible," see Erik Lars Sageng, "Colin MacLaurin and the Foundations of the Method of Fluxions" (Ph.D. diss., Princeton University, 1979), 283–297 (quotation from p. 289).

70. Colson, "Perpetual Comment," 143–144.

and extensive" than geometric analysis; and, it was applicable to the "more abstruse Geometrical Speculations."[71]

As Colson's spirited defense of algebra took him astray of the focus of *The Analyst*, so the metaphysics of "impossible quantities" that he developed at the end of his commentary bore a closer resemblance to Berkeley's reflections on algebra than on geometry and calculus. Colson argued that Newton's "moments, vanishing quantities, infinitely little quantities, and the like," even when considered independently of geometric underpinnings, ought to be taken as legitimate mathematical objects. His argument was daring and somewhat instrumentalist. "But it has been pretended," he began, "that the Mind cannot conceive quantity to be so far diminish'd, and such quantities as these are represented as impossible." Without taking sides on the question of impossibility here, he wrote that he could "not perceive, even if this impossibility were granted, that the Argumentation would be at all affected by it, or that the Conclusions would be the less certain." In short, mathematical demonstration could proceed on impossible quantities. By implication, the reasoning, not the concepts to which it was applied, was what mattered. Of course, if the concepts were inconsistent in any way, the reasoning would be vitiated. Without stating the latter point but arguing in the tradition of Barrow, Newton, and Boyle, he observed that: "The impossibility of Conception may arise from the narrowness and imperfection of our Faculties, and not from any inconsistency in the nature of the thing." These preliminaries laid, he proclaimed boldly:

> So ... we need not be very solicitous about the positive nature of these quantities, which are so volatile, subtile, and fugitive, as to escape our Imagination; nor need we be much in pain, by what name they are to be call'd; but we may confine ourselves wholly to the use of them, and to discover their properties.

To support this approach to calculus, he noted that geometers traditionally made "impossible and absurd Suppositions, which is the same thing as to introduce impossible quantities, and by their means to discover truth." As hinted here, his view was not totally instrumentalist. Impossible quantities were "not introduced for their own sake, but only as so many intermediate steps" leading to "the knowledge of other quantities, which are real, intelligible, and required to be known."[72]

Near the very end of his commentary, he explicitly linked the metaphysical problems of calculus and algebra. He argued in favor of mathematicians' admitting a new kind of quantity, which he suggested might be called "impossible quantity" and which was to encompass the imagi-

71. Ibid., 144.   72. Ibid., 335–337.

nary quantities of algebra as well as the moments and infinitely little quantities of calculus. He noted first that quantities involving $\sqrt{-1}$, which he called "impossible and imaginary Quantities," were "as inconceivable and as impossible" as the questionable quantities of calculus. But in algebra imaginary quantities served two vital functions. They indicated that some problems were impossible; in other cases, where imaginary quantities appeared and were subsequently "duly eliminated," they led to "sound and good" conclusions, as in the solution of irreducible cubic equations. Imaginary quantities, he stressed, "in all these and many other instances that might be produced, are so far from infecting or destroying the truth of these Conclusions, that they are the necessary means and helps of discovering it." Similarly, he argued, the impossible quantities of calculus, "being themselves duly eliminated and excluded, . . . may leave us finite, possible, and intelligible Equations, or Relations of Quantities."[73]

In his commentary, then, Colson both gave a geometric response to *The Analyst* and promoted an analytic or algebraic approach to calculus. Proving no truer to Newton than to Berkeley, he also implicitly departed from Newton's public stand on quantities involving $\sqrt{-1}$, and somewhat misrepresented that stand. Whereas in *Universal Arithmetick* Newton had not solved irreducible cubic equations algebraically because of his uneasiness with passing through impossible quantities to real solutions, Colson offered such solutions as evidence of the necessity of admitting imaginary quantities. Stretching the truth, he depicted his use of impossible quantities as "following the Example of our sagacious and illustrious Author, who of all others has the greatest right to be our Precedent in these matters." Speaking for the eighteenth-century British mathematicians with modern leanings, he finally observed that "'Tis enlarging the number of general Principles and Methods, which will always greatly contribute to the Advancement of true Science."[74] Generality and applicability were now asserted over strictness or foundational purism.

10

At the very highest mathematical level, that of the Lucasian professorship, mid-eighteenth-century Cambridge University was thus strongly committed to the arithmetico-algebraic focus of Newtonian mathematics. Moreover, in the two decades that had transpired since Newton's

73. Ibid., 337–338.   74. Ibid., 338–339.

death, Cambridge mathematicians, led by Saunderson and Colson, had taken English algebra somewhat beyond the cautious confines imposed upon the subject in *Universal Arithmetick*. The writings of Saunderson spoke of a growing pragmatic détente with expressions involving $\sqrt{-1}$; those of Colson spoke of an attempted philosophical coming to terms with such expressions. In his commentary on *The Method of Fluxions* Colson had gone so far as to urge that the idea of quantity be enlarged to include a new category of "impossible quantities" – in fact, "species of impossible quantities, if they must needs be thought and call'd so."[75] Moreover, Saunderson had implicitly dismissed the claim of synthetic geometry to sole standing as the paradigm of human thought; he had convinced his Cambridge friends that algebra could be "considered as the Logical Institutes of the Mathematician." Practically, he had challenged algebraists to pursue the demonstrative development of their subject. In passing, Colson, too, had reminded mathematicians that analysis was the pursuit of necessary connections; he had also proclaimed that algebraic analysis had made geometric analysis obsolete. Appropriately, Colson's last major work as Lucasian professor was a translation of Maria Gaetana Agnesi's *Analytical Institutions* from Italian into English. The translation, which was not published during his lifetime, was designed to expose the English-reading public, men and women alike, to the far-ranging analytic mathematics being cultivated by continental mathematicians of the eighteenth century.[76]

With the established algebraic tradition at Cambridge, it was almost to be expected that upon Colson's death the leading candidates for the Lucasian professorship would present strong algebraic credentials. And the Lucasian election of 1760 pitted Edward Waring, an exceedingly abstract algebraist for his time, against Francis Maseres, an opponent of the negative numbers whose major research, however, centered on raising algebra to the same logical standing as geometry. Waring was elected the sixth Lucasian professor and fifth successor of Isaac Barrow, who roughly a hundred years earlier had dismissed algebra as "yet . . . no Science."

75. Ibid., 338.
76. Maria Gaetana Agnesi, *Analytical Institutions, in Four Books,* trans. John Colson, vol. 1 (London, 1801). This work was published posthumously under the direction of John Hellins.

Epilogue

The contest for the Lucasian professorship in 1760 symbolized the state of English mathematics at the beginning of the second half of the eighteenth century, in more ways than one. Unlike Saunderson and Colson, Maseres and Waring were products of Cambridge University. Their mathematics was shaped, albeit with different results, by the Cambridge mathematical traditions in place by the midcentury. Like many Cambridge mathematical scholars of their period, both men were influenced by Saunderson's *Elements of Algebra*, which "was long the standard treatise on the subject."[1] In his major mathematical work, *A Dissertation on the Use of the Negative Sign in Algebra* of 1758, Maseres praised certain aspects of Saunderson's *Elements* even as he severely criticized the book's liberal use of negative quantities.[2] In 1760 Waring reported that everyone at Cambridge was reading Saunderson and cited the latter as the "Authority" for some of his own mathematical manipulations.[3]

Maseres and Waring were not ordinary Cambridge graduates. They were high wranglers, fourth and senior (or first) wranglers, respectively. That is, the election of 1760 was the first Lucasian election to be contested by Cambridge graduates who had distinguished themselves in the mathematical honors course that solidified at the university around the mid-eighteenth century. This course was a natural outgrowth of the university's emphasis on mathematics as a logic. If mathematics was really the best instrument to exercise and train the human mind, and

1. Augustus De Morgan, *A Budget of Paradoxes,* 2d ed., ed. David Eugene Smith, 2 vols. (Chicago: Open Court, 1915), 1:377.
2. For Maseres's remarks on "Saunderson's excellent discourse" on ratios, see Francis Maseres, *A Dissertation on the Use of the Negative Sign in Algebra* (London: Samuel Richardson, 1758), 6.
3. Edward Waring, *A Reply to a Pamphlet Entitled Observations on the First Chapter of a Book Called "Miscellanea Analytica"* (Cambridge, 1760), 27, 25.

Cambridge was charged by the Elizabethan statutes to teach logic to all second- and third-year undergraduates, the argument ran, then the undergraduate curriculum should center on mathematics. By the early eighteenth century the thesis that mathematics, especially geometry, was a paradigm of logic and, as such, an essential component of a liberal education had been incorporated into popular Cambridge literature.[4] For example, in his *Advice to a Young Student* of 1730, Daniel Waterland, the master of Magdalene College (which would later train Waring), declared that "The study of the Mathematicks also will help more towards it [the improvement of the reasoning powers] than any Rules of Logick."[5]

It was the institution of the Senate House examination, later known as the mathematical tripos, before the middle of the eighteenth century that provided the incentive needed to convince the colleges of Cambridge to focus undergraduate studies on mathematics. By 1750 this examination, which was named for the building in which it was given, was a university-wide, mathematically oriented examination taken by undergraduates in their fourth year of study.[6] Since university honors were awarded on the basis of performance on this examination, it quickly set academic priorities across the university. Each year seniors and colleges alike vied for the coveted top positions on the ranked list of examinees. From 1747 onward, as the examination assumed increasing status, this list was printed; from 1753 on, the list individually ranked each honors student in one of three classes, the top class being that of the wranglers (usually a dozen or so in number), with the first enjoying the special title of "senior wrangler."[7]

The examination required some mathematics of all undergraduates, and considerable mathematics of those aiming at honors. Around the mid-eighteenth century, the "non-reading" students, or those avoiding the honors section of the examination, were supposed to master "two

---

4. John Gascoigne, "Mathematics and Meritocracy: The Emergence of the Cambridge Mathematical Tripos," *Social Studies of Science* 14 (1984): 547–584, esp. 570–572.

5. Daniel Waterland, *Advice to a Young Student* (London, 1730), 20.

6. For a descriptive history of the Senate House examination, see W. W. Rouse Ball, *A History of the Study of Mathematics at Cambridge* (Cambridge: Cambridge University Press, 1889), 187–219.

7. The other honors classes were the senior optimes and the junior optimes, with the rest of the students being lumped together as pollmen. The term "wrangler" further substantiates the substitution of mathematics for logic at Cambridge, since logicians had frequently been called "wranglers" (Gascoigne, "Mathematics and Meritocracy," 571–572).

books of Euclid's Geometry, Simple and Quadratic Equations, [and] the early parts of Paley's Moral Philosophy," along with a smattering of elementary Latin, in order to secure a pass or "poll" on the examination.[8] As indicated by John Jebb in 1772, students reading for honors needed to know much more, including "the extraction of roots, the arithmetic of surds, the invention of divisors, the resolution of quadratic, cubic, and biquadratic equations; together with the doctrine of fluxions."[9]

By the years Maseres and Waring went up to Cambridge, 1748 and 1753, respectively, then, talented students like themselves were expected to concentrate on mathematics. Moreover, according to the message given to students and faculty alike, mathematics was associated with logic, with training the mind, and perhaps less so with advancing the frontiers of human knowledge. Oughtred's dream of making mathematics an academic subject had been more than realized: mathematics was not just one of many academic subjects at Cambridge; it was now the premier subject. But the dream had been realized at some cost, as the concern for logical strictness became an integral part of the British practice of mathematics.

The focus of the undergraduate honors curriculum, mathematics had to live up to its billing as the paradigm of human reasoning. Eighteenth-century Cambridge mathematicians could not have easily ignored the foundations of mathematics, even if they had desired to do so. To a large extent, *The Analyst* controversy revolved around the claim of geometry and hence calculus to standing as a logic. As we have seen, Colson wrote a long commentary on Newton's doctrine of fluxions; Saunderson struggled to explain the negative numbers; both Saunderson and Colson worried in print about imaginary quantities; and Saunderson made the discussion of the methods of mathematics, especially algebra, the hallmark of his professorship.

Well might Saunderson have devoted so much time to the foundational issues of algebra, for, in the eyes of some of his contemporaries, algebra was less a logic than geometry. Despite his strong stand in favor of algebra and the analytic method and despite the algebraic proclivities

---

8. T. G. Bonney, *A Chapter in the Life History of an Old University* (London, 1882), 12–13; quoted in Martha McMackin Garland, *Cambridge before Darwin: The Ideal of a Liberal Education, 1800–1860* (Cambridge: Cambridge University Press, 1980), 3. For moral philosophy, both Paley's *Moral Philosophy* and Locke's *Essay Concerning Human Understanding* were recommended.

9. Quoted in Ball, *Study of Mathematics at Cambridge*, 192.

of his immediate Lucasian successors, it was not at all clear to the larger Cambridge community of the late eighteenth century that algebra trained the human mind nearly as well as geometry. In 1774 John Lawson, writing anonymously, defended the unique pedagogical values of geometry. A graduate of Sidney Sussex College, Cambridge, in 1745, subsequently a mathematical lecturer and tutor at the college, and an admirer of Simson, Lawson emphasized that the study of geometry "tend[ed] greatly to establish a close thinking, and a methodical and just argumentation, when we apply ourselves to any other subject whatsoever, physical or moral, oeconomical or civil." Despite such pedagogical value, he complained, geometry "seems of late to have met with less regard than it's dignity and usefulness demand. Our schools and universities, our philosophical societies, and philomaths of all degrees, seem to have been very assiduous of late in paying their devoirs to the younger sister *Algebra*." For "those who apply themselves to the Mathematics only for the ends abovementioned, namely, to inure themselves to a method of close thinking and just reasoning," he added, "*Geometry* is the proper field."[10]

Cambridge scholars who worried that algebra fell short of geometry as a logic, and their Scottish counterparts, had two choices: the more conservative of their lot could retreat to the geometric focus of Newtonian mathematics; the others could focus their work on algebra in an attempt to reform the subject. Here again the Lucasian election of 1760 caught the mathematical pulse of Cambridge. It pitted the ultimately victorious Waring, who took algebra in its abstrusest directions,[11] against Maseres, a major opponent of the negative numbers who sought to elevate algebra to the same standing as geometry through a more restricted definition of the negatives.

Ironically, even as he had advised that philosophical forays were risky for mathematicians, Saunderson's own explanations of the negative and imaginary quantities seemed to help foment the late-eighteenth-century

10. [J. Lawson], *A Dissertation on the Geometrical Analysis of the Antients, with a Collection of Theorems and Problems, without Solutions, for the Exercise of Young Students* (Canterbury, 1774), ii–iv. In his popular *Rudiments of Mathematics* of 1785, William Ludlam endorsed Euclid's *Elements* as "the best book of logic extant" (quoted in Gascoigne, "Mathematics and Meritocracy," 570).
11. See *Meditationes Algebraicae: An English Translation of the Work of Edward Waring*, ed. and trans. Dennis Weeks (Providence, R.I.: American Mathematical Society, 1991).

attack on these numbers. Both Maseres and William Frend – who into the early nineteenth century stood side by side as the two major and influential English opponents of the negative (and imaginary) numbers – attacked specifically the writings of Saunderson as well as MacLaurin and even Newton. Both criticized Newton's definition of a negative quantity as a quantity less than nothing and Saunderson's reiteration of that definition.[12] Moreover, Saunderson's dismissal (as "narrow minds") of students who questioned the negative numbers had apparently affected Maseres to the quick. As late as 1800 he was still berating Saunderson for his treatment of the negative and imaginary numbers and, not incidentally, for his treatment of students who dared to ask for algebraic clarifications: "These absurd doctrines [of negative numbers and roots] seem to have delighted Dr. Saunderson, who frequently insults and rails at such of his readers as shall find a difficulty in understanding them, as being persons of narrow minds and incorrigible dullness and stupidity."[13]

In his *Dissertation on the Use of the Negative Sign in Algebra,* which he published in 1758 to bolster his anticipated candidacy for the Lucasian professorship, Maseres attempted to construct an algebra in which "the Negative Sign" was considered in no "other light than as the mark of the subtraction of a lesser quantity from a greater."[14] A high wrangler who viewed mathematics largely as a pedagogical tool, he had been struck by the inconsistency of setting mathematics (including algebra) up as a logical paradigm, on the one hand, and requiring undergraduates to accept the negative numbers as quantities less than nothing, on the other. In his book, he rejected the isolated negative and hence imaginary numbers. Largely following the deductive approach to algebra that Saunderson had begun to sketch, he reconstructed algebra as the science of universal arithmetic in the strictest sense. " 'Tis hoped therefore," he wrote at the end of his explanation of the different standing of algebra versus geometry as a logic, "that an attempt to raise it [algebra] to a level with Geometry, in respect both of perspicuity of conception, and accuracy of reasoning, will not prove an unacceptable present to such

12. See, e.g., Francis Maseres, *Observations on Mr. Raphson's Method of Resolving Affected Equations of All Degrees by Approximation,* appended to William Frend, *The Principles of Algebra* (London, 1796), 465–466.
13. Francis Maseres, *Tracts on the Resolution of Affected Algebraick Equations by Dr. Halley's, Mr. Raphson's, and Sir Isaac Newton's Methods of Approximation* (London: J. Davis, 1800), lviii.
14. Maseres, *Dissertation,* i.

lovers of the Mathematical Sciences as are as yet but in the beginning of their studies."[15]

During the same period that Maseres and Frend were criticizing the negative numbers and the handling of these numbers by Newton, Mac-Laurin, and Saunderson, the Scottish mathematician John Playfair directly questioned the imaginary numbers and their handling by MacLaurin. As we have seen, although contented with his explanation of the negative numbers through an appeal to the relation of contrariety, MacLaurin had appeared uneasy about the imaginary numbers. In his *Treatise of Fluxions* he had dissociated the "evidence" of fluxions from "any arts of this kind,"[16] in particular, imaginary quantities. It was in 1778, after Scottish mathematics had focused on geometry for the thirty years following MacLaurin's death, that Playfair, then joint professor of mathematics at Edinburgh, published his influential paper, "On the Arithmetic of Impossible Quantities." In this paper he carefully framed the problem of "imaginary expressions" (a term he borrowed from MacLaurin). As he reminded his readers, such expressions were "no more than marks of impossibility" but were yet sometimes "put to denote real quantities" and, as such, subjected to arithmetic operations. What really needed to be explained, he continued, was how the imaginaries, when so operated on, led to "just conclusions." Mimicking Berkeley's accusations against Newton and Leibniz, he accused MacLaurin and Jean Bernoulli of "being more intent on applying their calculus [of imaginaries], than on explaining the grounds of it." Explicitly referring to and rejecting MacLaurin's appeal to the compensation of errors, he then offered his own defense of the imaginaries.[17] The details of the defense, however, mattered less than the paper's role in stimulating concentrated British reflection on the problem of the imaginary numbers.

Scholars have discussed the writings of Maseres, Frend, Playfair, and a few other British mathematical thinkers as contributing to the crucial, turn-of-the-century discussion of the problem of the negative and imaginary numbers that helped to set the stage for the British development of modern symbolical algebra in the early decades of the nineteenth century.[18] Although the majority of turn-of-the-century British mathemati-

15. Ibid., iii.

16. Colin MacLaurin, *Treatise of Fluxions, in Two Books,* 2 vols. (Edinburgh: T. W. and T. Ruddimans, 1742), 2:578.

17. John Playfair, "On the Arithmetic of Impossible Quantities," *Philosophical Transactions* 68 (1778): 318–343, on 320–322.

18. See, e.g., Ernst Nagel, "'Impossible Numbers': A Chapter in the History of Modern Logic," *Studies in the History of Ideas* 3 (1935): 429–474, esp.

cal thinkers did not join Maseres and Frend in rejecting isolated negative numbers, some appreciated their criticisms of these numbers and began to rethink not only the negative and imaginary numbers but algebra itself. The present history of early modern British algebra shows, of course, that concern for the foundations of algebra was nothing new to the late eighteenth century. What was new was the urgency assigned the problem of the negative and imaginary numbers. From the late eighteenth century into the early nineteenth century these numbers were discussed in journal articles as well as textbooks, by seemingly more British mathematical thinkers than ever before, and with a new persistence and candor.

The pressures to do something about these numbers – not only pedagogical but also mathematical and philosophical pressures – had never been as great. Time and precedent were turning the "venturous" into nearly mainstream mathematics. Concurrently, the failure of Newton, MacLaurin, and Saunderson, in turn, to provide universally acceptable definitions of the negative and imaginary numbers, when traditional definitions were still assigned a premium within mathematics, strongly suggested that no such definitions or easy justifications of these numbers were forthcoming. Maseres and Frend made this point, in so many words, as they quoted the relevant differing and prolix statements of their mathematical predecessors. Either rejection of these numbers or a rethinking of algebra seemed in order. On an individual level, Maseres and Frend were vocal and influential players in the attack on the negative and imaginary numbers. Both were Unitarians in a period when religious dissent still carried a high price; their intellectual skepticism freed them to criticize Newton, his algebra, and his followers in public, to an extent and with a persistence from which men who were closer to the British establishment shied away. Frend turned the religious argument of Berkeley's *Analyst* upside down: if as a Unitarian he could reject any religious idea that he did not understand, as a mathematician he could certainly oppose any mathematical idea that eluded clear conception.[19]

438–441; Helena M. Pycior, "George Peacock and the British Origins of Symbolical Algebra," *Historia Mathematica* 8 (1981): 23–45, esp. 27–31; idem, "Internalism, Externalism, and Beyond: 19th-Century British Algebra," *Historia Mathematica* 11 (1984): 424–441, on 430; and David Sherry, "The Logic of Impossible Quantities," *Studies in History and Philosophy of Science* 22 (1991): 37–62, esp. 41–47.

19. On Frend's linking of his Unitarianism with rejection of the negative numbers, see Helena M. Pycior, "British Abstract Algebra: Development and Early Reception," in *History in Mathematics Education,* ed. Ivor Grattan-Guinness (Paris: Belin, 1987), 152–168, on 155.

Thus seen as more urgent than ever before, the problem of the negative and imaginary numbers proved fertile at the turn of the century. The problem preoccupied such lesser lights of British mathematics as Maseres and Frend as well as such key mathematical thinkers as Robert Woodhouse, George Peacock, and Augustus De Morgan. At least partially from their reflections on the negative and imaginary numbers emerged a new understanding of algebra as an abstract science of arbitrary signs and symbols, the interpretation of which followed rather than preceded their manipulation. Algebra thus entered a new stage: no longer early modern or even post-Newtonian, it became "symbolical algebra."

The story of how and why the original British symbolical algebraists – Charles Babbage, Peacock, J. F. W. Herschel, and the pioneering Woodhouse – and their later associates, including Augustus De Morgan and Duncan F. Gregory, formulated their abstract approach to algebra has been only partially reconstructed.[20] In certain fundamental ways, these mathematicians studied algebra under different conditions than their British predecessors. In the late eighteenth and early nineteenth centuries, British mathematics was more open to continental influences than at any prior point in the post-Newtonian period.[21] Cross-fertilization of British mathematics with continental mathematics certainly played some role in the emergence of symbolical algebra.[22] As early modern algebra

20. For a general history leading up to symbolical algebra, see Luboš Nový, *Origins of Modern Algebra,* trans. J. Tauer (Prague: Academia, 1973); for a brief analysis of the development of symbolical algebra (with bibliography), see Pycior, "British Abstract Algebra: Development and Early Reception"; and for reflections on the current state of the history of symbolical algebra (with many references), see Menachem Fisch, "'The Emergency Which Has Arrived': The Problematic History of Nineteenth-Century British Algebra – A Programmatic Outline," *British Journal for the History of Science* 27 (1994): 247–276.

21. Whereas earlier historians of mathematics largely agreed that eighteenth-century British mathematicians were isolated from continental mathematics, the extent of this isolation is now a subject of historical reexamination. See, e.g., Niccolò Guicciardini, *The Development of Newtonian Calculus in Britain, 1700–1800* (Cambridge: Cambridge University Press, 1989), vii–viii.

22. For a claim of a fundamental role for continental mathematics, see Elaine Koppelman, "The Calculus of Operations and the Rise of Abstract Algebra," *Archive for History of Exact Sciences* 8, no. 3 (1971): 155–242. Koppelman argued that "the work in algebra was a direct response of the English to a specific aspect of the work of Continental analysts [the calculus of operations] which became accessible to them" (ibid., 156).

became established in English mathematics during the Civil War, so British symbolical algebra was created in the troubled years following the French Revolution. More internally to mathematics, just before the turn of the nineteenth century Caspar Wessel discovered a geometric representation of the complex numbers. One of five similar treatments of the complex numbers to be developed through the first three decades of the nineteenth century,[23] Wessel's work gave a hitherto-unknown reality to the imaginary numbers. The geometric representation of the complex numbers permitted the symbolical algebraists to use $\sqrt{-1}$ as a meaningless symbol with the understanding that its geometric interpretation could follow and, to an extent, vindicate its manipulation.

In addition to these and other new conditions for the practice of algebra at the beginning of the nineteenth century, earlier speculations on algebra and its foundations – the stuff of the present book – contributed to the emergence of symbolical algebra. The deductive approach to algebra of the symbolical algebraists followed Saunderson's emphasis on the demonstrative nature of analysis. There seems to have been a natural progression from Saunderson's methodological reflections to Maseres's goal of "setting down almost every step of the [algebraic] reasoning, as Euclid has done in his Elements" and, finally, to Peacock's ideal of "conferring upon Algebra the character of a demonstrative science," expressed in the first sentence of the preface to his *Treatise on Algebra* of 1830.[24]

Furthermore, Woodhouse found some philosophical backing for his abstract approach to mathematics in the writings of Locke and Berkeley. In developing (what he called) his "philosophy of analysis," Babbage followed Dugald Stewart, who (as a philosopher) brought him under Berkeley's influence.[25] Generally, by the late eighteenth century British scholars were more favorably disposed toward Berkeley's philosophy than their predecessors of the earlier part of the century. Time had healed the scars coming from *The Analyst* controversy, making it possible for British mathematicians to read and appreciate the complete Berkeley.

23. On early work on the geometric representation of the complex numbers, see Michael J. Crowe, *A History of Vector Analysis: The Evolution of the Idea of a Vectorial System* (Notre Dame, Ind.: University of Notre Dame Press, 1967), 5–12.

24. Maseres, *Dissertation*, i; George Peacock, *Treatise on Algebra* (Cambridge, 1830), v.

25. See Pycior, "Internalism, Externalism, and Beyond," 434–438, and Sherry, "Logic of Impossible Quantities," 47–54. On the connection between Peacock and Locke, see M.-J. Durand, "Genèse de l'algèbre symbolique en Angleterre: Une influence possible de J. Locke," *Revue d'Histoire des Sciences et de leurs Applications* 43 (1990): 129–180. On Babbage's role in

There are also promising connections between Babbage and the early modern English algebraists. Babbage studied the history of early modern algebra enough to reflect on some of its high points in his important essays on "The Philosophy of Analysis." He dabbled in cryptography and, for a period during his early years, he focused on the pursuit of a universal language. According to Buxton, he "studied with much attention the investigations of his predecessors, and especially Bishop Wilkins in this field of enquiry."[26] And Babbage had special praise for the *Arithmetica infinitorum* of that venturous, algebraic explorer, John Wallis.[27]

The development of algebra from the post-Newtonian stage to that of British symbolical algebra is, however, another story. Its reconstruction will probably prove as episodic and conditioned by the sociocultural matrices of its protagonists as has the development of algebra from the early modern through post-Newtonian stages.

the creation of symbolical algebra, see J. M. Dubbey, "Babbage, Peacock and Modern Algebra," *Historia Mathematica* 4 (1977): 295–302, and idem, *The Mathematical Work of Charles Babbage* (Cambridge: Cambridge University Press, 1978), esp. 93–130; on Woodhouse's role, see Harvey W. Becher, "Woodhouse, Babbage, Peacock, and Modern Algebra," *Historia Mathematica* 7 (1980): 389–400.

26. H. W. Buxton, *Memoir of the Life and Labours of the Late Charles Babbage Esq. F.R.S.*, ed. Anthony Hyman (Cambridge, Mass.: MIT Press, 1988), 347.

27. See, e.g., Dubbey, *Mathematical Work of Babbage,* 112–113.

Index

arithmetic (*cont.*)
  Wallis, John, arithmetic formulation of
    algebra
arithmetic and geometry
  analytic geometry, 33–35, 78–79, 148,
    202–203
  arithmetic subordinate to geometry, 5, 6,
    151, 153, 155, 156–160, 166
  arithmetic superior to geometry, 124,
    143, 155, 158, 215, 256, 260
  geometry superior to arithmetic, 5, 142,
    143, 166, 231
Arminianism, 149, 165
Attercliffe Academy, 277, 281
Aubrey, John, 67, 90, 137–138
Aylesbury, Sir Thomas, 56n59

Babbage, Charles, 314–316
Bacon, Francis, 4–5, 46, 121, 136–137
Barrow, Isaac
  algebra, dismissal of, 6, 155, 162, 165,
    306
  Apollonius's *Conics*, 153, 165
  arithmetic subordinate to geometry (*see
    under* arithmetic and geometry)
  education and career, 148–152, 168,
    169, 182
  Euclid's *Elements*, 151–154, 170
  geometry as logic, 4, 6, 66, 163–165
  humanist, 151–152, 154, 164
  limits of human reason, 175
  Mathematical Lectures, 154–165, 175,
    212–213
  metaphysical numbers, 161
  philosophy of geometry, 155–157, 281–
    282
  and Plato, 155, 157, 161, 163, 166
  religious beliefs and ministry, 149–151,
    165, 237–238, 262
  surds, 159—160
  universal ideas, 155–156
  and Wallis, 155, 157–158, 159–160, 207
  see also under Berkeley, George; Collins,
    John; Newton, Isaac
Bentley, Richard, 278
Berkeley, George
  Alciphron, or the Minute Philosopher, 7,
    211, 212, 227–230, 231
  algebra, support for, 209–210, 213, 222
  algebra and geometry, ranking of, 231
  antiabstractionism, 6, 209, 218–219,
    225–227, 228, 236, 240–241, 252,
    254, 256
  Arithmetica, 213, 218
  and Barrow, 136, 157, 209, 212–213,
    216–218, 220, 222–223

calculus as part of geometry, 232–233
De motu, 221
*Defence of Free-thinking in Mathemat-
  ics,* 232, 241
education and career, 210–213
Essay towards a New Theory of Vision,
  211, 212, 217, 218, 224, 240
general arguments, 220–221
and Hobbes, 139, 209, 212, 220, 222–
  223
immaterialism, 218–220, 224, 252, 254
influence on British mathematics, 6, 210,
  230–231, 236, 240–241, 315
language and signs, 6–7, 8, 209–210,
  213–214, 222–224, 226–230
and Locke, 212, 214–217, 218–219,
  220, 222, 225, 226
Miscellanea mathematica, 213, 216, 218
and Newton, 212–213, 222, 232, 234,
  235, 312
Philosophical Commentaries, 212–213,
  216–218, 219, 220, 222–224, 233
philosophy of arithmetic and algebra, 6,
  135–136, 209, 216–218, 219, 222–
  227, 228–231, 240–241, 242
philosophy of geometry, 6, 209, 219–
  222, 281–282
primary qualities, 217, 224–225
*Treatise Concerning the Principles of Hu-
  man Knowledge,* 211, 212, 218–219,
  220–221, 224–227, 228, 252
and Wallis, 209, 212, 222–223, 230,
  272
see also The Analyst; MacLaurin, Colin,
  and Berkeley
Bernoulli, Johann (Jean), 247, 312
bifocal Newtonian mathematical tradition
  and Berkeley, 6, 210, 240–241
  at Cambridge, 5–6, 276, 305, 309–310
  and Colson, 302–304
  and MacLaurin, 6, 207, 242–243, 256–
    257, 273–274, 276
  from Newton's works, 5–6, 170–171,
    207–208
  and Saunderson, 276, 281, 283, 292,
    300–301
  and Simson, 6, 242, 274–275
binomial theorem, 132–133, 171
binomials
  cube roots of surd, 109–110, 177–178,
    183
  expansion of, 52
  see also irreducible cubic equation, ex-
    traction of cube roots
Boethius, 55
Bombelli, Rafael, 23–24, 108, 127

Index